Governing food security

Governing food security

Law, politics and the right to food

edited by:
Otto Hospes
Irene Hadiprayitno

Wageningen Academic
Publishers

Also available in the 'European Institute for Food Law series':

European Food Law Handbook
Bernd van der Meulen and Menno van der Velde
ISBN 978-90-8686-082-1
www.WageningenAcademic.com/foodlaw

Fed up with the right to food?
The Netherlands' policies and practices regarding the human right to adequate food
edited by: Otto Hospes and Bernd van der Meulen
ISBN 978-90-8686-107-1
www.WageningenAcademic.com/righttofood

Reconciling food law to competitiveness
Report on the regulatory environment of the European food and dairy sector
Bernd van der Meulen
ISBN 978-90-8686-098-2
www.WageningenAcademic.com/reconciling

ISBN: 978-90-8686-157-6
e-ISBN: 978-90-8686-713-4
DOI: 10.3920/978-90-8686-713-4

ISSN 1871-3483

First published, 2010

Table of contents

About the authors

G.E. (Gerard) Breeman studied public administration in Rotterdam and got his PhD from Leiden University. He works at the Public Administration and Policy Group at Wageningen University. His main research focus is about how to make policies trustworthy. He also publishes on agenda-setting, the politics of attention, and European Union policies. Since 2009 he has also become a research fellow at the Montesquieu Institute in The Hague.

J.W.M. (Han) van Dijk holds a special chair of Law and Governance in Africa, financed jointly by the African Studies Centre in Leiden and Wageningen University. His teaching and research focuses on the socio-legal aspects of natural resource management, land tenure, the analysis of development policy, food security, resource conflicts, impact of climate change, governance studies, nomadic pastoralism, livelihood studies, fisheries management, methodology of evaluation research and interdisciplinarity.

A. (Asbjørn) Eide is former director and presently senior fellow of the Norwegian Center for Human Rights at the University of Oslo. He was previously a director of the International Peace Research Institute in Oslo, a former secretary-general of the International Peace Research Association, and the author of numerous books, contributions to books, articles in periodicals and studies for the United Nations, including the seminal study on the right to adequate food as a human rights (1987) which has affected the subsequent literature and monitoring of economic, social and cultural rights. He has been Torgny Segerstedt Professor at the University of Gothenburg, Sweden and is presently visiting professor at the University of Lund. For 20 years, until 2004, he was expert member of the United Nations Sub-Commission on Promotion and Protection of Human Rights and from 1995 to 2004 Chairman of the United Nations working group on the Rights of Minorities. For some years until June 2006 he was also member and President of the Advisory Committee on the Council of Europe's Framework Convention for the Protection of National Minorities.

H.J.M. (Henri) Goverde is Associate Professor of Public Administration at Radboud University Nijmegen. He was Endowed Professor of Political Science at Wageningen University (1995-2009). Goverde chaired the Research Committee on Political Power of the International Political Science Association (1994-2008), which he serves now as honorary member. Recently, he co-edited a special issue about 'Power and Space' for the Journal of Power (2009, Vol. 2, No 2) and a book about 'Governance in de Groen-Blauwe Ruimte' (2009).

I.I. (Irene) Hadiprayitno is a Postdoc Researcher working on the Human Right to Food at Wageningen University. Her research interest focuses on the interactions between legal and non-legal aspects that define and give meaning to human rights and shape the efforts towards realisation. She has done extensive research on the dialectics of human rights and development in Indonesia and has published widely on this topic as well as the right to development and the right to food.

O. (Otto) Hospes is an associate professor Law and Governance at the Department of Social Sciences of Wageningen University. He is doing research on law and governance of sustainable palm oil and soy production, biofuels and food security, and human rights. He has visited, conducted evaluations, and has done research in many countries, including Indonesia, India, Sri Lanka, Zambia, Zimbabwe, Tanzania, Kenya and Brazil.

F.M. (Michiel) Köhne is Assistant Professor of Law and Governance at Wageningen University. His field of research and teaching concerns the role of legal pluralism in rural development with a focus on non-state law. He has been working for a long time on indigenous law and the role of indigenous organisations in the governance of land use. At present he is focusing on private initiatives of regulation, often through labelling, of the production of tropical commodities such as coffee, hard wood and palm oil.

B.M.J. (Bernd) van der Meulen is Professor of Law and Governance at Wageningen University. He is also chairman of the Dutch Food Law Association and director of the European Institute for Food Law. He has published widely in the field of food law including the human rights dimension. See: www.law.wur.nl.

M.R. (Peggy) Grossman is Bock Chair and Professor of Agricultural Law in the Department of Agricultural and Consumer Economics, University of Illinois, USA. She is the author of numerous US and European law review articles and book chapters on agricultural and environmental law topics. Grossman is editor of 'Agriculture and the Polluter Pays Principle' (BIICL, 2009) and co-editor of 'Agriculture and International Trade: Law, Policy and the WTO' (CABI Publishing, 2003)). Professor Grossman has received three Fulbright Senior Scholar Awards and a German Marshall Fund Research Fellowship to support her research in Europe. She has served as President of the American Agricultural Law Association and received the AALA Distinguished Service Award. She received the AALA Professional Scholarship Award in 2006 and 2008. The European Council for Agricultural Law awarded her the Silver Medal in 1999. Since 1986, Professor Grossman has been a frequent visitor to the Law and Governance Group at Wageningen University, The Netherlands. In Spring 2008, she was Visiting Professor at the University of Copenhagen (Denmark) and at Newcastle Law School (England).

D. (Dik) Roth is a social anthropologist and Assistant Professor at the Law and Governance Group of Wageningen University. Among his scientific interests are the anthropology of law, governance and management of natural resources, and flood policy and politics. He has done extensive research on land reform, local irrigation management, migration and regional autonomy in Indonesia, and has widely published on these and other topics. With Jeroen Warner, he has published previously on virtual water and on flood policy in the Netherlands.

A. (Anna) Szajkowska is a lawyer, currently writing her PhD dissertation on the principle of risk analysis in European Union food safety law at the Law and Governance Group at Wageningen University. Her research interests include EU internal market law, the relationship between EU law and the national legal systems, risk regulation, and the World Trade Organization regime. She has published in the field of food law, including articles in journals such as the Common Market Law Review and Food Policy.

C.J.A.M. (Catrien) Termeer is chair of the public administration and policy group at Wageningen University. She used to work at the Erasmus University of Rotterdam; the Technical University of Delft; the Ministry of Agriculture, Nature and Food and at Sioo, a Centre for Organisational Change and Learning. She graduated at the University of Wageningen in Landscape Planning.

F.M.C. (Frank) Vlemminx is a part-time Assistant Professor of Constitutional Law at Tilburg University teaching human rights to bachelor and master students and focusing primarily on the International Covenant on Economic, Social and Cultural rights, the European Convention on Human Rights and the direct applicability of international human rights law. He is engaged in research on the same subject matter. He is also an amateur archaeologist specialising in Middle Palaeolithic and Late Palaeolithic stone artefacts.

J.F. (Jeroen) Warner is Assistant Professor of Disaster Studies at Wageningen University. He also moonlights as a researcher on water policy and as a translator of English. He has published two books and a range of articles and contributions on risk, security and multi-stakeholder participation in water management. His latest book, 'The politics of water', which he co-edited with Kai Wegerich, will be published by Routledge any moment. The present article with Dik Roth is the latest instalment in an enduring research collaboration on the politics of food and floods.

B.F.W. (Bart) Wernaart teaches law and ethics at Fontys University of applied sciences in Eindhoven and 's-Hertogenbosch. He is currently working on his PhD research on the right to adequate food at Wageningen University. He is also a professional musician (drums, mallets and percussion), composer and conductor.

M.G. (Melanie) Wiber is Professor of Anthropology at the University of New Brunswick, Canada. She has published widely in the fields of legal and economic anthropology and is currently engaged in research on new forms of property, including fishing quota and the privatisation of coastal commons.

Chapter 1
Introduction

Otto Hospes, Han van Dijk and Bernd van der Meulen

1. Food security

Despite its simple label, food security is an immensely complex issue. This book addresses the issue in the context of law, politics and the right to food. The chapters, written by members and international partners of the Law and Governance Group of Wageningen University, address different dimensions of food security at different levels. While it is self-evident that laws do not feed people and governments only do so to a limited extent, the organisational arrangements of law and governance regimes of a public or private nature may contribute to the prevailing food security situation in positive or negative ways. To unravel how food security at different levels and in different places is governed by law, politics and the right to food, this book presents and combines different legal, political, sociological and anthropological approaches.

The concept of food security has been defined in different but similar ways. The FAO 'Voluntary Guidelines to support the progressive realisation of the right to adequate food in the context of national food security' apply the following formula:

> 'Food security exists when all people, at all times, have physical and economic access to sufficient, safe and nutritious food to meet their dietary needs and food preferences for an active and healthy life. The four pillars of food security are availability, stability of supply, access and utilization' (FAO, 2004: 5).

The majority of the people in the world have enough to eat, sometimes to such an extent that it is taken for granted. A wide range of factors and actors may have contributed to this state of food sufficiency, like agricultural production, efficient management of natural resources, climate conditions, peace and political stability, and well-designed policies and concerted efforts of governments, companies and civil society. However, a devastating and increasing number of people suffer from food insecurity in one or more aspects. According to the Food and Agriculture Organization of the United Nations, the number of undernourished was about 800 million in 1996 (the year of the World Food Summit) and has since risen to more than 1 billion in 2009 (FAO, 2006, 2009). At the same time, deteriorating diets and fast food have turned obesity into a truly global problem, no longer limited to Western countries but rapidly becoming a problem in China, Brazil and even urban Sub-Saharan Africa. Finally, food safety crises and concerns in different parts of the world show that access to safe food cannot be taken for granted at all.

The contributions to this book address different aspects of the relationship between law and governance and food (in)security. It will be shown that different situations lead to different priorities and different approaches.

2. Adequate food

The major legal expression of food security is the human right to adequate food. The Economic and Social Council of the United Nations defines the right to adequate food in terms that are quite similar to the ones used by FAO to describe food security; except that the Council adopts a human rights language and speaks of 'every man, woman and child' instead of 'all people':

> 'The right to adequate food is realized when every man, woman and child, alone or in community with others, has physical and economic access at all times to adequate food or means for its procurement.'

The parallel between food security and the right to adequate food is particularly relevant as the concept of 'adequate' has been further elaborated along and beyond the lines quoted above on food security. On the one hand the availability at all times is relevant, on the other hand adequacy is understood to mean: sufficient to satisfy dietary needs, free from adverse substances and acceptable in a given culture. It will be seen that, depending on the circumstances, policy emphasis is on one of these aspects.

In September 2000 the United Nations Commission on Human Rights appointed a Special Rapporteur on the right to food. His job is to ensure that governments are meeting their obligations with regard to the right to food of all people.[1] The Special Rapporteur elaborated a doctrine stating that the right to food does not require that governments feed the whole population. It does, however, include the obligation for state authorities not to interfere with the population's means and abilities to feed themselves. In this doctrine four different obligations have been identified: (1) to respect (non-interference), (2) to protect (from the interference by third parties), (3) to promote (support self-realisation) and (4) to fulfil (provide in case of emergency). These four obligations imply that we should not focus on food security as a condition or end situation but on the ways in which state actors directly or indirectly hamper people's ability to produce or acquire adequate food. This includes political and economic interventions that may be not directly related to food or food production itself. Power relations influencing the distribution of land, military insecurity and political oppression stopping people from producing

[1] The Special Rapporteur makes an annual report on his work to the Commission on Human Rights in Geneva in April and, at the request of the Commission, he also makes an annual report to the UN General Assembly in November each year. He also conducts country missions to assess food security conditions in terms of the right to food in different countries and regions of the world. Available at: http://www.righttofood.org/.

food are a case in point. Pollution affecting the safety of food and food production are another case in point. Therefore, in the view of the Special Rapporteur the right to food should be accepted – at the very least as far as its obligations to respect are concerned – as enforceable. This also means that individuals should have access to the (national) courts to defend their right to food in case national authorities unduly restrict it.

The implementation of the right to food in the Member States of the UN is further supported by general comments of the Committee on Economic, Social and Cultural Rights (UN) and by the Voluntary Guidelines to support the progressive realisation of the right to adequate food in the context of national food security (FAO, 2005).

3. Food policy

In most cases the reception of the UN's views on the human right to food by the Member States has so far been lukewarm at best. Nevertheless, its practical complement of 'food security', defined both in a quantitative way as having access to sufficient food as well as in a qualitative way as having access to safe food, has been a central policy concept for decades. The sustainable availability of sufficient food of good quality is an issue that plays a role in different guises all over the world. In the South the *provisioning* of the poor with food of sufficient quantity and quality has been a core objective of development policy at international as well as national level. In the North issues of food *safety* have re-emerged on the policy agenda with the introduction of genetically modified organisms (GMOs), mad cow disease (BSE: Bovine Spongiform Encephalopathy), dioxin chicken, and avian influenza. Moreover, the growing number of obese children and adults is raising concern, not only in the United States, but also in Europe and recently in the Newly Emerging Economies of Latin America and Asia, where growing affluence coincides with a growing incidence of obesity.

With respect to the South there is increasing concern and embarrassment over the growing number of people who are not food secure and do not have the political and economic means to claim a larger share of the 'cake'. With the formulation of the Millennium Development Goals this issue has climbed the policy agenda, though with the current pace of implementation of the proposed policies, it has become highly unlikely that the target of halving the number of hungry people by 2015, especially in Africa, will be met. In addition the quality of food and the possible presence of all kind of pollutants (pesticide residues but also mycotoxins, etc.) are also areas of growing concern.

The contributions to this volume highlight a number of aspects of this crucial policy agenda mainly from the perspective of regulation. Two general beliefs support the importance of regulation: that people have a tendency to organise their lives according to certain rules and that actions of people can be influenced (but not determined) by manipulating those rules or by superimposing other rules through regulation mechanisms. In this way, policy is about regulation, not just rules embodied by written legislation, but also by the various institutions and power networks that govern human behaviour. In our case this is about the production of food, the distribution of the means to produce food and the redistribution of the food that is produced. These domains are linked with regulatory mechanisms embedded in the law and the structures of governance of our societies at the global, international, national and even local level.

Food security is a deeply normative concept. It is regarded as a basic human right by some, something inherently connected with the minimum levels of human dignity, so that no concession or discussion about the content of the concept is acceptable or even possible. But, taking the concept only for its normative dimensions is a risky undertaking from a scientific point of view. While focusing on the normative dimensions of food security we may lose sight of the ways in which people, governments, non-state actors, and international organisations try to ensure food security as well as the structural constraints for improving food production, access to food and income, and social justice. Comparing the current situation with the desired situation is one thing. Analysing why this desired situation will not occur automatically, while the great majority of the world's population and its leaders are endorsing these goals, is quite another. In this perspective it might be wiser to start from its opposite, food *in*security, and begin to analyse where it starts and to identify the deeper lying causes of the problems sketched above.

Those responsible for food security in the world hold a wide variety of opinions and have developed a wide variety of frameworks for tackling the issue of food security. These frameworks are not only congruous with specific scientific approaches to food security, but also act as frameworks for orienting policies. They address what has been labelled in Article 11 of the International Covenant on Social, Economic and Cultural Rights as the 'improvement of methods of production, conservation and distribution of food by making full use of technical and scientific knowledge, by disseminating knowledge of the principles of nutrition and by developing or reforming agrarian systems in such a way as to achieve the most efficient development and utilisation of natural resources'. These imperatives can be analysed from a variety of perspectives: (1) the neo-Malthusian perspective focusing on food production and population growth; (2) neo-liberal economics, focusing on production and distribution and the functioning of markets for production, distribution and the means of production; (3) new institutional economics, focusing on transaction costs and the functioning of institutions in distributing means, production and the distribution of production; and (4) political economy approaches that focus not so much on production as such, but on the political dimensions and the structural constraints the disadvantaged have to face in acquiring the means of production and income to ensure the production and procurement of sufficient food to survive.

Each of these perspectives has a specific approach to food security. The focus of the neo-Malthusian perspective is on quantitative measures of food production in relation to population numbers and takes food self-sufficiency at a specific geographical level as its central concern. Technology is seen as the prime instrument to achieve a better balance between population and production. Neo-liberal economics is also focused on production, but regards market reform as the main instrument to achieve an efficient distribution of food and to ensure food security and safety. Efficient markets are seen as the best means to allocate the means of production, and food itself. Pioneers in the field of new institutional economics are Williamson (1998) and North (2005), conceptualising institutions as layered and consisting of formal and informal rules with different time-horizons. Lastly, political economy approaches look at the issue of food security in terms of access. According to the Nobel Prize laureate and economist Amartya Sen (1984), food insecurity is not a problem of food production and availability but of people's limited ability to mobilise resources, relationships and rules to access food. He also postulated a relationship between the political organisation of society and extreme conditions of food insecurity: 'No famine has ever taken place in the history of the world in a functioning democracy' (Sen, 1999), arguing that a *functioning* democracy offers institutional checks and balances to prevent deteriorating food insecurity from turning into famine.

Each of these approaches has a specific view on regulation and the policies that are required to better regulate food security. Even free market economies need forms of regulation, if only to ensure property rights, so that rights regarding food can be transferred from one person to another. Regulation on the one hand is achieved through all kinds of legal frameworks at various levels. However, these serve only as guidelines with which people and institutions must direct and co-ordinate their actions to achieve their goals. The latter aspect we consider as 'governance'. It is at the intersection of these two that the regulation of food security is 'produced'.

When we take a historical view on policies to promote food security, sovereignty and food safety, three main developments can be distinguished. In the first place food security (and other closely related policy domains like social security and other forms of security) was the almost exclusive domain of the nation-state in the 1960s and 1970s. It was not only in the affluent North that interventionist states emerged. The newly independent colonies in Africa and Asia also drafted and attempted to implement ambitious policies with respect to economic growth, development and the safeguarding of food supply. In a few cases these policies were quite successful but in most countries they were not, especially those in Africa. Clearly, development policies had to be revised.

By the end of the 1970's the first signs became apparent that responsibility for food security was to be delegated to other non-state actors. In the South the international financial institutions pleaded for downsizing of the state after debts mounted considerably. For the next few decades global governance in the form of structural adjustments programmes (SAPs) and market liberalisation imposed by international financial institutions, backed by the major donors was a major determinant of food policy. These measures were meant to reduce the burden of public spending on the economies, and open up markets to provide an impetus for economic growth, so that poverty would be reduced. This led to the roll-back of the state, and the promotion of free market economics. This so-called 'Washington consensus' in macro-economic policies has been mollified somewhat by the realisation that reductions in public spending also have negative consequences for the poor, since they depend on public services and will not be able to escape from poverty and food insecurity unassisted (Gore, 2000).

In an attempt to remedy the consequences of the roll-back of the state, development projects and later on Poverty Reduction Strategies were developed to achieve better co-ordination and planning of poverty reduction. The new architecture of aid was based on these plans and combined with Sector Wide Approaches (SWAPs) for specific economic and service sectors, such as agriculture, health and education, 'sound' macro-economic policies, market liberalisation, and if necessary poverty alleviation measures. However, there is an inherent risk in these policies that donors will focus on sectors where SWAPs work well and therefore will overemphasise social sectors at the expense of growth policies and cross-cutting themes like rural development.

It was also hoped that market reform and the privatisation of all kinds of state enterprises would lead to a buoyant private sector, bringing about fast economic growth automatically resulting in poverty reduction. The results of these policies have been mixed (Bardhan and Mokherjee, 2006). NGOs were considered to be better equipped to do the basic work of poverty alleviation and economic development at the level of the poor populations, because they were considered to be closer to the needs and ambitions of the poor.

For a number of countries this did not work at all. Privatisation led to corruption and local capacity was not there to create economic development through the private sector in many regions. Trade barriers, fluctuations in global market prices, and quality standards of importing countries continued to pose an obstacle to export-led growth. On the contrary, local markets were flooded with cheap rice from Thailand, maize and wheat from the United States, onions from Europe, and frozen beef from Argentina. Nothing was done for the subsistence sector and internal markets to make these more competitive, whereas probably more than 90% of all agricultural produce never gets beyond the border of a district in marginal areas. Though there are also winners in this process of economic globalisation the losers are often extremely vulnerable to changes in relative prices for commodities and wages (Basu, 2006; Nissanke and Thorbecke, 2006).

This delegation of responsibility to the private sector, and in case of emergencies to international and multilateral relief agencies, led in the North to a downsizing of the welfare state because taxpayers refused to keep paying for the enormous cash outlays needed to ensure social security for everybody. Implicitly food security was also made a matter of private concern. Though the state remained active in the domain of ensuring the production of sufficient food, this retreat manifested itself for example in the rise of the rate of malnutrition among low-income groups in the United States. Only when food security is or seems to be touching on matters of national security does the nation state take a renewed interest in the issue.

In Europe there has been a move towards delegating concerns over food security and food safety to the private sector and relying on self-regulation by the food industry. This orientation altered dramatically when the aforementioned food safety crises brought serious shortcomings to light. From that moment on the European Union resumed responsibility for food safety and increased its hold both on industry and the Member States (Van der Meulen and Van der Velde, 2008).

The collective result of all these changes has been that food security and safety have increasingly been taken out of the hands of the people concerned and entrusted more to some private or international institution. In the North regulatory powers are moving upward to the European Union and the FDA (U.S. Food and Drug Administration), bureaucratic organisations who seem almost immune to democratic control. Assessments of food safety are first and foremost based on all kinds of scientific knowledge determining 'acceptable levels' of 'unsafety'. With this move we have evolved into what the German sociologist Beck (1992) has labelled a 'risk society', where risk has become the object of political struggle and reflection about society itself. This scientific determination of Acceptable Daily Intakes (ADI's) also becomes increasingly an instrument to regulate world trade relations, as it has become one of the means to fence off Northern markets from Southern products in the era of free trade. With the state retreating from welfare, the effort to acquire sufficient food has also become a matter of private concern of the Northern poor. Even in a country like the Netherlands, with a social security system, food distribution banks have re-appeared to provide the poor with food (Hospes and Van der Meulen, 2009).

In the South a similar evolution of policy-making with respect to food security and food safety has taken place. Food safety has never been of much concern within developing nations, with the enormous urgency of food production lagging behind population growth in many African and some Asian countries. Often even the most basic control on food safety is lacking, with severe and chronic health problems as a result, manifesting itself in the occurrence of intestinal diseases, intestinal worms and so forth both in chronic forms and as epidemics (Rajaratnam *et al.*, 2010).

Here too the state retreated. With the wave of privatisations in the 1980's within the framework of Structural Adjustment Programmes, the complete infrastructure to regulate the production and distribution of food in the form of marketing boards, strategic stocks, extension services, parastatal development organisations and so on was handed over to the market. This whole process was initiated and controlled from outside, that is, by international financial institutions such as the International Monetary Fund and World Bank, who argued that states were inefficient bureaucracies and that markets would do a better job, providing for economic growth and promoting food production. The motto was getting the prices rights and then the food production and distribution sectors would do the work by themselves.

The results of these policies have been mixed at best. Malnutrition rates in many countries are still very high, especially in Africa. In many countries per capita food production is declining instead of rising (Sanchez, 2002). In addition, weakened states are threatened from the inside by corruption, mismanagement, underfunding and neo-patrimonial power structures and from the outside by contestants for power in the form of rebel movements and organised crime. In a number of countries this has led to major humanitarian crises and a total collapse of state power. In these instances, food security has become the responsibility of the international community in the form of UN organisations and the international aid community. Often these institutions are incapable of doing their job, because of a lack of funding, unsafety or the sheer inability to get the goods delivered under the conditions.

4. Law and governance approaches to food security and food policy

4.1 Law-in-motion approaches

Over recent years legal approaches to food security have gained a lot of attention and momentum. By 1966 the Human Right to Food had already been formulated within the framework of the International Covenant on Economic, Social and Cultural Rights. However, Article 11 concerning the Right to Food, remained dormant until the late 1990s. Under the impetus of Asbjørn Eide, then appointed Special Rapporteur on the Right to Food, a General Comment was adopted in 1999 by the UN Economic and Social Council that specified state obligations with regard to the right to food (see Chapter 2 and 4 in this volume). He and his successors Jean Ziegler and Olivier de Schutter have been developing the right to food and have succeeded in getting the right to food higher on the policy agenda. One of the key Millennium Development Goals adopted by the United Nations in 2000, has been to reduce by half the proportion of people who suffer from hunger (see Chapter 3 and 4 in this volume).

The discussion about the right to food has not been limited to more or less dramatic situations of food insecurity in the South, in which poverty, droughts and political conflict are involved. The right to food is also gaining attention in the North. Food insecurity is a growing problem in more affluent societies as poverty increases due to budget cuts to social benefit payments. In the Netherlands food distribution through food banks has started in recent years. In the United States of America malnutrition among children is on the rise. At the same time governments of these societies are reluctant to consider these problems in terms of the right to food (see Hospes and Van der Meulen, 2009) and to provide access to court (Chapter 5 in this volume).

There has also been an increasing emphasis in the North on the safety and health aspects of food. The enormous and growing number of people suffering from obesity due to over-consumption of fat and other high-calorie foods is a major societal problem in rich countries. Growing health concerns with regard to the multiple additives in processed food and pesticide residues and pollutants present in food have given an enormous boost to food safety regulation at the national and supranational level. Risk has become a major topic in discussions about food safety and the quality of food. Science-based legislation is seen as a major instrument to control risks for food safety, yet 'other legitimate factors' (Chapter 9 in this volume) and non-transparency of technocratic decision-making (Chapter 7 in this volume) complement or challenge the role of science in food safety policy.

As a result of processes of globalisation, the right to food has become a truly global issue connecting the North and the South and calling for a review of human rights law. The general rule of human rights law is that each state has the primary responsibility to respect and ensure human rights within its territory. However, due to globalisation, human rights scholars propose to shift attention from human rights as a 'domestic obligation' of a state to obligations of state *and* non-state actors across national boundaries towards citizens of other countries. Considering the impact of biofuels on food security, for example, Eide (2008) has proposed to 'reverse the order' and focus on 'extra-territorial obligations' as 'problems are often caused by actors having a global reach' (Hospes, 2009).

The legal history of the right to food shows that the making of law is meant to increasingly guarantee access to food for all but is at the same characterised by controversy, resistance and shifting focus (Hospes and Van der Meulen, 2009). Legal scientists face a double challenge here: on the one hand they wish to strengthen the internal consistency of the body of law on the right to food and to elaborate the normative content of the right to food in a systematic way; on the other hand they seek to adjust the logic of rights and obligations to new societal trends and to develop new paradigms for law-making (Chapter 3 and 4 in this volume). The making of law is subject to both centripetal and centrifugal forces. To face this double challenge and these forces, a view on law as static, homogeneous, self-explanatory and a quickly-unpack-and-use-tool for engineering food security, is misleading and not very useful. Law is in motion. Legal scientists both study and contribute to law-in-motion. This is done by seeking to understand, improve and specify law within existing legal orders and legal-scientific parameters, and by proposing new paradigms, principles and ordering mechanisms to address new social, economic, political and technological trends, innovations and events.

A disciplinary field with a long tradition of studying law-in-motion is the anthropology of law. The basic premise of anthropology of law is that law is produced and reproduced by human beings and in human interaction. Legal scientists, judges and politicians are not seen as the sole or sovereign producers of law. As a result, anthropology of law not only pays a lot of attention to the social fields or interactions in which human beings use and give meaning to law but also assumes plurality of laws or plural legal orders. The theoretical-methodological concept of the 'semi-autonomous social field' has greatly contributed to the conceptualisation of law in society and law-in-motion. The concept was launched by Moore (1973) to explain that groups and networks of people are subject to external rules, that is, rules emanating from outside their social field, but also have the ability to twist, squeeze or strategically use these external rules, and more: they have the ability to generate their own rules. Inspired by the now seminal article of Moore, anthropologists started to study the use of law by human beings in plural legal orders. For this purpose new concepts and approaches were coined, like 'forum shopping' (Von Benda-Beckmann, 1981), 'legal pluralism' (Von Benda-Beckmann, 2001; Von Benda-Beckmann *et al.*, 2009) and 'interlegality' (Merry, 2006; Sousa Santos, 2003). The concept of forum shopping has been launched to study how state and non-state actors make selective use of different laws and legal systems ('forum shopping') to win disputes or to gain legitimate access to critical resources, like land, water, food, technology, etc. The concepts of 'legal pluralism' and 'interlegality' are not meant as classificatory devices or to refer to the obvious co-existence of different legal systems or orders. Rather, they are meant to explore the relationships, tensions or symbiosis between these systems and orders manifested in social fields and human interaction.

Using and further developing these concepts, an anthropology of law approach can very well complement legal approaches to food security and food policy by providing insights into how different actors use different laws and legal systems to secure access to food. The legal sources in negotiating access to food and food policy are manifold: they vary from informal to formal rules and procedures; from customs and case law to national legislation and international treaties; from state law to codes of conduct adopted by companies, NGO project law and local law of communities. Similarly, an anthropology of law approach can help to explain the failure of international or national food policy by showing how such a policy is characterised by plural legal orders and caught by a hidden one. By analyzing social fields, processes and interactions, an anthropology of law approach can explain why, for instance, a human rights-based approach to food fails in practice. Recently, this unorthodox yet revealing approach has caught the eye of the International Council on Human Rights (2009).

4.2 Governance approaches

The concept of governance has evolved as one of the most popular concepts in the worlds of social science and policy-making alike. Whilst Van Kersbergen and Van Waarden (2004) expected that the concept of governance would evolve as 'a bridge' between disciplines, Nuijten and others (2004) have shown that the concept has been defined and used by different disciplines in diverging ways. If there is one thing that is clear from the glossary on governance as put together by Diedrichs and Wessels (2006), it is that governance is not a bridging or unifying concept but a garbage-can or container concept hiding different and contrasting assumptions on what governs and what are the possibilities to steer social change.

The conviction, realisation or experience that 'the state' can no longer be expected to provide effective and legitimate solutions to many of the local, national and global problems of today, including those on food security, has been a major boost for state and non-state actors alike to seek new 'governance' arrangements. UN officials, captains of industry, international NGOs and local communities have called for new partnerships, multi-stakeholder consultations, institutional reform and new forms of representation and decision-making (Annan, 2005; Stoker, 1998).

In spite of this openness, quite a few retrogressive and normative approaches have characterised both the launch and analysis of 'new' governance arrangements. Though the concept of governance was to emphasise the importance of different state and non-state actors to jointly and newly 'organise values and voices', for some reason some actors were often considered more important than others. When at the end of the day, it is the government that decides or is supposed to decide on new laws, principles or criteria, the question arises what is jointly developed and new in the governance arrangement. In this connection the phrase from 'government to governance' has been quite misleading, or even a magic spell (see Chapter 12 in this volume).

Something similar can be said about the organisation of private-private forms of governance in which NGOs and companies are supposed to jointly develop new principles and criteria. If the space for or ability of NGOs to access and influence decision-making processes is limited, the question arises to what extent multi-stakeholder consultations are or can be inclusive at all (see Chapter 14 in this volume). This is not to say that NGOs are the 'good ones' and companies the 'bad ones'. The thing is to constantly question what is supposedly 'good' governance (Nuijten *et al.*, 2004). NGOs also defend own, strategic interests (Clifford, 2009). In quite a few cases it is not very clear who or whose interests are actually represented or voiced by NGOs. Houtzager (2010) qualifies this as 'assumed representation'.

Strangely enough, the observation that there is not a single actor that has the ability to offer legitimate and effective solutions for today's problems (Annan, 2005; Stoker, 1998) has invoked easy glorification of new forms of governance – and governance as such – as the key solution to societal problems. The focus on 'getting the governance right' and the quest for 'models of governance' as part of new social engineering has turned attention away from exploring what actually happens and has happened in the name of governance.

To redress this, a combination of theory-driven, yet empirically-focused approaches is needed that enables analysis of different dimensions of governance as processes of organising frames, values and voices. Frames, values and voices are concepts that refer to cognitive, normative and power dimensions of governing. Governance is not something that is solely about frames, values or politics but all of them together: the way in which a policy or institutional arrangement is framed more or less explicitly refers to the use of values and expresses strategic interests and abilities of actors to trade images and manage meaning; governance is not about the promotion of normative monoculture but rather about the management of normative crowdedness in situations where actors have different frames of realities, problems and solutions, and have different interests, abilities and time-horizons to shape things to their own benefit; politics is an integral part of governance but has often been a forgotten or neglected dimension. Even in a more sophisticated conceptualisation of the concept of governance (for instance, Treib *et al.*, 2007), the political dimension and more in particular the concept of power, is underdeveloped. Yet, analysis of the different abilities of actors to mobilise resources, relationships and rights is critical to understand processes of governance as processes of inclusion and exclusion of values, voices and interests.

The first step in undoing the promotion and analysis of governance from normative bias is to treat governance as a discourse (Chapters 7, 12 and 14 in this volume). The promotion of any governance model can be seen as a discursive strategy, whether multi-stakeholder consultation or expert-based food policy, whether market-based or human-rights based food security. To unpack a discourse as a hegemonic, institutionalised way of labelling, talking and thinking, different types of analysis can be used: historical-institutional analysis of food politics that underlie food policies (Chapter 7 in this volume); sociological analysis of trust networks and policy communities that form the social backbone of old or new food policy (Chapter 8 in this volume); legal analysis to identify the legal basis of a dominant discourse but also the legal space for a counter-discourse or alternative mode of governance (Chapter 9 in this volume); and legal-anthropological analysis of different but connected social spaces in which a discourse is cultivated and contested (Chapter 14 in this volume).

A second step and element of a combination of theory-driven, yet empirically-focused approaches to governance of food security, is to emphasise that governance takes places in plural legal orders and requires a thorough understanding of interlegality and institutional complexity. Plural legal orders refer not only to regulation and rights of food security at local, national and international level, or the different, country-specific legal treatment of access, availability, safety and cultural appropriateness as dimensions of 'adequate food' (see and compare the first two parts of this volume). Such plurality also refers to laws and regulations in other fields or domains that affect food security, such as agriculture and environment (Grossman in this volume), water (Chapter 12 in this volume) and feed (Chapter 14 in this volume). The term of 'interlegality' (Sousa Santos, 2002) can be used to emphasise and explore how these laws and regulations 'clash, mingle, hybridise, and interact with one another' at several levels: 'between national legalities and among legalities not necessarily centered on any nation state' (Merry, 2006).

Institutions consist of 'both informal constraints (sanctions, taboos, customs, traditions and codes of conduct) and formal rules (constitutions, laws, property rights)'(North, 1991, 2005). Institutional complexity is not merely about the distinction of formal and informal rules as categories of institutions but more precisely about how these different rules are related to each other and to behavioural patterns or social practices at the level of individuals and organisation. The property of institutions as a mix of rules and practices is expressed in the definition of institutions by Uphoff (1993: 614) as 'complexes of norms and behaviour that persist over time by serving collectively valued purposes'. Institutions thus not only refer to, for instance, the normative content of a human rights covenant or an article of a general food law, but *at the same time* to the 'routinised' or 'cultural' ways in which such a covenant, law or article is interpreted and used in a specific social field or political arena. Governance is about organising institutional complexity.

A third step and approach is to theorise the concept of power with a view to analysing the ability of actors to govern and to exert power over others. Wolf (1999) distinguishes four modes of power: personal power, distributive power, organisational or tactical power, and structural power (Visser, 2009). The first mode refers to the capabilities and skills of an individual that can be improved and accumulated and do not necessary negatively influence another person. The second mode approaches power as a zero-sum game, which means that an increase in power of one person leads to a decrease of power of another person. The third mode of tactical power refers to someone's ability to influence the setting, the agenda and arena, in which power struggles take place by creating legal, institutional and organisational barriers to make voices heard or unheard. The fourth mode of power is about power that organises and orchestrates the setting itself. Structural power is not the property of a person but operates outside the direct control of any actor as part of networks of people. Wolf's treatment of culture as an ingredient of structural power resembles the notion of institutions as 'cultural regimes' and 'machines for thinking' (Douglas, 1987).

Different actor-structure approaches to power can be used to conceptually deepen and integrate these different modes of power and to further develop the notion of 'ability'. Ribot and Peluso (2003: 154) propose ''access analysis' to understand why some people or institutions benefit from resources, whether or not they have rights to them'. They define access as a 'bundle of powers' – embodied in and exercised through various mechanisms, processes and social relations – that affect people's ability to benefit from rights and resources. Sen (1984: 497) has developed the concept of 'entitlements' to refer to 'the set of alternative commodity bundles that a person can command in a society using the totality of rights and opportunities that he or she faces'. The concepts of 'access' and 'entitlement' can help to understand why and how different actors succeed or fail to feed themselves, not only in situations of scarcity, drought or political instability but also in situations of abundance, favourable climate conditions and political stability.

The concepts of disciplinary and governmental power of Foucault can be seen as advanced combinations of Wolfe's notions of tactical and structural power. Foucault's conceptualisation of power is based on an analysis of the development of modern states. He argues that the development of modern states is characterised by the transformation of sovereign power into disciplinary and governmental power. Sovereign power emphasises the repressive character of power, the imposition of law and the legitimacy of the central body's sovereignty. Disciplinary power is the pressure of a web of authorities that is watching you (hierarchical observation of a central authority) and is making you behave in a 'normal' way (through participation in armies, schools, churches, voluntary organisations, etc.). Governmental power is a kind of aggregate form of disciplinary power. Governmental power, or governmentality, is the ensemble of institutions and knowledge that allow the execution of power over a population, a political system, an economy.

The exertion of governmental power implies the use of 'political technologies' that make individuals believe that it is their own idea or choice to conform to laws, values or norms of the state or society, or even that they have constructed these laws, values or norms themselves. To qualify these self-governing capacities, Foucault speaks of 'technologies of the self'. These technologies refer to techniques through which individuals transform and construct their thoughts, conducts and beings, but at the same these techniques are integrated into structures of coercion and domination (Foucault (1993) in Lemke, 2002: 4).

Governmentality refers to the disciplining and ordering power of discourse, institutions and knowledge. The concept is not based on the assumption that power struggles are a zero-sum game. Though governmentality is about understanding ordering processes and the sophisticated use of power, it can very well be used to unravel policy processes (like those geared towards food security but also other ones that are of functional significance for food security) as processes of inclusion and exclusion. And this is important for understanding why food insecurity prevails and proliferates *and* for seeking ways to confront, if not embarrass, different state and non-state actors with the increasing number of people suffering from food insecurity and the mechanisms that underlie these processes.

5. This book

The central perspective adopted in this book is that a wide diversity of laws, policy-making processes and rights at different levels and in different domains enables and constrains food security of individuals, communities and countries in different ways. These laws, policy-making processes and rights offer both the scene and tools for power struggles of state and non-state actors on food and natural resources. As a consequence of the plurality of legal orders and multi-level politics of food and natural resources, food security cannot be treated at one level, in one time or in one domain only. To understand local food security, we not only have to study local laws but also the use and importance of global laws in national or local resource struggles and the transboundary 'side-effects' or global impact of national legislation. Food security is a 'glocal' problem. Though commonly defined as a 'situation', food security is not static and given, which means that we have to study food security through time and see how individuals, communities and countries re-address their situation in response to health hazards, ecological change, political turmoil, institutional voids and (other) material and political struggles. Finally, we cannot understand or promote food security without understanding natural resource complexes and people's struggles to secure or change access to natural resources, whether in the form of material, virtual or intellectual property.

To understand how laws, politics and rights govern food security, we need to develop and combine theories, concepts and methodologies from legal, political, anthropological and sociological sciences. To identify practical or strategic ways to (indirectly) contribute to food security, insights from these different approaches could be of great value, in different ways. They vary from improving justiciability to building trust, from seeking ways to address non-scientific concerns to creating room for plurality of lifestyles and norms, from unmasking dominant discourse to understanding or strengthening abilities or arrangements to cope with vulnerability.

This book consists of three parts. The first part offers a critical and constructive assessment of the contribution of human rights law to food security. The second part is about the symbiotic and tense relationships between law, politics and science in securing food safety in the Netherlands and the EU. The third part is about transnational law, resource complexes and food security. This part addresses the challenges and controversies of the making of global or national laws (in the field of food production, property or sustainability) for food security elsewhere.

5.1 Developing human rights law for food security

The contributors to the first part are Wernaart (Chapter 2), Van der Meulen (Chapter 3), Eide (Chapter 4), Vlemminx (Chapter 5) and Hadiprayitno (Chapter 6). They present the many different sources of the right to food and explain how this human right has been caught up in dead-end reasoning. They also propose challenging paradigms and nuanced perspectives to modernise human rights law for better use in a globalising world.

Wernaart presents an overview of the institutions engaged at UN level with the right to food and the many documents they have produced. His argument is that the right to food has evolved and is still evolving from many sources or 'plural wells'. He expresses the fear that proliferation may hamper the achievement of consensus on the meaning of the right to food and therewith also its acceptance in legal practice.

Van der Meulen provides a synthesis of legal history and distinguishes three paradigms to classify different political and scholarly approaches to human rights doctrine. The first he labels as 'the paradigm of generations'. In this approach to human rights with its roots in the Cold War, a sharp distinction is being made between two generations of human rights: the first generation are civil and political rights and the second generation are social, economic and cultural rights. The first generation human rights have been understood to create negative or passive duties for states, whereas the second generation rights have been considered as positive or active duties. The former are seen as binding law enforceable in the courts; the latter as policy guidelines leaving a margin of discretion to the states that is untouchable for the courts. Vlemminx shows how application of the paradigm of generations in Dutch practice renders the right to food virtually meaningless.

Van der Meulen refers to the second paradigm as 'the holistic paradigm'. This approach rejects the subdivision of human rights into two different types. Instead, its forerunners – like Eide – emphasise that it is not the rights that differ, but that different types of state obligations are connected to all human rights: the obligation to respect, the obligation to protect and the obligation to fulfil. Eide deepens his development of his theory of state obligations under human rights law and proposes a nuanced perspective on the third obligation. The holistic paradigm helps to clear up many prevailing misunderstandings with regard to the right to food and to human rights in general. This paradigm could also help the Netherlands in turning its poor legal cuisine (Vlemminx) into haute cuisine on the right to food.

Hadiprayitno analyzes the legal cuisine of Indonesia with regard to food security. She reveals that there is an abundance of legal provisions dealing with food security and food availability in this country. At the same time, she observes that hunger is not a rare incident in her case study of Papua, Indonesia. The contributor offers two conditions or mechanisms to explain this paradox. First, in spite of the abundance of legal provisions, there are very few possibilities for malnourished people to acquire access to justice and to treat hunger as a violation of the human right to food. Second, the ruling government is eager to express its sympathy with food insecure people but rather hesitant to consider hunger as more than the result of disappointing harvests.

Finally, Van der Meulen speculates that a third paradigm may emerge which he calls 'real'. Some aspects in the development of human rights doctrine evade classification in either of the first two mentioned paradigms. These regard relationships differently from those between people and their states and include horizontal relationships between people and external responsibilities of states.

5.2 Law, science and politics in securing food safety

The contributors of the second part are Goverde (Chapter 7), Breeman and Termeer (Chapter 8) and Szajkowska (Chapter 9). Their common observation is that food safety has become the major dimension of food security or simply the dominant, overarching concept for food issues in the Netherlands and the EU. Their analysis starts with the observation that science has been increasingly awarded a central role in food safety law and regulation of food chains in the Netherlands and the EU. Having noted this, the contributors articulate the limits of science as a basis for political decision-making and speaking truth on food safety and the importance of non-science factors in risk management, food politics and governance. Every author does so in a unique way.

Goverde provides a historical and institutional explanation of the dominant role of science in Dutch and EU food safety policy. In his view this role can first of all be explained by a deeply rooted and shared belief of public and private scientific networks in the objectivity and political neutrality of scientific knowledge. The second explanation is that these networks operate as a closed community that enjoy the equally non-transparent support of technocratic civil servants, who together decide on food as a public concern without involving the general public. Using insights from political sciences and ethics, Goverde provides an alternative to the hegemonic modernist approach in Dutch food politics and the EU General Food Law. With a view to enhancing democratic food politics, he calls for more transparency concerning the intermediary role of science and more room for discussing alternative food -and life-styles.

Breeman and Termeer characterise the post-Second World War history of agricultural policies in Europe in terms of a paradigm shift from food security to food quality. In the reconstruction period, most European agricultural policies aimed to secure food supply. In that period, farmers and food production chains were organised with a view to bulk production. This meant that predominantly actors who were able to produce bulk food were included in the food policy community. However, since the 1980's the aims of food production chains have changed in Western Europe. Chain management has evolved as a dominant institutional tool to safeguard food quality rather than to secure food supply. Based on public policy theory, Breeman and Termeer explain that rules-based governance and science-based food standards will damage interpersonal trust within policy sub-communities or networks. The contributors fear that actors who used to be essential for securing our food supply will be excluded from the policy community or production network. In addition, they present different adverse effects of chain co-ordination, like organisational blindness, selective activation and exclusion. They expect that these effects may lead to loss in production and innovation capacity.

Szajkowska shows that the EU General Food Law is very much based on a technocratic approach to food safety and risk assessment but reflects a more democratic approach, based on deliberative-constitutive theory, when it comes to risk management. The GFL states that 'other legitimate factors' (such as cultural, economic, social or ethical issues) come into play during the risk management stage. On the basis of a review of court cases, the contributor concludes that the lack of indication in the GFL of how 'other legitimate factors' should be incorporated in the decision-making process and to what extent measures based on these factors can deviate from the conclusions of scientific risk management, poses a major challenge to food safety governance at the EU level. On the one hand, this provides an opportunity to include socio-economic considerations and to promote diversity of the internal market. On the other hand, it is necessary to ensure the credibility of science-based risk governance of food safety and to clearly define how science and a wide set of other legitimate factors interact in policy-making in this field.

5.3 Transnational law, resource complexes and food security

The contributors to the fourth part are Grossman (Chapter 10), Wiber (Chapter 11), Roth and Warner (Chapter 12), Kohne (Chapter 13) and Hospes (Chapter 14). They explore the complex relationships between transnational law on the one hand and global, national or local food security on the other. Transnational law is all the law – local, national or international – that performs or has influence across state lines. The (growing) transnational flows of food and ingredients for food production (like water and soy) have turned nearly all food, agriculture and property law into transnational law. A major implication is that access and availability of food cannot be analyzed apart from multi-level governance of natural resources (like land, water, forests) and global agricultural commodities (like soy, palm oil, etc.). The contributors explain what legal challenges and political controversies have arisen as a result.

Against the background of increasing global food demand and widespread food insecurity at the national level in some 70 countries in the next decade, Grossman argues that the legal protection of land for agricultural production in the US is critical for global food security. The agricultural sector in the US not only ensures food security for the US but, through exports and food aid, also contributes significantly to the world food supply. The supply of arable land in the US, however, is under pressure from competing uses like urbanisation and infrastructure. The use for food production is under pressure from demand for alternative use production, like feed and biofuels. Grossman provides insights into the programmes at federal and state level that aim to protect farmland from irreversible removal from agricultural production and to conserve agricultural land quality.

The contribution of Wiber is about the jungle of international organisations and bodies of law directed at patenting staple crops and its implications for the food security of millions of the world's poor. The author argues that no other debate at the global level more clearly demonstrates the agricultural and trade realities that divide the North from the South. At the same time she shows that no other debate is more confusing: the debate is characterised by misleading 'moral signposts', poor conceptualisation of the notion of 'property' and the inability of science to produce clear and uncontested insights into the (lack of) utility of GM foods in fighting world hunger. The author argues that the proliferation of international organisations in the field of intellectual property rights offers space for political manoeuvring of interest groups and room for powerful trading nations and corporate actors to define the rules of the game. Farmers in the North and the South are rarely involved in multi-lateral or bilateral trade negotiations. Small farmers, particularly in the South, have little chance to generate income from intellectual property protection without major restructuring of land and resource ownership. The author fears that the global legal battle on intellectual property of genetic plant varieties will generate devastating outcomes in food security at grass roots level.

The angle that is used by Roth and Warner to discuss relationships between transnational law and food security is the policy concept of virtual water: the water needed for crop production. They explain that this concept has been embraced by policy-makers, water managers and economists to increase efficiency of water use. The basic idea is to adapt global food trade to existing variability in water scarcity: concentrating production of water-intensive food crops and products with low water productivity in water-abundant countries while turning water-scarce countries into food importers and producers of less water-intensive crops. Roth and Warner observe that the concept of virtual water has moved the global debate on water into calm waters. At the same time, they believe that the neat figures on virtual water are misleading: the accounts and accountants shift our attention away from political decision-making in the (re-)allocation of resources and from everyday but complex governance of land and water resources. The contributors fear that the food security and livelihood of many rural and urban people will be threatened if and when these issues are ignored.

Köhne describes the brokerage role of a community-based indigenous territorial organisation for food security in a Bolivian indigenous territory. He explains that the importance of this organisation for food security is due to its position in the centre of a web of power relations connecting villagers, state institutions and international NGOs. The latter two types of actors need villagers and their labour power to make development projects work, whereas villagers need the power of state institutions and international NGOs to provide access to land, water, knowledge and investments.

The contribution of Hospes is about transnational law on 'responsible soy' and its importance for feed security, described as 'food security of the rich'. At international meetings of the Round Table on Responsible Soy (RTRS) in Latin America and Europe, soy producers, agro-industry, banks and NGOs have been negotiating and formulating global principles and criteria for responsible soy. The contributor explains that this transnational law is meant to minimise the negative effects of soy expansion (like de-forestation, loss of bio-diversity and human rights violations) in Brazil and other soy-producing countries but is also contested and used for other purposes. To illustrate this, Hospes provides an analysis of transnational law on responsible soy as part and parcel of social and discursive struggles between NGOs and companies at different places and parts of global soy chains in Brazil and the Netherlands. In these struggles, the power to speak the truth is often more decisive than 'facts'. Hospes shows that multinational companies have been very eager and effective in trading images to secure their trade of goods and herewith the supply of soy as a major ingredient for the animal feed and meat industry in Europe.

References

Annan, K., 2005. In larger freedom: towards security, development and human rights for all. Report of the Secretary-General of the United Nations for decision by Heads of State and Government.

Bardhan, P.K. and Mokherjee D. (eds.), 2006. Decentralisation and Local Governance in Developing Countries. MIT Press, Cambridge, MA, USA.

Basu, K., 2006. Globalization, Poverty and Inequality: What is the relationship? What can be done? World Development 34: 1361-1373.

Beck, U. 1992. Risk Society: Towards a New Modernity. Sage, New Delhi, India.

Clifford, B. (ed.), 2010. The International Struggle for New Human Rights. University of Pennsylvania Press, Philadelphia, PA, USA.

Diedrich, U. and Wessels, W., 2006. Newgov: New Modes of Governance. Next stage of cluster glossary of governance. Integrated project: citizens and governance in the knowledge-based society, co-funded by the European Commission with the 6th Framework Programme (2002-2006).

Douglas, M., 1987. How institutions think. Routledge and Kegan, London, UK.

Eide, A., 2008. The Right to Food and the Impact of Liquid Biofuels (Agrofuels), FAO Right to Food Studies, FAO, Rome, Italy.

FAO (Food and Agriculture Organization of the United Nations), 2004. Voluntary guidelines to support the progressive realization of the right to food in the context of national food security. FAO, Rome, Italy.

FAO (Food and Agriculture Organization of the United Nations), 2005. Voluntary guidelines to support the progressive realisation of the right to adequate food in the context of national food security. FAO, Rome, Italy.

FAO (Food and Agriculture Organization of the United Nations), 2006, The state of food security in the world 2006. FAO, Rome, Italy.

FAO (Food and Agriculture Organization of the United Nations), 2009. Undernourishment in 2009 by region. Available at: http://www.fao.org/hunger/hunger-graphics/en/. Accessed June 21st of 2010.

Hospes, O. and Van der Meulen, B.M.J., 2009. Fed up with the right to food: the Netherlands' policies and practices regarding the human right to adequate food. Wageningen Academic Publishers, Wageningen, the Netherlands.

Hospes, O., 2009. Regulating biofuels in the name of sustainability or the right to food? In: Hospes, O. and Van der Meulen, B.M.J. (eds.). Fed up with the right to food: the Netherlands' policies and practices regarding the human right to adequate food. Wageningen Academic Publishers, Wageningen, the Netherlands, pp. 121-134.

Houtzager, P.P. and Lavalle, A.G., 2010. Civil Society's Claims to Political Representation'. Brazil Studies Comparative International Development 45: 1-29.

International Council on Human Rights, 2009. When legal worlds overlap: human rights, state and non-state law. Versoix, Switzerland.

Lemke, T., 2002. Foucault, Governmentality, and Critique. Rethinking Marxism 14, 3: 1-17.

Merry, S.E., 2006. Anthropology and international law. Annual Review of Anthropology 35: 99-116.

Moore, S.F., 1973. Law and social change: the semi-autonomous social field as an appropriate subject of study. Law and Society Review 7: 719-746.

Nissanke, M. and Thornbecke, E. (eds.), 2008. Globalization and the Poor in Asia: Can Shared Growth be Sustained? Palgrave Macmillan, Hampshire, UK.

North, D.C., 1991. Institutions. Journal of Economic Perspectives 5: 97-112.

North, D.C., 2005. Understanding the Process of Institutional Change. Princeton University Press, Princeton, NJ, USA.

Nuijten, M., Anders, G., Van Gastel, J., Van der Haar, G., Van Nijnatten, C. and Warner, J., 2004. 'Governance in action: some theoretical and practical reflections on a key concept'. In: Kalb, D., Pantsers, W and Siebers, H. (eds.), Globalization and development: themes and concepts in current research. Kluwer Academic, Dordrecht, the Netherlands.

Rajaratnam, J.K. Rajaratnam, J.K., Marcus, J.R., Flaxman, A.D., Wang, H., Levin-Rector, A., Dwyer, L., Costa, M., Lopez, A.D. and Murray C.J., 2010. Neonatal, postneonatal, childhood, and under-5 mortality for 187 countries, 1970-2010: a systematic analysis of progress towards Millennium Development Goal 4. The Lancet 375: 1988-2008.

Ribot, J.C. and Peluso, N.L., 2003. A theory of access. Rural Sociology 68: 153-181.

Sanchez, P.A., 2002. Ecology: Soil Fertility and Hunger in Africa. Science 295: 2019-2020.

Sen, A., 1984. Resources, Values and Development. Basil Blackwell, Oxford, UK.

Sen, A., 1999. Democracy as freedom. Oxford University Press, London, UK.

Sousa Santos, B., 2002. Toward a new legal common sense. Northwestern University Press, Evansten, IL, USA.

Stoker, G., 1998. Governance as theory: five propositions. International Journal of Social Sciences 50: 17-28.

Treib, O., Bähr, H. and Falkner, G. 2007. Modes of governance: towards a conceptual clarification. Journal of European Public Policy 14: 1-20.

Uphoff, N., 1993. Grassroots organizations and NGOs in rural development: opportunities with diminishing states and expanding markets. World Development 4: 607-622.

Van Kersbergen, K. and Van Waarden, F., 2004. Governance' as a bridge between disciplines: Cross-disciplinary inspiration regarding shifts in governance and problems of governability, accountability and legitimacy. European Journal of Political Research 43: 143-171.

Van der Meulen, B.M.J. and Van der Velde, M., 2008. European food law handbook. Wageningen Academic Publishers, Wageningen, the Netherlands.

Visser, M., 2009. Totalising Power of Empowerment – Liberating Strength of Power: Analysis of Chambers' and Putnam's conceptualisations of power and empowerment in the context of participation, based on Foucault's Insights on Power. Master thesis, Law and Governance Group of Wageningen University, Wageningen, the Netherlands.

Von Benda-Beckmann, K., 1981. Forum Shopping and Shopping Forums: Dispute Processing in a Minangkabau Village in West Sumatra. Journal of Legal Pluralism 19: 117-159.

Von Benda-Beckmann, F., 2001. Legal pluralism and social justice in economic and political development, IDS bulletin 32: 46-56.

Von Benda-Beckmann, F., Von Benda-Beckmann, K. and Eckert, J. (eds.), 2009. Rules of law and laws of ruling: on the governance of law. Ashgate Publishing, Surrey, UK.

Williamson, O.E., 1998. Transaction Cost Economics: How it works, where it is headed. The Economist 146: 23-58.

Wolf, E., 1999. Envisioning Power: Ideologies of Dominance and Crisis. University of California Press, Berkeley and Los Angeles, USA.

Part 1

Developing human rights law for food security

Chapter 2
The plural wells of the right to food

Bart Wernaart

1. Introduction

The right to adequate food is by no means a tranquil possession with one single undisputed content. It has been addressed uncountable times with the utmost of urgency. Being one of the prerequisites for a humane life with at least a right to a minimum of existence, this right is embedded in the international human rights system that evolved in the period after the Second World War. The right to food is interrelated with complex matters that are not easily solved. Therefore, it appeared to be difficult to give meaning to the content of the right to food, and to help Member States of the United Nations to implement this right in a suitable way in their national legal systems. It is no surprise then, that a large web of international institutions, each functioning within their own competences and mandates, are involved in the process of further developing the right to food. This is done in discussions between stakeholders during summits and gathering. It is done by negotiating and finally adopting legislative documents that stimulate the development of this right one step further. It is done by desk and field research to learn about the nature of this human right, about how this right is (or is not) realised in distressing times, and about what could be done to further clarify and develop this right. It is also done by (financial) contributions, all kinds of aid and education. Finally, it is done by measuring progress that has been made, by trying to make actors responsible for the realization of the right to food, and even accountable for when they fail to meet their obligations. The right to food streams throughout the whole international human rights system, from an ever increasing number of wells. This chapter aims to provide an inventory of those wells of the right to food.

Therefore, Section 2 will give a general impression on how the concept of the right to food has been developed throughout the years. Section 3 will demonstrate how the right to food appears in a broader sense in international documents. Section 4 and 5 will present the reader with an overview of international documents that refer to the right to food and international institutions that have helped develop this right. Finally, some concluding remarks will be made in Section 6.

2. Development of the concept of 'right to food' over time

This part sketches the general developments regarding the right to food since World War II. The institutions that were (and are) actively involved in the development of the right to food are described and placed in their functional context. The section intends to provide structure to the complex system in which the right to food has developed so far, and is therefore by no means intended to be an exhausting overview.

After the Second World War, the United Nations was founded with the main goal of maintaining peace and security. The UN charter may be seen as its constitution, establishing the primary organs of the UN – The General Assembly, the Security Council and the Economic and Social Council – and providing for procedures that enabled these primary organs to install secondary organs. The Commission on Human Rights was created at the first meeting of the Economic and Social Council, mainly as an organ of advice to help the ECOSOC fulfil its task.[2] It spent its first three years of existence writing the Universal Declaration of Human Rights (UDHR)which was adopted by the general Assembly on the 10th of December in 1948.[3] Article 25 in the UDHR mentions the human right to an adequate living standard, including adequate food. Originally, the intention was to create a worldwide human rights system, consisting of three parts: a declaration, a binding treaty and a monitoring/ accountability mechanism. Eventually, due to different opinions between Eastern and Western countries on the one hand and developing and developed countries on the other, the choice was made to draft two covenants instead of one. One covenant contained civil and political rights (ICCPR), and the other economic, social and cultural rights (ICESCR).[4] Article 11 of the International Covenant on Civil and Political Rights (ICCPR) mentions the human right to an adequate standard of living, including food. The third step of the international Human Rights system was integrated in these two covenants: the State Parties of the ICESCR have the obligation to submit reports on the measures which they have adopted and the progress made in achieving the observance of the ICESCR rights.[5] The Committee on Economic, Social and Cultural Rights was established in 1985,[6] in order to monitor the implementation of the ICESCR. This UN body receives and considers the submitted country reports. To stimulate the further implementation of the ICESCR in the national legal systems of its Members, the Commission regularly writes General Comments to the treaty, clarifying in more detail the content of its provisions and advising on how to implement the Covenant. In 2008, the General Assembly adopted an optional protocol to the ICESCR,[7] containing a procedure for individual communications and inter-state

[2] Economic and Social Council Resolution 5 (I), 16 February 1946. The decision was based on Article 68 UN Charter: 'The Economic and Social Council shall set up commissions in economic and social fields and for the promotion of human rights, and such other commissions as may be required for the performance of its functions.'

[3] General Assembly Resolution 217 A (III), 10 December 1948.

[4] International Covenant on Economic, Social and Cultural Rights, adopted by General Assembly Resolution 2200A (XXI) of 16 December 1966.

[5] International Covenant on Economic, Social and Cultural Rights, adopted by General Assembly Resolution 2200A (XXI) of 16 December 1966, part IV.

[6] Economic and Social Council Resolution 1985/17, 28 May 1985.

[7] A/RES/63/117, 10 December 2008, General Assembly Resolution.

communications. The committee on economic, social and cultural rights is the competent body to receive those communications. At the time of writing, the protocol had 30 signatories.

In 1974, The Declaration on the Eradication of Hunger and Malnutrition was written as a result of the World Food Conference. Increasing food production and more adequately sharing resources were considered to be key elements in order to combat hunger, a common responsibility of the international community.

Due to the failure of reaching the goals of the World Food Conference, the FAO organised the World Food Summit in 1996, bringing together close to 10,000 representatives at the highest level from 185 countries. Without doubt, the most important ambition of the World Food Summit was to reduce the number of undernourished people to half their level no later than 2015.[8] An ambition that was later reaffirmed in the Millennium Declaration, adopted by the UN General Assembly in 2000.[9]

Objective 7.4. of the World Food Summit intends '[to] clarify the content of the right to adequate food and the fundamental right of everyone to be free from hunger, ... and to give particular attention to implementation and full and progressive realisation of this right as a means of achieving food security for all.' In response, the Committee on Economic, Social and Cultural Rights wrote their General Comment 12 on the right to food. In this General Comment, the Committee defines the term 'adequate food' more precisely, and points out the different types of obligations for Member States resulting from the right to food: the duty to respect, protect and fulfil.[10] In addition, the High Commissioner for Human Rights convened expert consultations on the right to food.[11]

Once again at the invitation of FAO, a World Food Summit was held in 2001. The Summit was originally not foreseen, but had 'been prompted by the concern that the target set in the 1996 Rome Declaration may not be achieved.'[12] The summit resulted in another declaration: the declaration of the World Food Summit, five years later. Paragraph 10 of the declaration mandated FAO to 'establish an International Working Group...to draw a set of voluntary guidelines to support the Member States' efforts to achieve the progressive realisation of the right to adequate food in than the context of national food security,' in order to strengthen the respect of all human rights and fundamental freedoms.

---

[8] Rome Declaration on World Food Security, 13 November 1996, paragraph 2.
[9] A/55/L.2, 18 September 2000, General Assembly Resolution, paragraph 19.
[10] E/C.12/1999/5, 12 May 1999, Committee on Economic, Social and Cultural Rights, General Comment 12.
[11] E/CN.4/1998/21, 15 January 1998; E/CN.4/1999/45, 20 January 1999; E/CN.4/2001/148, 30 March 2001.
[12] H.E. Carlo Azeglio Ciampi, President of the Italian Republic, inaugural ceremony, 10 June 2002.

During its 123rd session, FAO installed the Intergovernmental Working Group (IGWG) as requested. In the period March 2003-September 2004, the IGWG held four sessions and an intersessional meeting, resulting in the final version of 'the voluntary guidelines to support the progressive realisation of the right to adequate food in the context of national food security.' The guidelines give practical guidance on 19 topics, based upon three underlying dimensions: adequacy, availability and accessibility.[13]

A third World Food Summit was held in November 2009: the World Summit on Food Security.[14] In the final declaration, the goal that was set during the first World Food Summit to halve the number of people who suffer from hunger or malnutrition by 2015 was reaffirmed,[15] and the declaration contained commitments and actions that would lead to food security. A worrying decline in interest seemed to emerge. The participating representatives did not meet the number and high level of representation that compared with the first two Summits. Their declaration does little more than to generally reaffirm existing targets and commitments.

Meanwhile, the Commission on Human Rights had mandated several working groups to investigate human rights situations. One of these mandates is the post of 'Special Rapporteur on the right to food,' held by Mr. Jean Ziegler in the period 2000-2008,[16] and since 2008 by Mr. Olivier de Schutter.[17] The Sub-Commission on the Promotion and Protection of Human Rights[18] also installed mandates on the right to food and the right to water. Mr. Asbjørn Eide[19] has functioned as Special Rapporteur on the right to food, and Mr. El Hadji Guissé as Special Rapporteur on the right to drinking water.[20] In their reports and activities, the Special Rapporteurs contributed to the development and realisation of the right to food.

The Commission on Human Rights was replaced by the Human Rights Council in 2006,[21] and the Sub-Commission on the Promotion and Protection of Human

13 E/C.12/1999/5, 12 May 1999, Committee on Economic, Social and Cultural Rights, General Comment 12, paragraph 8.
14 FAO Doc. WSFS 2009/2, Rome, 16-18 November 2009, Declaration of the World Food Summit on Food Security.
15 FAO Doc. WSFS 2009/2, Rome, 16-18 November 2009, Declaration of the World Food Summit on Food Security, objective 7.1.
16 Commission on Human Rights Resolution 2000/10, 17 April 2000, paragraph 10.
17 Mr. Olivier de Schutter was appointed Special Rapporteur on the right to food on March 26, 2008, by the Human Rights Council.
18 Before also named: Sub-Commission on the Prevention on Discrimination and Protection of Minorities.
19 E/CN.4/Sub.2/1999/12, 28 June 1999, Asbjørn Eide, updated study on the right to food, in accordance with Sub-Commission decision 1998/106.
20 Mr. El Hadji Guissé, E/CN.4/Sub.2/1998/7, 10 June 1998; E/CN.4/Sub2/2004/20, 14 July 2004; E/CN.4/Sub.2/2005/25, 11 July 2005.
21 A/res/60/251, 3 April 2006, paragraph 6, General Assembly Resolution.

Rights was replaced by the Advisory Council in 2007.[22] All previous resolutions adopted by the UN organs concerning the right to food were reaffirmed,[23] including the mandate of the Special Rapporteur on the right to food.[24]

To conclude, it appears that three pillars can very roughly be distinguished within the UN context (Figure 1) that contribute to the further development of the right to food since World War II. The first pillar seems to be the general human rights treaties and the ICESCR-related committee on economic, social and cultural rights. The second pillar that could be distinguished is the work done within the FAO context: the adoption of the World Food Summit declarations and the work that resulted from these summits, such as the adoption of the voluntary guidelines. The third pillar would be the work done by the Special Rapporteurs. Needless to say, the delineation between these three pillars is somewhat arbitrary as the proposed pillars are closely interrelated, but distinguishing them may provide a better understanding of the way the right to food has been developed so far.

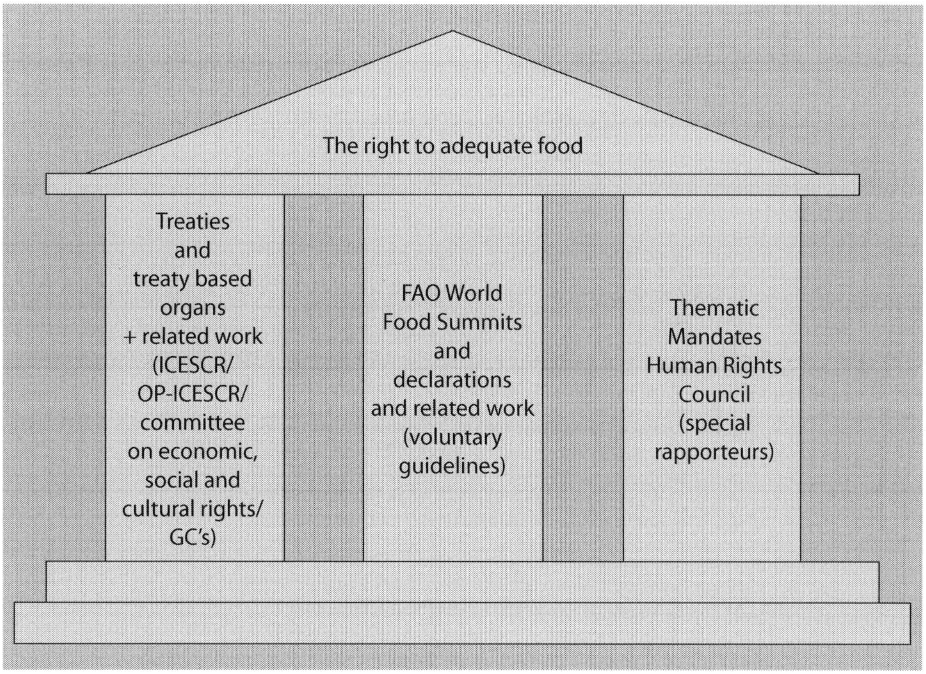

Figure 1. The right to food and its three pillars.

[22] Human Rights Council Resolution 5/1, 18 June 2007 and 6/102, 27 September 2007.
[23] Human Rights Council Resolution 7/14, 27 March 2008, the right to food.
[24] Human Rights Council Resolution 6/2, 27 September 2007, paragraph 2, mandate of the Special Rapporteur on the right to food. For a previous description of the mandate as written by the Human Rights Council, see Resolutions 2000/10, 17 April 2000, paragraph 8-12, and 2001/25, 20 April 2001, paragraph 11.

3. The right to food in a broader perspective

The right to food is mostly discussed in the context of article 11 ICESCR.[25] This does not mean however that the right to food is not mentioned elsewhere. A survey of international documents with regard to human rights or human rights related issues shows that the right to food is also mentioned as an independent right in documents that aim at the protection of a particular group of individuals (Section 3.1) and is explicitly or implicitly mentioned in the context of other human rights or human rights related issues (Section 3.2). This section shows that despite the fact that the content of the right to food was mainly discussed in the light of 11 ICESCR, the right to food has a far broader scope than this particular article.

3.1 The right to food mentioned as an independent right

When looking for provisions that recognise the right to food, it seems a logical step to start searching for provisions in legal documents that literally contain the phrase 'right to food'.

As demonstrated above, Article 25 of the Universal Declaration of Human Rights[26] and Article 11 of the International Covenant on Economic, Social and Cultural Rights[27] are well-known examples in which the right to food for every individual is recognised. The content of Article 11 ICESCR in particular, is explained in more detail by soft law, such as General Comment 12[28] and the voluntary guidelines.[29]

Some international provisions mention the right to food for a specific group of persons that need additional protection. For instance, Article 27 of the International Covenant for the Rights of the Child,[30] Article 12(2) of the Convention on the Elimination of All Forms of Discrimination against Women,[31] and Article 28 of the Convention on Persons with Disabilities[32] mention the right to food for respectively children, pregnant women and disabled persons. In accordance with international practice, the treaties just mentioned were preceded by declarations covering the same subject, but lacking the status of a juridical binding document.

---

[25] See Chapter 3 by Van der Meulen (in this volume).

[26] General Assembly Resolution 217 A (III), 10 December 1948, Article 25.

[27] International Covenant on Economic, Social and Cultural Rights, adopted by General Assembly Resolution 2200A (XXI) of 16 December 1966, Article 11.

[28] E/C.12/1999/5, 12 May 1999, Committee on Economic, Social and Cultural Rights, General Comment 12.

[29] The voluntary guidelines to support the progressive realization of the right to adequate food in the context of national food security, adopted by the 127th session of the FAO council, November 2004.

[30] Convention on the Rights of the Child, adopted by General Assembly Resolution 44/25, 20 November 1989, Article 27.

[31] The International Convention on the Elimination of All Forms of Discrimination Against Women, adopted by General Assembly Resolution 34/180, 18 December, 1979, Article 12(2).

[32] The Convention on the Rights of Persons with Disabilities, adopted by General Assembly Resolution 61/106, 13 December 2006, Article 28.

In some texts the freedom from want is mentioned in the preamble, expressing that this freedom is one of the considerations behind the writing of that text. The Universal Declaration on Human Rights, and both the International Covenant on Civil and Political Rights and the International Covenant on Economic, Social and Cultural rights are such documents.

3.2 The right to food implicitly or explicitly mentioned in a context

The right to food is closely related to other human rights and is therefore frequently mentioned in their context, by stressing the importance of adequate food in order to support the realisation of the other human right. Five human rights often refer to food related issues, and implicitly or explicitly refer to the right to food due to its interrelationship: the right to self-determination, the right to health care, the right to life, non-discrimination provisions and the right to social security. The right to food is also implicitly or explicitly mentioned in a thematic context.

3.2.1 The right to self-determination

The right to self-determination can in some circumstances be seen as a prerequisite for the realisation of all other human rights:

> 'the right to self-determination is of particular importance because its realization is an essential condition for the effective guarantee and observance of individual human rights and for the promotion and strengthening of those rights. It is for that reason that States set forth the right of self-determination in a provision of positive law in both Covenants and placed this provision as article 1 apart from and before all of the other rights in the two Covenants.'[33]

In order to realise the right to food, State Parties must comply with the provisions addressing the right to self-determination. The right to self-determination is mentioned in the context of land access in rural areas,[34] or access to resources in poor fishing communities.[35] Access to resources is reported to be particularly

[33] The Human Rights Committee, General Comment 12, 13 March 1984, paragraph 1.

[34] A/57/356, 27 August 2002, the Special Rapporteur on the right to food in his report to the General Assembly, chapter III.

[35] A/59/385, 27 September 2004, the Special Rapporteur on the right to food in his report to the General Assembly, chapter IV.

problematic for indigenous people[36] and woman.[37] They often experience difficulties in gaining access to resources that are necessary for food production, mostly as a result of discriminatory circumstances, which demonstrates the interrelationship of the right to self-determination not only with the right to food, but also with non-discrimination provisions.

3.2.2 Non-discrimination provisions

The right to food may be an important issue in case of the persistent discrimination of a particular group of people. The Committee on Economic, Social and Cultural Rights pointed out with regard to the ICESC that 'discrimination in access to food… constitutes a violation of the Covenant.'[38] Provisions that forbid discrimination may in this regard strengthen the right to food. As demonstrated in Section 3.1, the right to food is explicitly mentioned in the context of non-discrimination treaties for particular groups of persons requiring extra protection, and must be considered in the context of discrimination. It has also been frequently mentioned in soft law, not necessarily literally as a 'right to food', but also in terms of 'equal access to food,' or 'equal access to resources.'[39]

3.2.3 The right to health or healthcare

The right to health or healthcare is of particular importance for weaker groups in a society. For these groups, access to adequate nutrition may be a prerequisite for the realisation of their right to health or healthcare, especially with regard to women and young children.[40] In this context the right to breast-feeding can be

[36] For instance: The UN Declaration on the Rights of Indigenous People, Adopted by General Assembly Resolution 61/295 on 13 September 2007; Human Rights Council Resolution 7/14, 27 March 2008, paragraph 12; A/RES/62/164, 13 March 2008, General Assembly Resolution, paragraph 12; Agenda 21, chapter 26: Recognising and strengthening the role of indigenous people and their communities; the voluntary guidelines to support the progressive realization of the right to adequate food in the context of national food security, adopted by the 127th session of the FAO council, November 2004, preamble, paragraph 8.1; A/60/2005, 12 September 2005, Report of the Special Rapporteur on the Right to Food to the General Assembly, Chapter III. See also Knuth (2009), especially paragraph 1.3.1.

[37] For instance: A/58/330, 28 August 2003, Report of the Special Rapporteur on the Right to Food to the General Assembly, especially paragraph 22. Beijing Declaration and Platform for Action, adopted by the Fourth Conference on Women, 15 September 1995, especially chapter IV A and F.

[38] E/C.12/1999/5, 12 May 1999, Committee on Economic, Social and Cultural Rights, General Comment 12, paragraph 18 and 19. See also: E/C.12/GC/20, 10 June 2009, Committee on Economic, Social and Cultural Rights, General Comment 20, in particular paragraph 6, 23, and 30.

[39] For instance: the International Convention on the Suppression and Punishment of the Crime of Apartheid, ratification by General Assembly Resolution 3068 (XXVIII) of 30 November 1973, entry into force 18 July 1976, Article II (b) and (c); A/CONF.166/9, 19 April 1995, Report of the World Summit for Social Development; Beijing Declaration and Platform for Action, the Fourth World Conference on Women, 15 September 1995.

[40] For instance: the International Convention on the Elimination of All Forms of Discrimination Against Women, adopted by General Assembly Resolution 34/180, 18 December, 1979, Article 12. Bejing Declaration and Platform for Action, adopted by the Fourth Conference on Women, 1995, especially chapter IV C. The Convention on the Rights of the Child, adopted by General Assembly Resolution 44/50, 20 November 1989, Article 24 (c).

mentioned.[41] The Committee on Economic, Social and Cultural Right explained that 'the right to health embraces a wide range of socio-economic factors that promote conditions in which people can lead a healthy life, and extends to the underlying determinants of health, such as food and nutrition, access to safe and potable water...'[42] In its view, the obligations of States include at least 'to ensure access to the minimum essential food which is nutritionally adequate and safe, to ensure freedom from hunger to everyone.'[43]

3.2.4 The right to life

In some circumstances, the right to food is a prerequisite to actually fulfilling the right to life. Although the ICCPR is not yet often viewed in the context of economic, social and cultural rights, Article 6 ICCPR is explained by the Human Rights Committee as a duty of Member States to 'take all possible measures to reduce infant mortality and to increase life expectancy, especially in adopting measures to eliminate malnutrition and epidemics.'[44] To withhold (access to) food with the purpose of destroying life can be seen as a violation of Article 6 ICCPR or Article II (c) of the Convention on the Prevention and Punishment of the Crime of Genocide.[45]

3.2.5 The right to social security

In some situations, the right to food is not mentioned literally, but may be connected so strongly to another human right, that realisation of that particular right may only be possible when the right to food has been realised at least to a certain minimum. When we take a closer look at the country reports submitted by the Member States of the ICESCR, it appears that while discussing Article 11 of this covenant, the level of social security is a way of expressing to what extent the Member State complies with the realisation of the right to an adequate standard of living, including the right to food.[46] The validity of such arguments is not self-evident. The right to social security is explicitly mentioned elsewhere in the same covenant as a human right in its own right. These two rights should not be lumped together too easily. Access to the benefits of a social security system does

[41] For instance: The European Social Charter, Article 8; The Convention on the Rights of the Child, adopted by General Assembly Resolution 44/50, 20 November 1989, Article 24 (e).

[42] E/C.12/2000/4, 11 August 2000, Committee on Economic, Social and Cultural Rights, General Comment 14, paragraph 4.

[43] E/C.12/2000/4, 11 August 2000, Committee on Economic, Social and Cultural Rights, General Comment 14, paragraph 43 (b).

[44] Human Rights Committee, 30 April 1982, General Comment 6.

[45] International Convention on the Prevention and Punishment of the Crime of Genocide adopted by General Assembly Resolution 260 A (III) on 9 December 1948.

[46] For instance, the Dutch government states in its third periodic report while discussing Article 11 that: 'the Netherlands has a comprehensive system of social benefits guaranteeing its citizens and adequate minimum income.' See: E/1994/104/add.30, 23 August 2005, implementation of the International Covenant on Economic, Social and Cultural Rights. Third periodic reports submitted by states parties under Articles 16 and 17 of the Covenant, Addendum, the Netherlands, paragraph 329.

not automatically lead to access to food. Therefore, the Committee on Economic, Social and Cultural Rights held that the right to social security must at least provide the beneficiaries with access to a minimum life standard, in accordance with Article 11 ICESCR.[47]

Another example of the interrelationship between the right to social security and the right to food is given in the Convention on the Rights of the Child. According to Article 26 ICRC the State Parties 'shall recognise for every child the right to benefit from social security...the benefits should...be granted, taking into account the resources and the circumstances of the child and persons having responsibility for the maintenance of the child....' Article 27 ICRC states that 'parents or others responsible for the child have the primary responsibility to secure, within their abilities and financial capacities, the conditions of living necessary for the child's development.' The State parties 'shall take appropriate measures to assist parents and others responsible for the child to implement this right and shall in case of need provide material assistance and support programmes, particularly with regard to nutrition, clothing and housing.' The realisation of the right to food is thus a primary responsibility for parents or others responsible for the child. The State Party has the obligation to assist those who are primary responsible for the child when they fail to meet their responsibilities. It is a small step to see this in the context of social security, when social security should be granted taking into account the resources and circumstances of the persons who are responsible for the child. In the Netherlands, Article 27 was understood to be closely related to Article 26 by the government that signed the Convention, and by some parliamentarians considered even complementary to each other.[48]

The right to social security is also mentioned in the context of discrimination: equal access to social security is required to ban discrimination against a particular group of people. The International Convention on the Elimination of All Forms of Racial Discrimination,[49] the Convention on the Elimination of All Forms of Discrimination against Women,[50] the Convention relating to the Status of Refugees,[51] and the International Convention on the Protection of the Rights of All Migrant Workers and

[47] E/C.12/GC/19, 4 February 2008, Committee on Economic, Social and Cultural Rights, General Comment 19. See in particular paragraph 18, 22, 28 and 59.
[48] Parliamentary Documents, II 1992-1993, 22855 (R1451), nr. 3, Article 26 and 27. See also meetings of the Parliament: the House of Commons, 81st meeting, 23 June 1994; the House of Commons, 84th meeting, 30 June 1994.
[49] The International Convention on the Elimination of All Forms of Racial Discrimination, adopted by General Assembly Resolution 2106 (XX) of 21 December 1965; Article 5 (d)(iv).
[50] The International Convention on the Elimination of All Forms of Discrimination Against Women, adopted by General Assembly Resolution 34/180, 18 December, 1979, article 11 (e).
[51] The Convention relating to the Status of Refugees, adopted by General Assembly Resolution 429 (V) of 14 December 1950, Article 24 (1)(b).

Members of Their Families[52] provide such examples. Maintaining the Committee's view that the right to social security at least means that an adequate standard of living is assured, would mean that these provisions recognising the right to social security could very well be regarded in the light of the right to food.

3.2.6 The right to food in a thematic context

The importance of the eradication of hunger is also mentioned in a thematic context, mostly in documents resulting from summits, congresses and other gatherings, addressing a specific theme. For example, Agenda 21 (a result of the Earth Summit in 1992) frequently mentions the importance of the eradication of hunger in order to combat poverty. The same goes for the Millennium Declaration (a result of the Millennium Summit in 2000) and its related documents.[53] The right to food may also be regarded in contexts such as HIV/AIDS, poverty in rural areas, local food security, land access/land reform, fishing livelihoods, the impact of transnational corporations, humanitarian law, 'hunger refugees' the use of agricultural resources for the production of biofuels, the use of food as an instrument of pressure, international trade, trade liberalisation and economic crises.[54]

4. Documents and institutions

This section provides the reader with a list of international documents that are related to the right to food. International documents that refer to the right to food in general (Section 4.1) and for specific groups (Section 4.2) and regional human right documents that address the right to food (Section 4.3) will be discussed.

4.1 International documents addressing the right to food

4.1.1 The UN Charter

After the events of the Second World War, the United Nations was called into being: a second attempt to create a worldwide organisation with the main goal to maintain peace and security.[55] The United Nations was founded by the UN

[52] The International Convention on the Protection of the Rights of All Migrant Workers and Members of Their Families, adopted by General Assembly Resolution 45/158 of 18 December 1990, Article 27.

[53] A/55/L.2, 18 September 2000, paragraph 19.

[54] The author suggests reading the reports of the Special Rapporteur on the right to food: the reports often give a sound overview of issues that are related to the right to food at the time of its writing. Mostly, the reports refer to the most actual congresses, summits and other gatherings in which the issue was discussed.

[55] The League of Nations was founded after the First World War in 1919 (The Covenant of the League of Nations), and its main intention was to promote international co-operation and to achieve international peace and security. The League of Nations did not become the success that was intended at its foundation, and silently ceased to function.

Charter[56] which entered into force on 24 October 1945. The importance of human rights is frequently mentioned.

The UN Charter can be seen as the 'constitution' of the United Nations. The primary organs of the UN – the General Assembly, the Security Council and the Economic and Social Council – have been established by the UN Charter.[57] Respect for human rights and fundamental freedoms are mentioned as a purpose of the UN, and two out of the three primary organs are explicitly mandated to contribute to this purpose. Subsidiary UN organs are installed by procedures mentioned in Articles 22, 57 and 68 UN Charter, or by a special treaty. All UN institutions and its activities can be traced back to the Charter.

4.1.2 The Universal Declaration on Human Rights

The Declaration on Human Rights was mainly written by the Commission on Human Rights,[58] and adopted by the General Assembly on 10 December 1948.[59] The freedom from want is mentioned in its preamble, and Article 25 UDHR contains the human right to an adequate standard of living, including food. The second section of this article provides for a special protection in this regard for motherhood and childhood, and declares that all children shall enjoy the same social protection. Furthermore, the UDHR contains non-discrimination provisions[60] and the right to life.[61] The Declaration is not legally binding, and was intended to be the first step towards a universal human rights system.

4.1.3 The International Covenants on Human Rights

The second step in the realisation of an international human rights system was the creation of a legally binding treaty containing the Human Rights that were declared in the UDHR. The choice was made to draft two covenants, one containing civil and political rights (ICCPR),[62] and the other containing economic, social and cultural rights (ICESCR).[63] One of the reasons for this decision was the difference

[56] The UN Charter was signed by the Member States of the UN on 26 June 1945 in San Francisco, USA. The Charter entered into force on 24 October 1945.

[57] See for the installation and competences of the General Assembly chapter IV UN Charter. See for the installation and competences of the Security Council chapter V UN Charter. See for the Economic and Social Council chapter X of the UN Charter.

[58] The drafting Committee consisted of a representation of the UN secretariat, Australia, Chile, France, the Philippines, the former Soviet Union, Ukrainian SSR, the United Kingdom, Uruguay and the former Yugoslavia. The chairperson was Eleanor Roosevelt.

[59] General Assembly Resolution 217 A (III), 10 December 1948.

[60] Universal Declaration on Human Rights, adopted by General Assembly Resolution 217 A (III) of 10 December 1948, Article 2 and 7.

[61] Universal Declaration on Human Rights, adopted by General Assembly Resolution 217 A (III) of 10 December 1948, Article 3.

[62] International Covenant on Civil and Political Rights, adopted by General Assembly Resolution 2200A (XXI) of 16 December 1966.

[63] International Covenant on Economic, Social and Cultural Rights, adopted by General Assembly Resolution 2200A (XXI) of 16 December 1966.

in ideological perception of human rights between 'East and West', strongly driven by the Cold War: Western countries, with a focus on capitalism, emphasised the importance of civil political rights, while the Eastern bloc countries, with a focus on socialism, stressed the importance of economic, social and cultural rights. Another reason was the differences in level of development between countries. The developing countries in particular expressed their concerns with regard to their capacity to realise the economic, social and cultural rights. This may also explain the different procedures for monitoring and accountability between the ICCPR and the ICESCR.

The third step of the global human right system was embedded in the Covenants themselves. Both contain a monitoring procedure.[64] Articles 16 and 17 ICESCR contain the ICESCR monitoring procedure.[65] The State Parties have the obligation to submit reports on the measures which they have adopted and the progress made in achieving the observance of the ICESCR rights.[66] The Committee on Economic, Social and Cultural Rights[67] receives and considers the submitted country reports. Unfortunately, not every Member State copes with the deadlines for these reports.[68]

The ICCPR contains a procedure for complaints by Member States.[69] An optional protocol contains a procedure for individual complaints.[70] No such procedures used to exist with regard to the ICESCR. Since 1990 however, attempts have been made to formulate an optional protocol that gives jurisdiction to the Committee on Economic, Social and Cultural Rights to receive individual communications and communications from fellow-Member States.[71] The Open-ended Working Group on the OP-ICESCR was installed in 2002,[72] the extended mandate established in

[64] See Article 40 ICCPR and Articles 16 and 17 ICESCR.
[65] This procedure was explained in more detail in: Committee on Economic, Social and Cultural Rights, General Comment 1, 24 February 1989.
[66] International Covenant on Economic, Social and Cultural Rights, adopted by General Assembly Resolution 2200A (XXI) of 16 December 1966, part IV.
[67] Established on 28 May 1985 by Economic and Social Council Resolution 1985/17.
[68] For instance, the Netherlands submitted their third report on 18 August 2005, but should have submitted the report no later than 30 June 1997.
[69] See chapter IV of the International Covenant on Civil and Political Rights, adopted by General Assembly Resolution 2200A (XXI) of 16 December 1966, in particular Article 41-43.
[70] Optional Protocol to the International Covenant on Civil and Political Rights, adopted by General Assembly Resolution 2200 A (XXI), 16 December 1966.
[71] Optional Protocol to the International Covenant on Economic, Social and Cultural Rights, adopted by General Assembly Resolution A/RES/63/117 on 10 December 2008, Article 1.
[72] Commission on Human Rights Resolution 2002/24, 22 April 2002.

2006[73] and the first draft of an optional protocol finished in 2007.[74] The General Assembly adopted the optional protocol to the ICESCR in 2008,[75] containing *inter alia* a procedure for submitting 'communications...by or on behalf of individuals or groups of individuals, under the jurisdiction of a State Party, claiming to be a victim of a violation of any of the...rights set forth in the Covenant by that State Party.'[76] There is also a second procedure for 'communications' from Member States 'that claim that another State Party is not fulfilling its obligations under the Covenant.'[77] At the time of writing, the protocol had 32 signatories, but not yet any accessions or ratifications.[78]

Article 11 of the ICESCR was written in co-operation with the FAO, who proposed sub-section (b), to provide for a legal basis for their 'freedom from hunger campaign' that was set up in 1960. It explains why sub (a) refers to 'the right to adequate food,' and sub (b) refers to 'freedom from hunger.'[79] During the negotiations that led to the final draft of the ICESCR in the early sixties, the content of Article 11 was debated heavily. For instance, the Dutch delegation stated that the article 'was too detailed, covered many matters which went beyond the competence of the Third Committee, was not consistent with the bald statements relating to the rights of housing and clothing, was more appropriate to a declaration than a legally-binding instrument.'[80]

Together with a group of other Member States, this government believed that food-related problems were so diverse that different approaches were required. They considered the formulation of the provisions in Article 11 too specific to be compatible with that requirement.

Article 11 ICESCR must be read in conjunction with Article 2 ICESCR to formulate state obligations as a result of the right to adequate food. It says that:

> 'Each State Party to the present Covenant undertakes to take steps, individually and through international assistance and co-operation, especially economic

[73] A/HRC/1/L.10, 29 June 2006, Human Rights Council Resolution 2006/3.
[74] A/HRC/6/WG.4/2, 23 April 2007, draft optional protocol to the International Covenant on Economic, Social and Cultural Rights; A/HRC/8/WG.4/2, 24 December 2007, revised draft optional protocol to the International Covenant on Economic, Social and Cultural Rights; A/HRC/8/WG.4/3, 25 March 2008, (second) revised draft optional protocol to the International Covenant on Economic, Social and Cultural Rights.
[75] A/RES/63/117, adopted on 10 December 2008, the 50th Anniversary of the UDHR.
[76] Optional Protocol to the International Covenant on Economic, Social and Cultural Rights, adopted by General Assembly Resolution A/RES/63/117 on 10 December 2008, Article 2-9.
[77] Optional Protocol to the International Covenant on Economic, Social and Cultural Rights, adopted by General Assembly Resolution A/RES/63/117 on 10 December 2008, Article 10-14.
[78] See for an official overview of signatories, accessions and ratifications of all UN human rights treaties: http:/treaties.un.org/Pages/Treaties.aspx?id = 4&subid = A&lang = en.
[79] See also Alston and Tomasevski (1984).
[80] A/C.3/SR.1266, 1963, para 57-63.

> and technical, to the maximum of its available resources, with a view to achieving progressively the full realization of the rights recognised in the present Covenant by all appropriate means, including particularly the adoption of legislative measures.'

The meaning of Article 2 ICESCR was further explained by the Committee on Economic, Social and Cultural Rights in its General Comment 3.[81]

4.1.4 The Declaration on the Eradication of Hunger and Malnutrition

The World Food Conference in 1974 was attended by 135 representatives of UN Member States, representatives of liberation movements, the UN secretary general and some UN organs and bodies, and a couple of global specialised agencies. In the resulting Declaration on the Eradication of Hunger and Malnutrition,[82] the urgency of providing a solution for malnutrition is mentioned, and the eradication of hunger is seen as a common responsibility of the entire international community. The declaration was partly based on the assumption that there are adequate resources to feed each human being, but men did not yet fully use or adequately share these resources. Therefore, each government should strive to increase and improve food production. The relationship between malnutrition and international trade is mentioned, and striving for a free market is proposed to help eradicate malnutrition in the context of world trade. Human rights or human rights treaties are not mentioned. The Declaration on the Eradication of Hunger and Malnutrition is still considered to be a milestone with regard to the development of the right to food, and was the first declaration that addressed the urgent need to ban hunger on an international level. The declaration is recalled by all UN institutes that write resolutions or other documents related to the right to food. The declaration was not the only outcome of the World Food Conference: it was accompanied by several more detailed resolutions containing strategies and guidelines for the realisation of the goals mentioned in the declaration.

The Committee on World Food Security that was originally established upon the recommendation of the World Food Conference was reformed prior to the World Summit on Food Security in 2009. Its latest goal is to function as the worldwide forum to discuss issues on food security.[83]

4.1.5 Rome Declaration on World Food Security and World Food Summit Plan of Action

At the invitation of FAO, the World Food Summit was held on 13-17 November 1996 in Rome, bringing together close to 10,000 representatives at the highest level from

[81] Committee on Economic, Social and Cultural Rights, General Comment 3, 14 December 1990, Art 2, section 1.

[82] Report of the World Food Conference, 5-16 November 1974, Rome.

[83] Committee on World Food Security, Thirty Fifth Session, 14, 15 and 17 October 2009, Rome, Agenda Item III, Reform of the committee on world food security, final version.

185 countries. The reason for organising the Summit was the failure to reach the goals that were set during the World Food Conference in 1974. The World Food Summit was the first global meeting of heads of states concerning the right to food since the World Food Conference in 1974. For five days, a debate took place in different settings. In the first place heads of states or high representatives gathered, resulting in the final version of the Rome Declaration and its Plan of Action.[84] In the second place there were parallel events that resulted in alternative declarations: the Statement by the NGO Forum to the World Food Summit, The Declaration of the International Youth Forum, and the Declaration Adopted by Consensus at the Parliamentarians' Day. Without doubt, the most important ambition of the World Food Summit was to reduce the number of undernourished people to half their present level no later than 2015.[85] An ambition that was reaffirmed in the Millennium Declaration, adopted by the UN General Assembly in 2000.[86]

There was not always consensus between the final declaration written by the heads of states and other high representatives and the declarations resulting from the parallel events. For instance, the attitude towards international trade in relation to the right to food was a heavily debated issue and led to different conclusions.[87]

Objective 7.4. of the Rome Declaration and its Plan of Action expressed the need 'To clarify the content of the right to adequate food and the fundamental right of everyone to be free from hunger, as stated in the International Covenant on Economic, Social and Cultural Rights and other relevant international and regional instruments, and to give particular attention to implementation and full and progressive realisation of this right as a means of achieving food security for all.' In response,[88] the Committee on Economic, Social and Cultural Rights wrote their General Comment 12 on the right to food, and the High Commissioner for Human Rights convened expert consultations on the right to food,[89] resulting in several reports that intend to respond to objective 7.4. of the Rome Declaration.[90]

[84] Rome Declaration on World Food Security and World Food Summit Plan of Action, 13 November 1996.
[85] Rome Declaration on World Food Security 13 November 1996, paragraph 2.
[86] A/55/L.2, 18 September 2000, General Assembly Resolution, paragraph 19.
[87] Compare the World Food Summit, Plan of Action, 13 November 1996, paragraph 37-38. The Declaration Adopted by Consensus at the Parliamentarians' Day, 15 November 1996, paragraph 9; The Statement by NGO Forum to the World Food Summit, 17 November 1996, key element 6; the Declaration of the International Youth Forum, 17 November 1996.
[88] And also at the invitation of the Commission on Human Rights (Resolutions 1998/23, 17 April 1998, paragraph 7, and 1999/24, 26 April 1999, paragraph 7).
[89] Also on invitation of the Commission on Human Rights: Commission on Human Rights Resolutions 1998/23, 17 April 1998, paragraph 5-6; 1999/24, 26 April 1999, paragraph 6-7; 2000/10, 17 April 2000, paragraph 7; 2001/25, 20 April 2001, paragraph14-15; 2002/25, paragraph 10. See also General Assembly Resolutions 56/155, 15 February 2002, paragraph 10; 57/226, 26 February 2003, paragraph 13; 58/186, 18 March 2004.
[90] E/CN.4/1999/45, 20 January 1999; E/CN.4/2000/48, 13 January 2000; E/CN.4/2001/148, 30 March 2001.

4.1.6 General Comment 12[91]

As already mentioned, General Comment 12 is a direct response to Objective 7.4. of the World Food Summit:

> 'To clarify the content of the right to adequate food and the fundamental right of everyone to be free from hunger, as stated in the International Covenant on Economic, Social and Cultural Rights and other relevant international and regional instruments, and to give particular attention to implementation and full and progressive realization of this right as a means of achieving food security for all.' It defines the term 'adequate food' as follows: 'The availability of food in a quantity and quality sufficient to satisfy the dietary needs of individuals, free from adverse substances, and acceptable within a given culture; The accessibility of such food in ways that are sustainable and that do not interfere with the enjoyment of other human rights.'

It also points out the different types of obligations for Member States resulting from the right to food: the duty to respect, protect and fulfil. The terminology is inspired by a doctrine that classifies States' obligations with regard to human rights. Special rapporteur on the right to food Asbjørn Eide[92] made a particularly significant contribution to this doctrine in his 'updated study on the right to food.'[93] The status of General Comments may be defined as: 'an authoritative legal interpretation clarifying the normative content of the right and the respective State obligations.'[94]

4.1.7 Other General Comments

General Comment 12 is at the time of writing part of a series of 20 General Comments, each of them further clarifying a part of the ICESCR. Some of these General Comments are somehow connected with the right to food. General Comment 1 further explains the reporting procedure of State Parties.[95] In General Comment 3 the Committee explained what the States Parties' obligations are under the Covenant.[96] The Committee clarified the meaning of the phrase 'to take steps' (Article 2 ICESCR) by explaining that 'while the full realisation of the relevant rights may be achieved progressively, steps towards that goal must be taken within a reasonably short time after the Covenant's entry into force for the States concerned. Such steps should be deliberate, concrete and targeted as clearly

[91] E/C.12/1999/5, 12 May 1999, Committee on Economic, Social and Cultural Rights, General Comment 12.

[92] In accordance with UN Sub-Commission on Prevention of Discrimination and Protection of Minorities decision 1998/106, 26 August 1996.

[93] E/CN.4/Sub.2/1999/12, 28 June 1999, Asbjørn Eide, updated study on the right to food, in accordance with Sub-Commission decision 1998/106, especially paragraph 44-57.

[94] See for instance with regard to General Comment 12: E/CN.4/2001/148, 30 March 2001, Report on the Third Expert Consultation on the Right to Food, paragraph 14.

[95] Committee on Economic, Social and Cultural Rights, General Comment 1, 24 February 1989.

[96] Committee on Economic, Social and Cultural Rights, General Comment 3, 14 December 1990.

as possible towards meeting the obligations recognised in the Covenant.'[97] The Committee also further explained what kind of (legal) measures could be taken:

> 'Among the measures which might be considered appropriate, in addition to legislation, is the provision of judicial remedies with respect to the rights which may, in accordance with the national legal system, be considered justiciable.'[98] With regard to the right to food, the Committee explained that it 'is of the view that a minimum core obligation to ensure the satisfaction of, at the very least, minimum essential levels of each of the rights is incumbent upon every State party. Thus, for example, a State party in which any significant number of individuals is deprived of essential foodstuffs...is, prima facie, failing to discharge its obligations under the Covenant. If the Covenant would be read in such a way as not to establish such a minimum core obligation, it would be largely deprived of its raison d'être.'[99]

In General Comment 4,[100] 6,[101] 8,[102] 14,[103] 16,[104] and 20,[105] the Right to Food is also mentioned in the context of the right to adequate housing, the rights of older persons, the relationship between economic sanctions and respect for economic social and cultural rights, equality between women and men, and discrimination. General Comment 15 deals with the meaning of the right to adequate water.[106]

General Comment 19[107] is concerned with the right to social security (Article 9 ICESCR). In this General Comment, social security is on many occasions linked to the access to adequate food:

---

[97] Committee on Economic, Social and Cultural Rights, General Comment 3, 14 December 1990, paragraph 2.
[98] Committee on Economic, Social and Cultural Rights, General Comment 3, 14 December 1990, paragraph 5.
[99] Committee on Economic, Social and Cultural Rights, General Comment 3, 14 December 1990, paragraph 10.
[100] Committee on Economic, Social and Cultural Rights, General Comment 4, 1991, 13 December 1991, paragraph 1 and 8 (b).
[101] Committee on Economic, Social and Cultural Rights, General Comment 6, 8 December 1995, paragraph 5, 32 and 35.
[102] E/C.12/1997/8, 12 December 1997, Committee on Economic, Social and Cultural Rights, General Comment 8, paragraph 3.
[103] E/C.12/2000/4, 11 August 2000, Committee on Economic, Social and Cultural Rights, General Comment 14, in particular paragraph 2, 3, 4, 11, 15, 36, 40, 43, 51 and 65.
[104] E/C.12/2005/4, 11 August 2005, Committee on Economic, Social and Cultural Rights, General Comment 16, paragraph 4, 28 and 29.
[105] E/C.12/GC/20, 10 June 2009, Committee on Economic, Social and Cultural Rights, General Comment 20, paragraph 6, 23, and 30.
[106] E/C.12/2002/11, 20 January 2003, Committee on Economic, Social and Cultural Rights, General Comment 15.
[107] E/C.12/GC/19, 4 February 2008, Committee on Economic, Social and Cultural Rights, General Comment 19.

> 'Family and Child benefits, including cash benefits and social services, should be provided to families, without discrimination on prohibited grounds, and would ordinary cover food, clothing, housing, water and sanitation, or other rights as appropriate,'[108] and: 'Benefits, whether in cash or in kind, must be adequate in amount and duration in order that everyone may realise his or her rights to family protection and assistance, and adequate standard of living and adequate access to health care, as contained in Articles 10, 11, and 12 of the Covenant.'[109]

The Committee pointed out that the right to social security plays an important role in supporting the realisation of *inter alia* Article 11 ICESCR.[110] And stated that:

> 'States parties have a core obligation to ensure the satisfaction of, at the very least, minimum essential levels of each of the rights enunciated in the Covenant. This requires the State party: To ensure access to a social security scheme that provides for a minimum essential level of benefits to all individuals and families that will enable them to acquire at least essential health care, basic shelter and housing, water and sanitation, foodstuffs, and the most basic forms of education.'[111]

4.1.8 Declaration of the World Food Summit: five years later

In 2001, a second World Food Summit was held, also hosted by FAO. The Summit was originally not foreseen, but had 'been prompted by the concern that the target set in the 1996 Rome Declaration may not be achieved.'[112]

In the Declaration of the World Food Summit: five years later, the participating States reaffirmed that trade was a key element in achieving food security, and maintained this view that was already pointed out in commitment four of the 1996 World Food Summit Plan of Action, by urging all Member States to implement

[108] E/C.12/GC/19, 4 February 2008, Committee on Economic, Social and Cultural Rights, General Comment 19, paragraph 18.

[109] E/C.12/GC/19, 4 February 2008, Committee on Economic, Social and Cultural Rights, General Comment 19, paragraph 22.

[110] E/C.12/GC/19, 4 February 2008, Committee on Economic, Social and Cultural Rights, General Comment 19, paragraph 28.

[111] E/C.12/GC/19, 4 February 2008, Committee on Economic, Social and Cultural Rights, General Comment 19, paragraph 59 (a).

[112] H.E. Carlo Azeglio Ciampi, President of the Italian Republic, inaugural ceremony, 10 June 2002. Since then, the urgency to meet the goals as set in the World Food Summits and the Millennium Declaration has been expressed several times by the Human Rights Council and its predecessor (for instance, Commission on Human Rights Resolutions 2004/19, 16 April 2004, paragraph 5; 2005/18, 14 April 2005, paragraph 6; Human Rights Council Resolution 7/14, 27 March 2008, paragraph 19). Also, the New York Declaration Against Hunger and Poverty, made on the 20 September 2004, urged the state parties of the Millennium Declaration to fulfil its promises.

the outcome of the WTO Doha Conference.[113] In the ministerial declaration of this Doha Round, the special interest of developing countries with regard to agriculture was recognised. Also, a fair and market-orientated trading system was seen as a solution to malnutrition.[114] In the declaration, the participating States recognised other UN activities and their outcomes with regard to the right to food. An important result of the declaration was the adoption of section 10, that mandated the FAO to 'establish an International Working Group...to draw up a set of voluntary guidelines to support the Member States' efforts to achieve the progressive realisation of the right to adequate food in the context of national food security,' in order to strengthen the respect of all human rights and fundamental freedoms.

Also here, parallel events took place: an NGO forum, a multistakeholder dialogue, a Parliamentarian's day and a private sector forum. The conclusions of these parallel events differed considerably. On hardly any point did consensus emerge. The role of international trade in relation to the right to food proved particularly controversial.[115] There seemed, however, to be unanimous support for the realisation of voluntary guidelines that would help to better implement the right to food. In one of his reports, the Special Rapporteur on the right to food expressed his disappointment with regard to the results of the 'World Food Summit: five years later.'[116]

4.1.9 The voluntary guidelines to support the progressive realisation of the right to adequate food in the context of national food security

During its 123rd session, the FAO installed the Intergovernmental Working Group (IGWG) as requested in section 10 of the declaration of the World Food Summit, five years later. The IGWG reported to the Committee on World Food Security, one of the specialised bodies of the FAO. In the period March 2003 - September 2004, the IGWG held four sessions and an intersessional meeting, in which the voluntary guidelines were developed. The sessions could be characterised as harsh negotiations in which slow progress was made due to the different interest

[113] Declaration of the World Food Summit: Five Years later, held in Rome, 10-13 June 2002, preamble and paragraph 12.

[114] WTO, ministerial conference, fourth session, Doha, 9-14 November 2001, Ministerial declaration adopted on 14 November 2001, paragraph 13.

[115] Compare Annex III to the report of the World Food Summit: five years later, Statement by the NGO forum, 2002; Annex III to the report of the World Food Summit: five years later, Statement as a result of the multistakeholder dialogue, 2002; Final Document of the Inter-Parliamentary Conference on 'Attaining the World Food Summit's Objectives through a Sustainable Development Strategy,' organised by the Inter-Parliamentary Union with the support of FAO and hosted by the Italian Parliament in Rome (Italy), from 29 November - 12 December 1998, paragraph 14 and 47; annex III to the report of the World Food Summit: five years later: Report on the Parliamentarians' day, 2002; Report of the World Food Summit, five years later; annex III: reports on the round tables, the multistakeholder dialogue, and the parallel events: statement by the private sector forum, by Mr. Augusto Bocchini.

[116] A/57/356, 27 August 2002, the Special Rapporteur on the right to food in his report to the General Assembly, Chapter II.

of the participants and the demand for consensus.[117] Finally, in October 2004, the Working Group agreed upon a final draft of the voluntary guidelines. The guidelines were adopted by the Council of the Food and Agriculture Organization of the United Nations in November 2004.[118]

The main purpose of the voluntary guidelines is to 'provide practical guidance to States in their implementation of the progressive realisation of the right to food in the context of national food security.'[119] The voluntary nature of the document is expressed in the preamble. A human rights based approach is strongly recommended,[120] and the legal entitlements of rights holders are supported. Furthermore, the guidelines point out that by signing the ICESCR, States have the obligation to respect, protect and promote and to take appropriate steps to achieve progressively the full realisation of the right to adequate food. Therefore, guideline 7 invites States to consider legal and policy frameworks in their domestic law that facilitate the progressive realisation of the right to food, or more directly, to consider the inclusion of provisions that directly implement the progressive realisation of the right to food, including administrative, quasi juridical and juridical mechanisms that provide remedies to especially members of vulnerable groups to be able to realise their right to food.[121] Although the voluntary character of the guidelines was made clear, the IGWG referred to other legislation that is not voluntary, such as the ICESCR, and pointed out that these voluntary guidelines were developed to support the progressive realisation of the right to adequate food, a human right recognised in non-voluntary treaties. The resulting nineteen guidelines are based upon three underlying dimensions: adequacy, availability and accessibility, a subdivision inspired by General Comment 12.[122]

4.1.10 Declaration of the World Food Summit on Food Security

A third World Food Summit was held in November 2009, resulting in a declaration.[123] The declaration holds four strategic objectives, including the objective to realise the target set during the first World Food Summit to halve the number of people

[117] A/59/385, 27 September 2004, the Special Rapporteur on the right to food in his report to the General Assembly, chapter III, in particular paragraph 27. See also Oshaug (2009), chapter six.

[118] E/CN.4/2005/131, 28 February 2005, Voluntary Guidelines to Support the Progressive Realization of the Right to Adequate Food in the Context of Food Security; annex.

[119] The voluntary guidelines to support the progressive realisation of the right to adequate food in the context of national food security, adopted by the 127th session of the FAO council, November 2004, foreword.

[120] A human rights approach is in the preamble of the guidelines, paragraph 19, explained as 'the need for individuals to be able to take part in national decision-making processes in freedom.'

[121] The voluntary guidelines to support the progressive realisation of the right to adequate food in the context of national food security, adopted by the 127th session of the FAO council, November 2004, paragraph 7.1-7.4.

[122] E/C.12/1999/5, 12 May 1999, Committee on Economic, Social and Cultural Rights, General Comment 12, paragraph 8.

[123] WSFS 2009/2, Rome, 16-18 November 2009, Declaration of the World Food Summit on Food Security.

who suffer from hunger or malnutrition by 2015,[124] and commitments and actions, based on five principles.[125] The practical results of the World Summit on Food Security remain to be seen. The interest of the international community seems to be waning. The participating representatives did not meet the number and high level of representation during the first two Summits. The final declaration generally does not seem to go beyond reaffirming already existing targets and commitments. Even the rather fierce debates that took place during the side events of the first two World Summits were absent during this summit: hardly any side event took place simultaneously to the summit itself, but rather in preparation for the event. There was no separate NGO-, Youth- or Parliamentarian's forum. In advance however, a Private Sector Forum took place in which representatives from the private sector debated the role of the private sector with respect to food security.[126] Also A High Level Expert Forum was held in Rome on 12-13 October 2009, that 'examined policy options that governments should consider adopting to ensure that the world population can be fed when it nears its peak of nearly 9.2 billion people in the middle of this century.'[127] The Committee on World Food Security met on 14, 15 and 17 October 2009, basically discussing its own reforms to be more effective as an international platform that discusses food security (FAO, 2009). Finally, the Non-Aligned Movement organised the second NAM First Ladies Summit on 15 November 2009, discussing women's access to resources. Its report was presented to the World Summit on food security.[128]

[124] Objective 7.1.

[125] Principle 1: invest in country-owned plans, aimed at channeling resources to well-designed and results-based programmes and partnerships. Principle 2: foster strategic coordination at national, regional and global level to improve governance, promote better allocation of resources, avoid duplication of efforts and identify response-gaps. Principle 3: strive for a comprehensive twin-track approach to food security that consists of (1) direct action to immediately tackle hunger for the most vulnerable and (2) medium- and long-term sustainable agricultural, food security, nutrition and rural development programmes to eliminate the root causes of hunger and poverty, including through the progressive realisation of the right to adequate food. Principle 4: ensure a strong role for the multilateral system by sustained improvements in efficiency, responsiveness, coordination and effectiveness of multilateral institutions. Principle 5: ensure sustained and substantial commitment by all partners to investment in agriculture and food security and nutrition, with provision of necessary resources in a timely and reliable fashion, aimed at multi-year plans and programmes.

[126] A list of participants can be found on the World Food Summit on Food Security website www.fao.orf/wsfs/world-summit/en/.

[127] Available at: www.fao.org/wsfs/forum2050/wsfs-forum/en/.

[128] The Second First Ladies Summit of the Non-Aligned Movement on Food Security and Women's Access to Resources, Rome, 15 November 2009.

4.2 International documents addressing the right to food for specific groups

4.2.1 The Convention on the Elimination of All Forms of Discrimination against Women

The Convention on the Elimination of All Forms of Discrimination against Women (CEDAW) was written to implement the principles set forth in the Declaration on the Elimination of Discrimination against Women. The Convention was adopted on 18 December 1979.[129] Its general purpose is obviously to ban all discrimination against women. One of the causes of malnutrition is the discrimination against women in favour of men/the breadwinners, when food is distributed. Especially in rural areas, women often function as the engine of food production on the one hand, but on the other hand turn out to be the last to profit from their work.[130] In the preamble, the Convention reminds its Members that State Parties to the International Conventions on human rights have the obligation to ensure the equal rights of men and women to economic, social and cultural rights. The Convention mentions furthermore the right to social security (Article 11 (1)(e)) and the right to protection of health (Article 11 (1)(f)). The right to food in case of pregnancy and the period afterwards is mentioned in Article 12 (1). The convention grants special protection to women who live in rural areas by mentioning the right to benefit directly from social security programmes (Article 14 (2)(c)), the right to have access to agricultural credit and loans marketing facilities, appropriate technology and equal treatment in land and agrarian reforms as well as in land resettlement schemes (Article 14 (2)(g)), and the right to enjoy adequate living conditions, particularly in relation to housing, sanitation, electricity and water supply, transport and communications (Article 14 (2)(h)).

The optional protocol to the Convention on the Elimination of Discrimination against Women contains a communications procedure. Individuals or groups of individuals (or persons on their behalf) may submit individual complaints to the Committee on the Elimination of Discrimination against Women.[131]

4.2.2 The Convention on the Rights of the Child

The Convention on the Rights of the Child was adopted by the General Assembly in November 1989[132] and entered into force on 2 September 1990. Predecessors of the Convention were the Geneva Declaration on the Rights of the Child (1924)[133]

[129] A/RES/34/180, 18 December 1979, General Assembly Resolution.

[130] FAO website, Gender and food security: Agriculture. Available at: www.fao.org.

[131] See especially Article 2 Optional Protocol to the International Convention on the Elimination of Discrimination against Women, adopted as Annex to General Assembly Resolution 54/4 of 6 October 1999.

[132] A/RES/44/25, 20 November 1989, General Assembly Resolution.

[133] Geneva Declaration of the Rights of the Child, adopted by the League of Nations on 26 September 1924.

and the Declaration of the Rights of the Child, adopted by the General Assembly in 1959.[134] The Convention on the Rights of the Child (CRC) explicitly mentions the right to food for children in Article 27:

> States shall take appropriate measures 'To combat disease and malnutrition, including within the framework of primary health care, through, inter alia, the application of readily available technology and through the provision of adequate nutritious foods and clean drinking-water, taking into consideration the dangers and risks of environmental pollution.'

Article 4 of the covenant mentions that with regard to the covenants economic, social and cultural rights, 'the State Parties shall undertake such measures to the maximum extent of their available resources and, where needed, within the framework of international co-operation.' The Convention also contains articles that may be indirectly linked to the right to food, such as a non-discrimination provision (Article 2), the principle of the best interests of the child (Article 3), the right to life (Article 6 (1)), the right to humanitarian assistance for refugees (Article 22), special protection for mentally or physically disabled children (Article 23), and the right to social security (Article 26). In the Convention, the right to the enjoyment of the highest attainable standard of health is recognised (Article 24). To do so, Member States shall 'pursue full implementation of this right and, in particular, shall take appropriate measures to combat disease and malnutrition... through the provision of adequate nutritious food and clean drinking water.'

The Convention on the Rights of the Child contains no complaint procedures. Two optional protocols to the convention were adopted by the General Assembly: the optional protocol on the Convention on the Rights of the Child on the involvement of children in armed conflict[135] and the optional protocol on the Convention on the Rights of the Child on the sale of children, child prostitution and child pornography.[136] Both protocols do not mention or refer to the right to food, but of course cover matters in which the right to food could be of importance.

4.2.3 The Convention on the Rights of Persons with Disabilities

In December 2996, the Convention on the Rights of Persons with Disabilities was adopted by the UN General Assembly.[137] The purpose of the Convention is 'to promote, protect and ensure the full and equal enjoyment of all human rights and fundamental freedoms by all persons with disabilities, and to promote respect for

---

[134] A/4354, 20 November 1959, General Assembly Resolution 1386 (XIV).

[135] A/RES/54/263, 25 May 2002, Optional protocol on the Convention on the Rights of the Child on the involvement of children in armed conflict.

[136] A/RES/54/263, 25 May 2002, Optional protocol on the Convention on the Rights of the Child on the sale of children, child prostitution and child pornography.

[137] A/RES/61/106, 24 January 2007, the Convention on the Rights of Persons with Disabilities.

their inherent dignity.'[138] Article 28 recognises the right to an adequate standard of living and social protection, including the right to adequate food. The same article mentions the right to equal access to clean water, and the right to access to social protection programmes and poverty reduction programmes. Also the article concerned with the right to healthcare mentions that state parties shall 'prevent discriminatory denial of health care services or food and fluids on the basis of disability' (Article 25 (f)). The importance of equal rights for disabled women and children is stressed.[139] The right to life, the right to the enjoyment of the highest attainable standard of health are also mentioned (Article 10 and 25). This convention contains no independent right to social security.

To ensure the implementation of the covenant, it contains a monitoring procedure: every fourth year Member States shall submit a report on measures taken to give effect to their obligations under the convention and on the progress made in that regard (Article 35). The Committee on the Rights of Persons with Disabilities, established in Article 34, regards these reports and makes suggestions and general recommendations on these reports (Article 36). The optional protocol to the convention on the rights of persons with disabilities provides for a procedure for the consideration of complaints from individuals or groups of individuals who feel the victim of a violation by a state party of the covenant.[140]

4.2.4 The Geneva Conventions and its protocols

Several humanitarian law instruments address the right to food in armed conflicts. In the four Geneva Conventions and their three protocols the human right to adequate food is recognised for the following groups of persons: medical personnel of a neutral country assisting one of the parties to a conflict,[141] prisoners of war in general,[142] prisoners of war who are being evacuated[143] or transferred,[144]

[138] A/RES/61/106, 24 January 2007, the Convention on the Rights of Persons with Disabilities, Article 1.
[139] A/RES/61/106, 24 January 2007, the Convention on the Rights of Persons with Disabilities, especially in its preamble and Articles 6 and 7.
[140] A/RES/61/106, 24 January 2007, Optional protocol to the Convention on the Rights of Persons with Disabilities.
[141] First Geneva Convention for the Amelioration of the Condition of the Wounded and Sick in Armed Forces in the Field, 12 August 1949, Article 32 (2) jo art 27.
[142] Third Geneva Convention Relative to the Treatment of Prisoners of War, 12 August 1949, Article 26, 28, 51.
[143] Third Geneva; Geneva Convention Relative to the Treatment of Prisoners of War, 12 August 1949, Article 20.
[144] Third Geneva; Geneva Convention Relative to the Treatment of Prisoners of War, 12 August, Article 46.

civilians,[145] detained civilians,[146] and persons whose liberty is restricted.[147] The starvation of civilians as means of pressure is forbidden in national and international armed conflicts,[148] as well as the deliberate destruction of foodstuff and drinking water.[149] Forced displacements of civilians leading to starvation are prohibited.[150] There are also international rules concerning the protection of humanitarian assistance in occupied territories[151] and during non-international armed conflicts.[152] Also (shipment) delivery of means of existence – including food – for prisoners of war or detained civilians should be allowed.[153] In case of the establishment of a neutralised zone, the delivery of food supply (amongst others) for wounded and sick combatants or non-combatants and civilians should be agreed upon amongst the conflicting parties.[154]

In addition, the starvation of civilians by depriving them of objects indispensible to their survival during an international armed conflict and impending relief supplies during an international armed conflict is considered to be a war crime under the Rome Statute of the International Criminal Court.[155]

---

[145] Fourth Geneva Convention relative to the Protection of Civilian Persons in Time of War, 12 August 1949, Article 55, 76, Protocol Additional to the Geneva Conventions of 12 August 1949, and relating to the Protection of Victims of International Armed Conflicts, 8 June 1977, Article 69.

[146] Fourth Geneva Convention relative to the Protection of Civilian Persons in Time of War, 12 August 1949, Article 87, 89, 100, 108, 127.

[147] Protocol Additional to the Geneva Conventions of 12 August 1949, and relating to the Protection of Victims of Non-International Armed Conflicts, 8 June 1977, Article 5.

[148] Protocol Additional to the Geneva Conventions of 12 August 1949, and relating to the Protection of Victims of International Armed Conflicts, 8 June 1977, Article 54, paragraph 1; Protocol Additional to the Geneva Conventions of 12 August 1949, and relating to the Protection of Victims of Non-International Armed Conflicts, 8 June 1977, Article 14.

[149] Protocol Additional to the Geneva Conventions of 12 August 1949, and relating to the Protection of Victims of International Armed Conflicts, 8 June 1977, Article 54, paragraph 2; Protocol Additional to the Geneva Conventions of 12 August 1949, and relating to the Protection of Victims of Non-International Armed Conflicts, 8 June 1977, Article 14.

[150] Fourth Geneva Convention relative to the Protection of Civilian Persons in Time of War, 12 August 1949, Article 49.

[151] Fourth Geneva Convention, Article 59, 12 August 1949; Protocol Additional to the Geneva Conventions of 12 August 1949, and relating to the Protection of Victims of International Armed Conflicts, 8 June 1977, Article 69, 70.

[152] Protocol Additional to the Geneva Conventions of 12 August 1949, and relating to the Protection of Victims of Non-International Armed Conflicts, 8 June 1977, Article 18.

[153] Third Geneva Convention Relative to the Treatment of Prisoners of War, 12 August 1949, Article 72; Fourth Geneva Convention relative to the Protection of Civilian Persons in Time of War, 12 August 1949, Article 108.

[154] Fourth Geneva Convention relative to the Protection of Civilian Persons in Time of War, 12 August 1949, Article 15.

[155] A/CONF.183/9, 17 July 1998, Rome Statute of the International Criminal Court, Article 7 (2)(b) and Article 8 (2)(b)(xxv).

Humanitarian law grants special protection to specific categories of persons, such as children, mothers and pregnant women[156] and prisoners of war,[157] due to the fact that they are considered not to be participating (anymore) in the armed conflict.

4.3 Regional human rights documents

4.3.1 The European Social Charter

The European Social Charter was adopted in Turin on 18 October 1961, and revised in the same city on 21 October 1991. The Charter contains no general provision that recognises the right to food, although in Article 8 the right for employed woman to have sufficient time to nurse (including breast-feeding) their infants is mentioned. The ESH is strongly based on ILO legislation, and therefore has a focus on labour law, clearly demonstrated in the fact that Article 1 recognises the right to work. The right to food is not mentioned in the Charter. Nevertheless, some articles may implicitly connect to the right to food. The right to social security is recognised (Article 12), as well as the right to safe and healthy working conditions (Article 3), the right of workers to a remuneration such as will give them and their families a decent standard of living (Article 4), the right to protection of health (Article 11), the right to social and medical assistance (Article 13) and the right to benefit from social welfare services (Article 14). The rights of children (Article 7), women (Article 8) the family (Article 16), and migrant workers (Article 19) receive additional protection in the Charter. The Social Charter has two additional protocols. One of them provides for a system of collective complaints. The European Committee of Social Rights (ECSR) *'decides on the conformity with the ESC of the law and practice of states,'*[158]and receives and decides on collective complaints.

4.3.2 The African Charter on Human and People's Rights[159]

The African Charter on Human and People's Rights was adopted on the initiative of the Organization of African Unity (OAU) in 1981 and entered into force in 1986. The charter established the African Commission on Human and People's Rights, which has a mandate to protect and promote human rights and advises on the interpretation of the African Charter on Human and Peoples' Rights. The charter does not literally recognise the right to adequate food, but does state in Article 21 that *'all peoples shall freely dispose of their wealth and natural resources.'*

---

[156] Fourth Geneva Convention relative to the Protection of Civilian Persons in Time of War, 12 August 1949, Art 23, 50, 89.

[157] Third Geneva Convention relative to the Treatment of Prisoners of War, 12 August 1949, Article 20, 26, 28, 46, 51 and 72.

[158] European Social Charter, European Committee on Social Rights, *Conclusions XVII-1, Volume 1*, Strasbourg: Council of Europe Publishing, December 2006, paragraph 2.

[159] African Charter on Human and Peoples' Rights, adopted by the eighteenth Assembly of Heads of State and Government, June 1981, Nairobi, Kenya.

A protocol to the Charter on Human Rights concerning the rights of women does however explicitly recognise the right to food security for women.[160] And in 1990, the OAP drew up the African Charter on the Rights and Welfare of the Child that explicitly mentions the right to adequate food in Articles 14 and 20.

The Charter itself originally contains only a reporting procedure.[161] But another additional protocol installs the African Court on Human and Peoples' Rights, that is mandated to make legally binding decisions[162] for violations of all human rights recognised in the African Charter and any other relevant human rights instrument ratified by the States concerned.[163] The additional protocol does not differentiate between civil and political rights on the one hand and economic, social and cultural rights on the other, a development which was praised by the Special Rapporteur on the right to food.[164]

In 2009 the OAP adopted the African Union Convention for the Protection and Assistance of Internally Displaced Persons in Africa (Kampala Convention).[165] In this convention the States Parties pledge themselves to provide internally displaced persons with adequate humanitarian assistance, including food and water.[166] Furthermore, members of armed groups 'shall be prohibited from denying internally displaced persons the right to live in satisfactory conditions of dignity, security, sanitation, food, water, health and shelter...'[167]

4.3.3 The Charter of the Organization of American States

In 1948, the Organization of American States (OAS) adopted the Charter of the Organization of American States. Article 34 of the charter states that proper nutrition and modernisation of rural life and land reforms are basic goals in order to support *inter alia* equality of opportunity, elimination of poverty and equal distribution of wealth and income.

---

160 Additional Protocol to the African Charter on Human and Peoples' Rights on the rights of Women in Africa, Adopted by the 2nd Ordinary Session of the Assembly of the Union, Maputo, 11 July 2003, Article 15.

161 African Charter on Human and Peoples' Rights, adopted by the eighteenth Assembly of Heads of State and Government, June 1981, Nairobi, Kenya, Article 62.

162 Additional Protocol to the African Charter on Human and Peoples' Rights on the establishment of an African Court on Human and Peoples' Rights, June 1998, Article 28-30.

163 Additional Protocol to the African Charter on Human and Peoples' Rights on the establishment of an African Court on Human and Peoples' Rights, June 1998, Article 3.

164 E/CN.4/2002/58, 10 January 2002, Report submitted to the Commission on Human Rights, paragraph 64.

165 African Union Convention for the Protection and Assistance of Internally Displaced Persons in Africa (Kampala Convention), adopted by the Special Summit of the Union held in Kampala, on 22 October 2009.

166 African Union Convention for the Protection and Assistance of Internally Displaced Persons in Africa (Kampala Convention), 22 October 2009, Article 9 (2) (b).

167 African Union Convention for the Protection and Assistance of Internally Displaced Persons in Africa (Kampala Convention). 22 October 2009, Article 7 (5)(c).

In 1969, the American Convention on Human Rights was adopted by the OAS, and entered into force on 1978. The Convention has a strong focus on Civil and Political Rights, and only obliges States to guarantee a progressive development of the economic, social and cultural rights (Article 26). The Inter American Court of Human Rights has a mandate to receive individual human rights complaints and can apply and interpret the Convention.

The San Salvador Additional Protocol to the American Convention on Human Rights, 1988, does however focus on economic, social and cultural rights, and recognises the right to adequate food in Article 12. Furthermore, the protocol has a non-discrimination provision (Article 3), and recognises the right to just, equitable, and satisfactory conditions of work (Article 7), the right to social security (Article 9) and the right to health (Article 10). The protocols include a monitoring procedure and a very limited procedure for complaints, both supervised by the Inter-American Commission on Human Rights.[168] The Member States have an obligation to enact domestic legislation in case the rights of the protocol are not yet guaranteed in the legal order of the Member State (Article 2), and to adopt measures 'for the purpose of achieving progressively and pursuant to their internal legislations, the full observance of the rights recognised in the Protocol' (Article 1).

4.3.4 The Cairo Declaration

The Cairo Declaration on Human Rights in Islam, drawn up by the Organization of the Islamic Conference (OIC), can be seen as a reaction to the UN human rights mechanisms that were based on the UDHR. A group of Muslim countries felt that they did not have sufficient input into the creation of the UDHR, and therefore felt the need to draw up their own declaration. The right to adequate food is recognised in case of armed conflict, for children, and for every individual in Articles 3, 7 and 17. The Declaration had a predecessor that also recognised the right to adequate food.[169] Both declarations have been heavily criticised for the fact that they affirm that Shari'ah law is the only legal source.

4.3.5 The Asian Human Rights Charter

The Asian Human Rights Charter, most recently updated on 30 March 1998, recognises the right to food in Article 7.1. The charter strongly suggests that States take appropriate measures to make sure each individual can realise its human rights.[170] It is recommended to achieve this by legislation: 'Human rights are violated by the state, civil society and business corporations. The legal protection for rights has to be extended against violations by all these groups.'[171] The Asian

[168] The San Salvador Additional Protocol to the American Convention on Human Rights, November 1988, Article 19.

[169] Universal Islamic Declaration of Human Rights, 21 Dhul Qaidah 1401, 19 September 1981, Article XVIII.

[170] The Asian Human Rights Charter, Updated on 30 March 1998, Article 15-16.

[171] The Asian Human Rights Charter, Updated on 30 March 1998, Article 16.

Commission on Human Rights is an institution that makes recommendations, and initiates studies.[172]

5. International institutions related to the human right to adequate food

The institutions mentioned below have contributed to the development of the human right to adequate food, and are all international organs linked in some way to the UN. The General Assembly, the Security Council and the Economic and Social Council – the primary organs of the UN – have been directly installed by the UN Charter. The others by the procedure mentioned in Articles 22, 57 and 68 UN Charter, or by a special treaty.

5.1 The General Assembly

The General Assembly is one of the three primary organs of the UN, and consists of all UN members.[173] The Assembly may discuss any questions or any matters within the scope of the UN Charter or relating to the powers and functions of any organs provided for in the UN Charter, and may make recommendations to the Member States of the UN or to the Security Council on such matters.[174] Human rights are explicitly mentioned as one of these matters in article 13 UN Charter.

To fulfil its task, the General Assembly can 'establish such subsidiary organs as it deems necessary for the performance of its functions.'[175] For instance, UNICEF, UNHCR, UNCTAD and UNDP were installed by this procedure. Furthermore, the Economic and Social Council and its secondary organs – such as the Human Rights Council and the Special Rapporteurs – make recommendations to the General Assembly. Each year the General Assembly adopts a regulation concerning the right to food in particular.[176] These resolutions basically confirm the recommendations of other UN institutions, especially from the ECOSOC and its secondary organs, for instance with regard to the realisation of the mandate of the Special Rapporteur on the Right to Food, the World Food Summit and its follow-up, and the voluntary guidelines. In these regulations, the General Assembly consequently reaffirms the right to food, and recalls in their preamble – among other things – the UDHR, the ICESCR, the Universal Declaration on the Eradication of Hunger and Malnutrition, the World Food Summit and its follow-up, and the voluntary guidelines. The General

[172] Available at: http://www.ahrchk.net.
[173] UN Charter, 26 June 1945, Article 9.
[174] UN Charter, 26 June 1945, Article 10.
[175] UN Charter, 26 June 1945, Article 22.
[176] The latest Resolutions are: General Assembly Resolutions A/RES/56/155, 15 February 2002; A/RES/57/226, 26 February 2003; A/RES/58/186, 18 March 2004; A/RES/59/202, 31 March 2005; A/RES/60/165, 2 March 2006; A/RES/61/163, 21 February 2007; A/RES/62/164, 13 March 2008; A/RES/63/187, 17 March 2009; A/RES/64/159, 10 March 2010.

Assembly played an important role with regard to the UN Human Rights bodies by deciding to replace the Commission on Human Rights with the Human Rights Council in 2006,[177] and the sub-Commission on the Promotion and Protection of Human Rights by the Advisory Council in 2007.[178]

5.2 The Security Council

The Security Council's primary responsibility is the maintenance of international peace and security.[179] The Council may consider famine a threat to peace and security, and its resolutions may therefore affect the human right to adequate food, mostly with regard to (the protection of) humanitarian assistance.[180] The resolutions of the Security Council are of a binding nature.

5.3 The Economic and Social Council

The Economic and Social Council (ECOSOC) may

> 'make recommendations with regard to the General Assembly, and make or initiate studies and reports with respect to international economic, social, cultural, educational, health, and related matters and may make recommendations with respect to any such matters to the General Assembly to the Members of the United Nations, and to the specialized agencies concerned. It may make recommendations for the purpose of promoting respect for, and observance of, human rights and fundamental freedoms for all. It may prepare draft conventions for submission to the General Assembly, with respect to matters falling within its competence....It may call, in accordance with the rules prescribed by the United Nations, international conferences on matters falling within its competence.'[181]

Furthermore, the ECOSOC is responsible for supervising the monitoring procedure as mentioned in part IV of the International Covenant on Economic, Social and Cultural Rights, and has for this reason installed the Committee on Economic, Social and Cultural Rights.[182] By its first meeting in 1946, the Economic and Social Council decided to install the former Commission on Human Rights, in accordance with Article 68 UN Charter.[183]

---

[177] A/RES/60/251, 3 April 2006, General Assembly Resolution.
[178] Human Rights Council Resolution 5/1, 18 June 2007 and 6/102, 27 September 2007.
[179] UN Charter, 26 June 1945, Article 24.
[180] For instance: S/RES/0688, 5 April 1991; S/RES/764, 13 July 1992; S/RES/794, 3 December 1992, paragraph 10; S/RES/1258, 6 August 1999; S/RES/1265, 17 September 1999; S/RES/1296, 19 April 2000; S/2009/23, 8 January 2009. Most Resolutions mention the free pass or armed protection of humanitarian help, including food supplies.
[181] UN Charter, 26 June 1945, Article 62.
[182] ECOSOC Resolution 1985/17, 28 May 1985.
[183] ECOSOC Resolution 5 (I), 16 February 1946.

5.4 The Committee on Economic, Social and Cultural Rights

The Committee on Economic, Social and Cultural Rights was established in 1985,[184] in order to monitor the implementation of the ICESCR. Its predecessor, the sessional working group on the implementation of the international covenant on economic, social and cultural rights,[185] was not very successful.

In order to facilitate implementation of the rights mentioned in the ICESCR, the Committee has tried to further clarify the ICESCR provisions through the years by writing General Comments to the Covenant. At the time of writing, the Committee wrote 20 General Comments.

The Committee is also charged with receiving and considering State Reports that are submitted in accordance with chapter IV of the ICESCR. Since the adoption of the Optional Protocol to the ICESCR, the Committee has also been competent in receiving and considering individual communications[186] or communications from Member States[187] that claim violations of ICESCR rights.

5.5 The Human Rights Council

The Commission on Human Rights was founded at the first meeting of the Economic and Social Rights in 1946. The Commission's primary mandate was to investigate human rights violations.[188] Furthermore, the drafts of the UDHR, ICCPR and the ICESCR were mainly written by the Commission. To assist in preventing or stopping human rights violations by investigation and recommendation, the Commission has set up several working groups, sometimes to initiate research to the situation of human right in a country in general (the so-called 'country mandates'), sometimes to investigate the situation of a specific human right or human right topic worldwide (the so-called 'thematic mandates'). Since 2000, the Special Rapporteur on the Human Right to Adequate Food has held one of these thematic mandates.[189]

In the last decade of its existence, the Commission on Human Rights was heavily criticised due to the membership of countries that had a doubtful status concerning human rights. Because of the persisting criticism of the Commission on Human Rights the General Assembly decided to replace this commission by the Human

[184] ECOSOC Resolution 1985/17, 28 May 1985.
[185] Established by ECOSOC Decision 1978/10, 3 May 1978.
[186] Optional Protocol to the International Covenant on Economic, Social and Cultural Rights, adopted by General Assembly Resolution A/RES/63/117 on 10 December 2008, Article 2-9.
[187] Optional Protocol to the International Covenant on Economic, Social and Cultural Rights, adopted by General Assembly Resolution A/RES/63/117 on 10 December 2008, Article 10-14.
[188] ECOSOC Resolution 1236 (XLII) 1967.
[189] Commission on Human Rights Resolution 2000/10, 17 April 2000, paragraph 10.

Rights Council on the 3rd of April 2006.[190] The attitude of its members towards human rights is taken into account more explicitly. The Human Rights Council reaffirmed all the previous resolutions adopted by the UN organs concerning the right to food,[191] and extended and reformulated the mandate of the Special Rapporteur on the right to food.[192]

5.6 The Advisory Council

The Sub-Commission on the Promotion and Protection of Human Rights[193] used to serve as a think-tank for the Commission on Human Rights. Twenty-six independent experts initiated research in the field of Human Rights and provided the Commission on Human Rights with advice. In order to fulfil its tasks, the Sub-Commission also installed thematic mandates on the right to food and the right to water. The research done by these Special Rapporteurs consisted mostly of desk-research. Mr. Asbjørn Eide[194] has functioned as Special Rapporteur on the right to food, and Mr. El Hadji Guissé as Special Rapporteur on the right to drinking water.[195] The General Assembly replaced the sub-Commission by the Advisory Council in 2007.[196] This body, consisting of 18 experts in the field of human rights, is the new think tank for the Human Rights Council.

5.7 The Special Rapporteur on the Right to Adequate Food

The Special Rapporteur on the Right to Adequate Food was installed by the Commission on Human Rights for two times a period of three years in 2001 and 2003.[197] The Human Rights Council extended this mandate for another three years in 2007.[198] The first mandate holder was Mr. Jean Ziegler (2001-2008).[199] The Human Rights Council wrote a new and updated description of his mandate, replacing the original one that was provided for by the Commission on Human Rights. The Special Rapporteur is mandated *inter alia* to promote and recommend on

---

190 A/res/60/251, 3 April 2006, General Assembly Resolution.

191 Human Rights Council Resolution 7/14, 27 March 2008, the right to food.

192 Human Rights Council, Resolution 6/2, 27 September 2007, mandate of the Special Rapporteur on the right to food.

193 Before also called: 'Sub-Commission on the Prevention on Discrimination and Protection of Minorities.'

194 E/CN.4/Sub.2/1999/12, 28 June 1999, Asbjørn Eide, updated study on the right to food, in accordance with Sub-Commission decision 1998/106.

195 Mr. El Hadji Guissé, E/CN.4/Sub.2/1998/7, 10 June 1998; E/CN.4/Sub2/2004/20, 14 July 2004; E/CN.4/Sub.2/2005/25, 11 July 2005.

196 Human Rights Council Resolution 5/1, 18 June 2007 and 6/102, 27 September 2007.

197 The Commission on Human Rights Resolutions 2000/10, 17 April 2000, paragraph 10, 20003/25, paragraph 8.

198 Human Rights Council, Resolution 6/2, 27 September 2007, paragraph 2.

199 Information about Mr. Jean Ziegler and his work as Special Rapporteur is available at: www.righttofood.org. The original mandate was formulated in The Commission on Human Rights Resolutions 2000/10, 17 April 2000, paragraph 11, and 2001/25, 20 April 2001, paragraph 8-12.

the full realisation of the right to food and to examine ways to overcome obstacles to that realization. In doing so, he should pay special attention to the disadvantaged position of women and children. Furthermore, the Special Rapporteur is mandated to present recommendations on how to halve the number of people who suffer from hunger by 2015. Finally, the Special Rapporteur should work closely with all relevant international actors, and continue participating in and contributing to relevant international conferences and events with the aim of promoting the realisation of the right to food.[200]

In the same resolution, the Human Rights Council welcomed the valuable work of Mr. Jean Ziegler, and invited him to end the fulfilment of his mandate in 2008.[201] In 2008, Mr. Olivier de Schutter was given the mandate of the Special Rapporteur on the Right to Food. At the time of writing he still holds this mandate. To fulfil his mandate, the Special Rapporteur presents annual reports to the Human Rights Council and to the General Assembly (the Third Committee) on his studies and activities. He also undertakes country visits to investigate the situation with regard to the right to food. Furthermore, he receives individual complaints concerning the right to food and discusses them with the state parties involved. Finally, he generally promotes the right to food.[202]

5.8 The High Commissioner for Human Rights

The post of High Commissioner for Human Rights was installed by the General Assembly in 1993.[203] The Office of the UN High Commissioner for Human Rights (OHCHR) functions as a co-ordinator of human right activities within the UN, and is mandated to promote and protect the enjoyment and full realisation of all rights established in the UN Charter and International Human Rights laws and treaties.[204] In the High Commissioner's Strategic Management Plan 2010-2011, the importance of the right to adequate food is mentioned frequently.[205] In response to the World Summit's objective 7.4, the High Commissioner for Human Rights convened expert consultations on the right to food,[206] resulting in reports that contributed to a better clarification of the meaning of its meaning.[207]

---

[200] Human Rights Council, Resolution 6/2, 27 September 2007, paragraph 2.

[201] Human Rights Council, Resolution 6/2, 27 September 2007, paragraph 1 and 5.

[202] Available at: www2.ohchr.org/English/issues/food/index.htm.

[203] A/RES/48/141, 7 January 1994, General Assembly Resolution. Information about the work of the High Commissioner for Human Rights is available at: www.ohchr.org.

[204] A/RES/48/141, 7 January 1994, General Assembly Resolution, paragraph 4.

[205] High Commissioner's Strategic Management Plan 2010-2011, available at: www.ohchr.org.

[206] Also on invitation of the Commission on Human Rights: Commission on Human Rights Resolutions 1997/8, 3 April 1997, paragraph 7; 1998/23, 17 April 1998, paragraph 5-6; 2000/10, 17 April 2000, paragraph 7; 2001/25, 20 April 2001, paragraph14-15.

[207] E/CN.4/1998/21, 15 January 1998; E/CN.4/1999/45, 20 January 1999; E/CN.4/2001/148, 30 March 2001.

5.9 The Food and Agricultural Organization

The FAO (Food and Agricultural Organization) of the United Nations was founded in 1945.[208] FAO's mandate is to raise levels of nutrition, improve agricultural productivity, better the lives of rural populations and contribute to the growth of the world economy.[209] FAO initiates study, promotion and assistance to fulfil its tasks.[210] The FAO organised the World Food Summits in 1996, 2002 and 2009. In the declaration of the World Food Summit: five years later, a request was formulated by the FAO to set up an International Working Group (IGWG) that would write a code of conduct for the implementation of the right to adequate food. As a result, the voluntary guidelines were adopted in 2003.[211] During its existence the FAO has financed various research projects concerning the right to food. The FAO also makes annual reports, sometimes in co-operation with other institutions, in which the status of global food security is monitored.

5.10 The World Health Organization

The World Health Organization (WHO) was founded in 1948 as a result of the World Health Conference (1946). Its primary task is to promote the highest attainable standard of health.[212] Nutrition is considered to be of great importance, for it is an *'input to and foundation for health and development.'*[213] WHO often co-operates with FAO and other UN institutions for research projects and worldwide activities. WHO frequently publishes its research conducted in the field of nutrition.[214]

5.11 The United Nations Children's Fund

The United Nations Children's Fund (UNICEF) focuses on the protection of human rights for children in general, and has developed several action programmes. Health, education, equality and protection are the cornerstones of UNICEF's activities. Nutrition programmes are of fundamental importance to achieve its goals. In its activities, UNICEF is led by a human rights based approach.[215]

---

208 FAO constitution, 1945, preamble.

209 FAO constitution, 1945, preamble.

210 FAO, constitution Article 1.

211 The voluntary guidelines to support the progressive realization of the right to adequate food in the context of national food security, adopted by the 127th session of the FAO council, November 2004.

212 Constitution WHO Article 1, adopted by the International Health Conference in 1946.

213 Constitution WHO Article 2 (i), adopted by the International Health Conference in 1946.

214 Available at: www.WHO.int/nutrition/publications/en/.

215 Available at: www.UNICEF.org.

5.12 The International Fund for Agricultural Development

The International Fund for Agricultural Development (IFAD)[216] is a fund, and does not act directly. IFAD supports third world countries with a rural poverty problem by funding projects under special low-cost conditions. IFAD also makes reports and drafts on a very regular basis with information and advice on how to cope with rural poverty problems.[217]

5.13 The World Food Programme

The World Food Programme was installed after an experimental period of three years on the advice of the FAO[218] with permission of the General Assembly.[219] Its purpose is to provide food in emergency situations and to stimulate economic and social development. Therefore, WFP teams work directly in famine environments and try to set up food programmes.[220]

5.14 The United Nations System Standing Committee on Nutrition

The United Nations System Standing Committee on Nutrition (SCN)[221] was founded in 1976, and is a co-ordinating organ for all UN institutes and organs and other NGO's that deal with nutrition-related issues. The institutions that are co-ordinated are the members of the SCN, including the FAO, WHO, UNICEF, UNHCR, ILO, UNDP and the UN itself. The annual meetings must result in a common approach to worldwide malnutrition. The SCN has also installed working groups that are involved with specific questions on malnutrition. SCN work often results in reports or advice intended to help the international community to solve nutrition problems.

6. Afterword

From many wells the right to food streams throughout the international human rights system with a final goal, which is to be consumed by humans. It would be interesting to analyse whether the water in these wells finally reaches the people it was intended for and why this does or does not succeed. This could be done by analysing the impact of the various international human rights systems on the

[216] Available at: www.IFAD.org.
[217] Since the publication of the Rural Poverty Report 2001, The Challenge of Ending Rural Poverty, IFAD frequently reports on rural poverty issues. The publications are available at: www.ifad.org.
[218] FAO Resolutions 1/61, 1961 and 4/64, 1965.
[219] General Assembly Resolution 1714 (XVI), 19 December 1961; General Assembly Resolution 2095 (XX), 20 December 1965.
[220] See for WFP's strategic objectives: WFP strategic plan 2008-2011, available at: www.wfp.org.
[221] The former ACC/SCN; available at: www.unscn.org.

national legal systems and how (and why) the right to food is interpreted in the national courts and politics.[222]

The right to food is mostly discussed with a focus on article 11 ICESCR. This contribution shows that there is much more under the sun worth analysing. Would different wells of the right to food also lead to different practices? For example: is there a difference in the effect of global human right treaties and regional equivalents? Do the treaties that offer protection to a discriminated group have a stronger effect due to their specific aim when compared to treaties that are addressed to people or states in general? Do the international attempts to clarify and unify the content of the right to food indeed lead to the intended results? Or might the diversity expressed in this international human rights system lead to the opposite effect instead of an increased understanding and unification of the right to food?

The development of the human right to food is a continuing process, which was clearly demonstrated during the time this chapter was written: the author had to adapt its content several times to keep up with the latest developments. Therefore, the author would like to recommend reading the reports of the Special Rapporteur on the right to food in order to remain informed about the most recent developments in this area. The Special Rapporteur publishes his reports frequently and keeps his readers well informed.[223]

References

Alston P. and K. Tomasevski (eds.), 1984. The right to food. SIM, Martinus Nijhoff Publishers, The Hague, the Netherlands, 228 pp.

FAO, 2009. Press Release: Global platform for food security revitalized. Committee on World Food Security, 35th Session, 14, 15 and 17 October 2009, Agenda Item III, Reform of the Committee on World Food Security, final version, Rome, Italy.

Knuth, L., 2009. The right to food and indigenous people, how can the right to food help indigenous people? FAO, Rome, Italy.

Oshaug, A., 2009 The Netherlands and the Making of the voluntary Guidelines on the Right to Food. In: Hospes, O and Van der Meulen, B. (eds.). Fed up with the right to food? Wageningen Academic Publishers, Wageningen, the Netherlands, 192 pp.

[222] The author is currently working on a PhD in this area.

[223] The reports are available at: www2.ohchr.org/english/issues/food/index.htm.

Chapter 3
The freedom to feed oneself: food in the struggle for paradigms in human rights law

Bernd van der Meulen

'The problem of this world is not a shortage of food but a shortage of justice.'[224]

1. Introduction

Discussions focusing on the human right to food have greatly advanced the academic understanding of human rights (Craven, 1995: 110-114). This chapter addresses the development of legal doctrine on human rights in general and the right to food in particular, focusing on the interaction between the two. Tenacious obstacles to the full realisation of the right to food gave rise to new approaches to this human right and may do so again.

2. Human rights

2.1 What are human rights?

Today in many legal orders, human rights are seen as basic norms representing fundamental values. They are found in international treaties of a fundamental nature and in national constitutions. They can be understood to define and protect the position of people within the given legal order, in particular the national state. Each and every person is entitled to a sphere of autonomy (freedom) within the legal order and to minimum conditions of dignified life. For this purpose human rights set limits to the domain of the state and impose requirements on the state as well. Although human rights are often elaborated and detailed in legislation of an ordinary nature, human rights stand out from this legislation in the sense that they set limits and put requirements on this legislation. Human rights provide a yardstick for the quality of the legal order and a safeguard against its deterioration.

2.2 Inalienable rights

Legal philosophy gives two different answers to the question from where human rights derive their fundamental nature. For some authors human rights exist only insofar as they are recognised by law. For others they represent values of a pre-legal nature. In this view human rights are part and parcel of human nature.

[224] Oxfam Novib, 30 October 2006, quoted in Dutch in Metro 31 October 2006: 'Het probleem van deze wereld is niet gebrek aan voedsel, maar gebrek aan rechtvaardigheid.'

They are inalienable, that is to say they cannot be given[225] or taken away, but remain with each person always. Law can only be law in a true sense when and if it respects human rights. This view is known as *natural law*. The philosophy of natural law presumes the existence of law independent from human will and creation (Luhmann, 1965: 38-52). By contrast the philosophy of *legal positivism* recognises as law only the rules set by the organs endowed with rule-making.

2.3 Human rights in the concept of law

Without taking a position on the discussion between natural law and legal positivism, this chapter focuses on human rights recognised in legal documents. Regardless of one's view on the *nature* of human rights, in *practice* to exercise their full force they need to have found form, structure and recognition within the legal order. In other words: positivation (Ishay, 2004; Lewis, 2003) is necessary for the realisation of human rights.

According to general legal theory, the heart of the concept of law is that society is organised in a peaceful manner, through regulation of human behaviour by general rules creating enforceable rights and duties. The factor of enforceability distinguishes rules of law from rules of morality, religion, philosophy and the like.

Legislative requirements that are not enforceable because they are not intended or not suitable (e.g. too vague) for that purpose, are sometimes referred to as *lex imperfecta*. Such imperfect law may have a symbolic value. Rules of law are derived from sources of law. Usually four sources are recognised: international treaties, legislation, custom and case law.

The rights of people that are recognised as human rights correspond to obligations of the state that holds power over the people concerned. Therefore human rights can only flourish in states adhering to the rule of law. The rule of law states that the power of the state is based on and limited by the law and that the law can be enforced against the state. For human rights to work in this way, it is vital that powers within or over the legal order are willing and able to uphold them against the state.

2.4 Treaties and legislation

The way sources of law function technically, differs between countries. This is in particular true for international treaties. International treaties are agreements

[225] By coincidence I am reviewing this line on my flight back from Champaign-Urbana University (Illinois) where among other things I gave a presentation on the human right to feed oneself. When flying in I signed an immigration form, waiving my right to appeal any decision taken by an immigration officer regarding my admissibility. So much for an 'inalienable' right of access to court (Article 12 ICCPR).

between states creating mutual rights and obligations between these states. If the aim of the treaty is to create rights and obligations for other entities as well – in the way human rights treaties do with regard to people – the national constitutional system dictates how this result is to be achieved. Basically two types of systems exist: monistic and dualistic systems. Monistic systems consider the national and the international legal order as one. In a monistic system international treaties can be applied the same way as national legislation. A dualistic system regards international treaties as exclusively binding upon the state. To have effect within the national legal order, the treaty provisions first have to be transposed into national legislation.

When, later in this chapter, the legal character of human rights is discussed, the focus is not on this issue. For monistic systems the question is if and how human rights provisions are applied regardless of whether they originate in a treaty or national constitution (or other law), and for dualistic systems, if and how these provisions in transposing national legislation are applied.

3. Dawn of modern human rights

3.1 Four freedoms

Although the notion that man has an inherent dignity which is worthy of protection, can be traced back for centuries (Ishay, 2004; Lewis, 2003) and from the era of enlightenment onward a variety of fundamental freedoms found protection in several declarations and constitutions, the Second World War is usually seen as the starting point of the development of human rights as we know them today. In particular the state of the union delivered by president Franklin Delano Roosevelt on 6 January 1941 is a landmark. In this address to Congress – now referred to as the Four Freedoms speech[226] – he called for preparedness for war, but he also presented an outlook on the world as it should be after the expected hostilities had subsided:

> 'In the future days, which we seek to make secure, we look forward to a world founded upon four essential human freedoms.
> The first is freedom of speech and expression – everywhere in the world.
> The second is freedom of every person to worship God in his own way everywhere in the world.

[226] The text is available in numerous places on the Internet for instance: http://www.libertynet.org/edcivic/fdr.html; http://www.roosevelt.nl/en/four_freedoms/; http://www.ourdocuments.gov/doc.php?flash=true&doc=70; http://www.quotedb.com/speeches/four-freedoms; http://www.u-s-history.com/pages/h1794.html; http://www.sagehistory.net/worldwar2/docs/FDR4Free.html. An audio version can be found at: http://www.americanrhetoric.com/speeches/fdrthefourfreedoms.htm.

> The third is freedom from want, which, translated into world terms, means economic understandings which will secure to every nation a healthy peacetime life for its inhabitants – everywhere in the world.
> The fourth is freedom from fear, which, translated into world terms, means a world-wide reduction of armaments to such a point and in such a thorough fashion that no nation will be in a position to commit an act of physical aggression against any neighbor – anywhere in the world.
> That is no vision of a distant millennium. It is a definite basis for a kind of world attainable in our own time and generation. That kind of world is the very antithesis of the so-called 'new order' of tyranny which the dictators seek to create with the crash of a bomb.
> To that new order we oppose the greater conception – the moral order. A good society is able to face schemes of world domination and foreign revolutions alike without fear.
> Since the beginning of our American history we have been engaged in change, in a perpetual, peaceful revolution, a revolution which goes on steadily, quietly, adjusting itself to changing conditions without the concentration camp or the quicklime in the ditch. The world order which we seek is the cooperation of free countries, working together in a friendly, civilized society.
> This nation has placed its destiny in the hands and heads and hearts of its millions of free men and women, and its faith in freedom under the guidance of God. Freedom means the supremacy of human rights everywhere. Our support goes to those who struggle to gain those rights and keep them. Our strength is our unity of purpose.
> To that high concept there can be no end save victory.'

3.2 The Universal Declaration of Human Rights

The Second World War did change the face of history. To avoid its atrocities from ever occurring again, a serious attempt has been made to bring forth from the ashes a better world. Immediately after the war, on the 24 of October 1945, a new global organisation, the 'United Nations' was called into being.[227] Article 1 of the United Nations Charter indicates the purposes of the UN. In the third paragraph it expresses the purpose:

> 'To achieve international co-operation in solving international problems of an economic, social, cultural, or humanitarian character, and in promoting and encouraging respect for human rights and for fundamental freedoms for all without distinction as to race, sex, language, or religion.'[228]

[227] For a more detailed account of the connection between the Second World War and the emergence of human rights and international organisation, see Kennedy (2005).
[228] On the crucial role of human rights in the framework of the UN, see also Articles 13(1), 55(c), 56, 62(2), 68 and 76(c) of the UN Charter.

Roosevelt did not live to see his vision come true. He passed away on 12 April 1945. His widow Anna Eleanor Roosevelt, however, played an important role. She chaired the UN commission that drew up the Universal Declaration of Human Rights (UDHR). The UDHR gave substance to the up till then rather imprecise notion of 'human rights and fundamental freedoms' by providing a catalogue of the rights and freedoms generally accepted as fundamental to the preservation of human freedom and dignity and the development of the human personality. It did not purport to create new obligations, or to broaden the commitments under the Charter. Indeed every effort was made at the time of the proclamation of the Declaration to deprive it of any legal or compulsory attribute and to safeguard it against such an attribution in the future (Moskowitz, 1958: 52). In other words: it was not law.

The Declaration was proclaimed by the General Assembly on 10 December 1948 'as a common standard of achievement for all peoples and all nations, to the end that every individual and every organ of society, keeping this Declaration constantly in mind, shall strive by teaching and education to promote respect for these rights and freedoms and by progressive measures, national and international, to secure their universal and effective recognition and observance, both among the peoples of Member States themselves and among the peoples of the territories under their jurisdiction' (Moskowitz, 1958: 25).

The structure chosen in the United Declaration of Human Rights (UNDHR) provided a model for later documents on human rights. On the one hand it expresses the individual rights, on the other it gives a clause to determine the limits of these rights. In the UDHR this limitations clause is uniform for all rights. It is given in Article 29:

> 'Article 29 UDHR
> 1. Everyone has duties to the community in which alone the free and full development of his personality is possible.
> 2. In the exercise of his rights and freedoms, everyone shall be subject only to such limitations as are determined by law solely for the purpose of securing due recognition and respect for the rights and freedoms of others and of meeting the just requirements of morality, public order and the general welfare in a democratic society.
> 3. These rights and freedoms may in no case be exercised contrary to the purposes and principles of the United Nations.'

Limitations to the exercise of human rights are only acceptable if they serve a specified worthy purpose. Furthermore, limitations may only be set by a heavy procedure predating the acts to which they apply ('determined by law'). In later limitations clauses and in case law the required connection between the purpose and the limitation of the exercise of a human right is further specified ('necessary') and a limit of proportionality is set to the impact of the limitation.

The Article that applies to the subject of this chapter, food security, is Article 25:

> 'Article 25
> 1. Everyone has the right to a standard of living adequate for the health and well-being of himself and of his family, *including food*, clothing, housing and medical care and necessary social services, and the right to security in the event of unemployment, sickness, disability, widowhood, old age or other lack of livelihood in circumstances beyond his control.'

The UDHR was intended as a first step towards an international bill of rights. A covenant on human rights with the legally binding force of a treaty was to be the second and decisive step towards achievement of that goal (Moskowitz, 1958: 58). However, the impact of the cold war within the UN was such that hopes to take this second step soon after the proclamation of the UDHR rapidly evaporated. The UN had to pass on the momentum in the development of human rights to regional organisations like the Council of Europe.[229]

4. Human rights in Europe

4.1 Council of Europe

Within the Council of Europe, the subject matter of the UDHR was split into two groups of human rights. In the language of the Four Freedoms speech they can be called the 'freedoms of' versus the 'freedoms from'. The 'freedoms of' have been labelled 'civil and political rights', 'the freedoms from' have been labelled 'social, economic and cultural rights'. In this distinction the influence of the cold war is felt. The civil and political rights like the freedom of speech and the freedom of worship embody the values of the first world, while state obligations concerning the welfare of the people, like the right to work and the protection from hunger, stand in a closer relation to the values of the socialist second world (Halász, 1966). Or, to put it more mildly; civil and political rights are at the core of liberal approaches to the rule of law (in German: *Liberaler Rechtsstaat*) and

[229] The Council of Europe is an international organisation active in co-operation in human rights and related fields of mutual interest of its Member States. The Council of Europe is distinct from the European Union and should not be confused with the EU Council. Information on the Council of Europe is available at: http://www.coe.int/.

social, economic and cultural rights of the social approaches to the rule of law (in German: *Sozialer Rechtsstaat*).[230]

Within the Council of Europe, on 4 November 1959, less than two years after the proclamation of the UDHR, the European Convention on the Protection of Human Rights and Fundamental Freedoms (ECHR) – also known as the Treaty of Rome[231] – was concluded. It took the Council of Europe another eleven years to agree on a European Social Charter (ESC).

The legal power of the ESC is minimal in comparison to the ECHR. The ECHR is endorsed by a supranational dispute resolution structure including a European Court of Human Rights where both Member States and individuals can bring complaints about infringements on human rights by Member States. The case law of the European Court of Human Rights has gained great authority. Even though the reasonings of the Court are sometimes criticised, convictions often have a considerable impact in the Member States concerned. The precedents set by the Court of Human Rights are by themselves sources of law adding weight and detail to the corpus of human rights law. In this way the provisions in the ECHR developed into hard law that can be successfully invoked by individuals – also – in cases before the national courts of the Member States. This will be illustrated by some examples.

4.2 The European Court of Human Rights

The role of the European Court of Human Rights is vital for at least three reasons:
1. it provides an instrument for individuals to uphold their rights;
2. it provides a forum where Member States can be held accountable for their (dis)respect of human rights to the people who's rights are at stake;
3. case law is in itself a source of law that further develops the human rights set out in the Convention.

In this section we will focus on the latter. The European Court of Human Rights set out to elaborate in its case law the human rights requirements laid out in the Convention.

[230] The expression 'rule of law' refers to the notion that in order to protect the people from abuse of power, public authorities should be bound by the law setting enforceable requirements and limits on the exercise of power. It is closely related to the separation of powers doctrine (*trias politica*), demanding that legislative, executive and judiciary powers be distributed over different organs in the state that are capable of keeping each other in check. The expression *Rechtsstaat* refers to a state where such conditions apply.

[231] Again to be distinguished from the Treaty of Rome that established the European Economic Community.

A landmark case is ECHR 26 April 1979 'Sunday Times'. The plaintiff was the London-based publisher of the Sunday Times newspaper. This newspaper had planned to publish an article concerning a case that was *sub judice* (pending before the court). The Attorney General issued a writ against Times Newspaper Limited in which he claimed an injunction to stop them from publishing. On 17 November 1972 the court granted the injunction. The injunction aimed to prevent the article from affecting and prejudicing the tribunal and the witnesses in the case concerned. After exhausting national remedies the plaintiff engaged in the procedures under the ECHR. This interference by a public authority in the exercise of the freedom of expression would entail a violation of Article 10 ECHR if it did not fall within one of the exceptions provided in the limitations clause in the second paragraph of this Article. The ECHR had to examine in particular if this interference was 'prescribed by law' and 'necessary in a democratic society'. The first question was answered to the affirmative. The unwritten (customary) common law requirements of due process were accepted as 'prescribed by law'.[232] The measure was, however, considered disproportionate and therefore not 'necessary in a democratic society' as required in the limitations clause. The interference with the freedom of expression was not justified, therefore the judicial authorities in the UK had been beholden to abstain from them. The freedom of speech acquired a strong position in the legal systems of the Member States of the European Council as well. In the Netherlands, for instance, spatial planning measures to ban TV-antennas (necessary to receive opinions) or neon lights (used to express political views) were successfully contested before the courts.[233]

Another landmark 'Ärtze für das Leben'[234] also concerns free speech. Austrian authorities had banned an anti-abortion demonstration for fear that opponents would cause riots. The Court of Human Rights ruled that in a case like this, authorities may not ban a demonstration but must use their police forces to protect the exercise of the freedom of speech from hostile audiences. This case had a similar impact on practice in the Member States.[235]

In a ruling of 1985 the Kingdom of the Netherlands was found to infringe the right of access to justice.[236] Under Dutch administrative law a system of legal protection existed that gave interested parties the possibility to appeal to the Crown from decisions taken by lower administrative authorities. The Crown is the Queen acting under the responsibility of the government, e.g. the Minister concerned. Decisions to be taken in Crown appeal were prepared by the Council

[232] In other words, the exercise of human rights may be limited not only by legislation, but also by other sources of law e.g. case law or customary law.
[233] ARRvS 10 October 1978, AA 1979, 477 (*Antenneverbod Leerdam*); HR 24 January 1967, NJ 67, 270 (*Nederland ontwapent*).
[234] ECHR 21 June 1988, A 139.
[235] For the Netherlands see Van der Meulen (1993).
[236] ECHR 23 October 1985 (No. 97), 8 E.H.R.R. 1 'Benthem v. Netherlands'.

of State; an independent advisory body to the government. The involvement of the government in deciding conflicts with administrative authorities was branded by the Court of Human Rights as an infringement of the right to a fair trial, as it did not give access to 'an independent and impartial tribunal' as required by Article 6 of the European Convention. As a consequence a major reform of Dutch administrative law had to be – and was – undertaken to ensure that access to an independent and impartial tribunal was provided for everyone who is entitled to this under the convention.

Other examples are closer to the subject matter of this chapter than free speech and access to court. The Öneryildiz case,[237] for example, deals with the right to life protected in Article 2 ECHR. On 28 April 1993 a methane explosion in a 'rubbish tip' caused a landslide burying some ten slum dwellings including one belonging to Mr. Öneryildiz. Thirty-nine people died, including nine members of the Öneryildiz family. The authorities had been warned of the risks in an expert report issued on 7 May 1991. The Court ruled that there had been a violation of the right to life. It concluded that the administrative authorities knew or ought to have known that the inhabitants of the slum areas at issue were faced with a real and immediate risk both to their physical integrity and their lives on account of the deficiencies of the municipal rubbish tip. The authorities failed to remedy those deficiencies and could not, moreover, be deemed to have done everything that could reasonably be expected of them within the scope of their powers under the regulations in force to prevent those risks materialising. Furthermore, they failed to comply with their duty to inform the inhabitants of the area of those risks, which might have enabled them – without diverting State resources to an unrealistic degree – to assess the serious dangers in continuing to live in the vicinity of the rubbish tip.

The Court went on to say that the obligations deriving from Article 2 do not end at the obligation to protect lives. Where lives have been lost in circumstances potentially engaging the responsibility of the State, that provision entails a duty for the State to ensure, by all means at its disposal, an adequate response – judicial or otherwise – so that the legislative and administrative framework set up to protect the right to life is properly implemented and any breaches of that right are repressed and punished. In the instant case, in a judgement of 4 April 1996, the Istanbul Criminal Court sentenced the two mayors held responsible for the accident to suspended fines of TRL 610,000 (an amount equivalent at the time to approximately 9.70 euros) for negligent omissions in the performance of their duties within the meaning of the Criminal Code. According to the European Court

[237] ECHR 19 June 2002, Öneryildiz v. Turkey (Application no. 48939/99). The case was referred to the Grand Chamber which delivered judgement on 30 November 2004. The Grand Chamber basically upholds the judgement quoted above laying more emphasis on the authorities' failure to prevent the accident and less on their failure to give warning.

of Human Rights, it cannot be said that the manner in which the Turkish criminal justice system operated in response to the tragedy secured the full accountability of State officials or authorities for their role in it and the effective implementation of provisions of domestic law guaranteeing respect for the right to life, in particular the deterrent function of the criminal law. In short, it concluded that there had also been a violation of Article 2 of the Convention in its procedural aspect, on account of the lack, in connection with a fatal accident provoked by the operation of a dangerous activity, of adequate protection 'by law' safeguarding the right to life and deterring similar life-endangering conduct in future.

If rulings of the European Court of Human Rights can necessitate Member States to rebuild their national legal infrastructure, as the Benthem example shows, the impact of the European Convention on Human Rights and Fundamental Freedoms is indeed tremendous. Human rights protected in this Convention provide individuals with legal entitlements which they can uphold against the state. The human rights as set out in the European Convention are being fleshed out and solidified in case law.

From the case law of the Court follows that not only may states not infringe human rights, they must also protect people from infringements by others, take action against those who do infringe other people's rights, provide information people need to protect themselves, and – finally – provide a system of law in which human rights can be fully exercised.

4.3 Progressive realisation

The above shows that the promotion of respect for human rights and freedoms 'by progressive measures, national and international, to secure their universal and effective recognition and observance, both among the peoples of Member States themselves and among the peoples of the territories under their jurisdiction' as envisaged at the proclamation of the UDHR has advanced within Member States of the Council of Europe to the point where civil and political rights are solid legal entitlements that interested parties can uphold in the courts of law.

4.4 The European Social Charter

The European Social Charter was not brought under the jurisdiction of the European Court of Human Rights. Some aspects mentioned in Article 25 UDHR are discernable in the ESC, such as the protection of health (Article 11), social security (Article 12-14 and 16), protection of families, mothers and children (Article 16-17) but food is not mentioned as a human right.

The rights included in the ESC never acquired the same legal impact as those included in the ECHR. It seems very likely that the absence of the power of the European Court of Human Rights greatly contributes to this state of affairs.

5. The UN Bill of Rights and the right to adequate food

5.1 Two covenants

Finally, the UN acquired its Bill of Rights. Despite the principle of indivisibility and interdependence of human rights that was asserted by the UDHR and reaffirmed time and again, the division of the single set of rights set forth in the UDHR, into two distinct categories was continued. The view of the states that were in favour of two separate covenants, in particular the United States and other Western states, prevailed over the view of the states that were in favour of creating one single document, including several countries adhering to the socialist line of thought (Arambulo, 1999: 17).

On 16 December 1966 two separate UN Treaties were adopted and opened for accession: the International Covenant on Civil and Political Rights (ICCPR) and the International Covenant on Economic, Social and Cultural Rights (ICESCR).

A right to adequate food is recognised in the ICESCR in Article 11:

> 'Article 11 ICESCR
> 1. The States Parties to the present Covenant recognize the right of everyone to an adequate standard of living for himself and his family, including *adequate food*, clothing and housing, and to the continuous improvement of living conditions. The States Parties will take appropriate steps to ensure the realization of this right, recognizing to this effect the essential importance of international co-operation based on free consent.
> 2. The States Parties to the present Covenant, recognizing the fundamental right of everyone *to be free from hunger*, shall take, individually and through international co-operation, the measures, including specific programmes, which are needed:
>   a. To improve methods of production, conservation and distribution of food by making full use of technical and scientific knowledge, by disseminating knowledge of the principles of nutrition and by developing or reforming agrarian systems in such a way as to achieve the most efficient development and utilization of natural resources;
>   b. Taking into account the problems of both food-importing and food-exporting countries, to ensure an equitable distribution of world food supplies in relation to need.'
>
> (italics added).

The limitations clause in the ICESCR is Article 4:

> 'The States Parties to the present Covenant recognize that, in the enjoyment of those rights provided by the State in conformity with the present Covenant, the State may subject such rights only to such limitations as are determined by law only in so far as this may be compatible with the nature of these rights and solely for the purpose of promoting the general welfare in a democratic society.'

In the clause we recognise the example set in the UDHR in that it makes requirements on the procedure ('determined by law') and the purpose ('promoting the general welfare in a democratic society') of limitations imposed on the enjoyment of human rights. The requirement of proportionality is expressed more clearly than in the UDHR ('only in so far as this may be compatible with the nature of these rights').

5.2 The right to adequate food

Article 11 ICESCR quoted above is the most general, and for this reason the most important, codification of the human right to adequate food.[238] We have seen above that it is absent in the Treaties of the Council of Europe.

The right to adequate food enjoys particular attention from several UN bodies.[239] The Sub-Commission on Prevention of Discrimination and Protection of Minorities appointed Asbjørn Eide as Special Rapporteur on the Right to Food. Later the Economic and Social Council appointed Jean Ziegler[240] as Special Rapporteur on the Right to Food. The reports of both these Special Rapporteurs and the General Comments by the Economic and Social Council greatly contributed to the development of understanding of the right to adequate food. They endeavour to make abstract notions like 'adequate food' sufficiently concrete for application to specific cases and to elaborate on the legal character of the right to food.

5.3 The concept of adequate food

The implementation of the right to food in the Member States is supported by general comments by the Committee on Economic, Social and Cultural Rights (UN) and by Voluntary Guidelines to support the progressive realisation of the right to adequate food in the context of national food security (FAO).

---

238 But it is not the only one. For a further elaboration on the sources of the right to food, see Chapter 2 of Wernaart (in this volume). Kearns (1998) argues that the right to food can be seen as customary law (as well).

239 For an overview of the UN bodies engaged in the right to food see Chapter 2 by Wernaart (in this volume).

240 Now succeeded by Olivier de Schutter, see Chapter 2 by Wernaart (in this volume).

In its 20th session the UN Economic and Social Council approved General Comment 12: the right to adequate food.[241] In this general comment the notion of adequate food in Article 11 ICESCR is further elaborated. The key considerations are:

'Adequacy and sustainability of food availability and access

7. The concept of adequacy is particularly significant in relation to the right to food since it serves to underline a number of factors which must be taken into account in determining whether particular foods or diets that are accessible can be considered the most appropriate under given circumstances for the purposes of article 11 of the Covenant. The notion of *sustainability* is intrinsically linked to the notion of adequate food or food security, implying food being accessible for both present and future generations. The precise meaning of 'adequacy' is to a large extent determined by prevailing social, economic, cultural, climatic, ecological and other conditions, while 'sustainability' incorporates the notion of long-term availability and accessibility.
8. The Committee considers that the core content of the right to adequate food implies:
   The availability of food in a quantity and quality sufficient to satisfy the dietary needs of individuals, free from adverse substances, and acceptable within a given culture;
   The accessibility of such food in ways that are sustainable and that do not interfere with the enjoyment of other human rights.
9. *Dietary needs* implies that the diet as a whole contains a mix of nutrients for physical and mental growth, development and maintenance, and physical activity that are in compliance with human physiological needs at all stages throughout the life cycle and according to gender and occupation. Measures may therefore need to be taken to maintain, adapt or strengthen dietary diversity and appropriate consumption and feeding patterns, including breast-feeding, while ensuring that changes in availability and access to food supply as a minimum do not negatively affect dietary composition and intake.
10. *Free from adverse substances* sets requirements for food safety and for a range of protective measures by both public and private means to prevent contamination of foodstuffs through adulteration and/or through bad environmental hygiene or inappropriate handling at different stages throughout the food chain; care must also be taken to identify and avoid or destroy naturally occurring toxins.
11. *Cultural or consumer acceptability* implies thc need also to take into account, as far as possible, perceived non nutrient-based values attached to food and food consumption and informed consumer concerns regarding the nature of accessible food supplies.

[241] E/C.12/1999/5, CESCR d.d. 12 May 1999.

12. *Availability* refers to the possibilities either for feeding oneself directly from productive land or other natural resources, or for well functioning distribution, processing and market systems that can move food from the site of production to where it is needed in accordance with demand.
13. *Accessibility* encompasses both economic and physical accessibility.' (italics added).

In short, the right to adequate food is understood as addressing issues of:
- nutrition;
- safety;
- cultural acceptability.

It is interesting to note that although the human right to food is usually regarded in the context of food security in the strict quantitative sense of availability of food, matters of food safety and food ethics can be approached from a human rights angle as well. The latter may come to bear not only with regard to food considered to be *kosher* or *hallal*, but also in the contemporary discussion on the acceptability to certain consumers of the application of certain techniques in food production like irradiation, the application of growth hormones and genetic modification.

5.4 The right to water

The concept of food in human rights law has always included drink. In its General Comment no. 15 the Economic and Social Council made explicit that the right to water is under the protection of Article 11 (and 12) ICESCR. The right to water encompasses both drinking water and 'access to water for the irrigation of food crops...particularly for subsistence farming and vulnerable peoples' (Ziegler, 2003; see also Ziegler, 2001). In 1998 the Sub-Commission on Prevention of Discrimination and Protection of Minorities appointed Mr. El Hadji Guissé as Special Rapporteur on the Right to Water. See Chapter 12 on water by Roth and Warner (in this volume).

5.5 Access to land

The Special Rapporteur Ziegler believes that access to land is another of the key elements necessary for eradicating hunger in the world. In his opinion, this means that policy options such as agrarian reform must play a key part in countries' food security strategies, in which access to land is fundamental (Ziegler, 2002). See Chapter 10 on access to land by Grossman.

6. The paradigm of generations

As is elaborated in more detail in Chapter 5 by Frank Vlemminx (in this volume), while the Netherlands reformed their legal infrastructure to comply with the requirements of the ECHR, it rejected applicability of the right to food in the national legal order out of hand without recourse to the limitations clause. In many other UN Member States the right to food does not fare much better. This striking difference with the approach by the European Court of Human Rights described above, is partly explained by the splitting up of the subject matter of the UDHR into civil and political rights on the one hand and social, economic and cultural rights on the other hand.

6.1 Social and economic rights

Unlike civil and political rights, the legal binding character of social, economic and cultural rights is under debate. The underlying argument is that civil and political rights require non-interference from public authorities. Non-interference is called a 'negative' obligation, that is an obligation to abstain from action. It does not require specific resources to abstain from torturing,[242] from persecuting political opponents, from curbing free speech,[243] from curtailing worship,[244] etc. The realisation of social, economic and cultural rights, on the other hand, requires positive action from the authorities or market parties. Houses must be built, schools equipped, the environment protected and food produced.

Each and every government can live up to negative obligations, but the extent to which positive obligations can be fulfilled largely depends on the availability of resources and the effectiveness of policies. For these and similar reasons, social, economic and cultural rights are perceived as policy guidelines rather than as provisions of law (Alkema, 1982; Bossuyt, 1975; Koekkoek and Konijnenbelt, 1982). This understanding is sometimes reflected in the wording of social, economic and cultural rights. Emphasis is then on the aims to be pursued by the authorities, not on entitlements of certain right holders.

6.2 Limits to the enjoyment of economic, social and cultural rights

It would be entirely consistent with the previous section to argue that, where there is no provision of law, there is no reason to label hunger or other non-realisations of economic, social or cultural rights as an 'infringement' or 'violation' of a human right, or even to apply a limitations clause.

Indeed the limited legal weight of economic, social and cultural rights is not defended on the basis of the limitations clause, but of Article 2 ICESCR. This article reads in its first paragraph:

[242] Article 7 ICCPR, Article 3 ECHR.

[243] Article 19 ICCPR, Article 10 ECHR.

[244] Article 18 and 27 ICCPR.

> 'Each State Party to the present Covenant undertakes to take steps, individually and through international assistance and co-operation, especially economic and technical, to the maximum of its available resources, with a view to achieving progressively the full realization of the rights recognized in the present Covenant by all appropriate means, including particularly the adoption of legislative measures.'

This provision can be read as saying that the obligations of the Member State are limited to making an effort to achieve progress. That is precisely how this provision was understood as late as 1990 by the UN Committee on Economic, Social and Cultural Rights itself. General Comments no. 3 addresses 'The nature of States parties obligations'. A crucial line is found in paragraph 9 of this general comment:

> 'The principal obligation of result reflected in article 2(1) is to take steps 'with a view to achieving progressively the full realization of the rights recognized' in the Covenant. The term 'progressive realization' is often used to describe the intent of this phrase. The concept of progressive realization constitutes a recognition of the fact that full realization of all economic, social and cultural rights will generally not be able to be achieved in a short period of time. In this sense the obligation differs significantly from that contained in article 2 of the International Covenant on Civil and Political Rights which embodies an immediate obligation to respect and ensure all of the relevant rights.'

6.3 Generations

The distinction that doctrine makes between civil and political rights on the one hand and economic, social and cultural rights on the other, is sometimes expressed in terms of generations (Alkema, 1982; Bossuyt, 1975; Koekkoek and Konijnenbelt, 1982). The former 'classic' human rights belong to the first generation, the latter to the second generation.[245] For this reason I think it is appropriate to use this notion of 'generations' to indicate the underlying paradigm.

7. The holistic paradigm

7.1 The Special Rapporteurs on the Right to Food

At the time when General Comment 3 was adopted, another approach to state obligations was already emerging. As mentioned above, Asbjørn Eide was appointed Special Rapporteur on the Right to Food by the Sub-Commission on the Prevention of Discrimination and Protection of Minorities (a sub-commission to the Commission on Human Rights). He delivered his first report in 1987. The report was updated

[245] Sometimes a third generation is mentioned encompassing collective rights of self-determination and development.

in 1998 and 1999. Jean Ziegler, in 2000 appointed Special Rapporteur on the Right to Food by the Economic and Social Council, published his first report in 2001. His analysis of the right to food follows very similar lines to Eide.

Eide contests the paradigm that views economic, social and cultural rights as mere policy aims as well as its underlying argument. His first report contained a detailed analysis of the nature of state obligations for human rights, noting that international human rights law, like other parts of international law, is legally binding for states; it is not a set of recommendations, but requirements that have to be implemented. Three levels of obligations of states were identified: the obligations to respect, protect and fulfil. Failure to perform any one of these three obligations, according to Eide constitutes a violation of the rights.[246]

I understand this reasoning to encompass the idea that there are always two sides to the human rights coin – even if this is not expressed clearly in all the documents. On the one hand there are the rights belonging to people as individuals and as groups, on the other hand there are obligations of the state. The Special Rapporteurs elaborated a doctrine stating that human rights are indivisible and that all human rights entail both negative and positive state obligations. For this reason I call this new paradigm 'holistic'.

For example, the freedom of speech requires more than non-interference. There must be an adequate infrastructure for a free press as well. The right to food does not require that governments feed the whole population. It includes the 'negative' obligation for authorities not to interfere with the population's means to feed themselves.

A fourth state obligation has been identified so that in this doctrine four different obligations are distinguished: an obligation to respect (non-interference),[247] to protect (from the interference by third parties),[248] to facilitate (promote; support self-realisation)[249] and to fulfil[250] (provide in case of emergency). It is interesting to note that the European Court of Human Rights in the Öneryildiz case shows consideration with the state, qualifying non-fulfilment as an infringement only in a situation where fulfilment was possible 'without diverting state resources to an unrealistic degree'.

[246] Eide's report elaborated on earlier publications (Alston and Eide, 1984) and must have been inspired by discussion in literature as well. See, for example, the contributions of Shue and Van Hoof to Alston *et al.* (1984) and earlier sources mentioned there, Eide among them.
[247] In the case law mentioned in this chapter: Sunday Times provides an example of the obligation to respect.
[248] In the case law mentioned in this chapter, *Ärtzte für das Leben* and Öneryildiz provide examples of the obligation to protect.
[249] In the case law mentioned in this chapter, Öneryildiz provides an example of the obligation to promote (by providing adequate information).
[250] In the case law mentioned in this chapter, Benthem provides an example of the obligation to fulfil.

See for a more detailed account of state obligations Chapter 4 by Eide (this volume).

7.2 General Comment no. 12

The new approach advocated by the Special Rapporteurs has been taken up in official UN policy documents. General Comment no. 12 on the right to food – mentioned above – has the following to say on the subject of state obligations:

> 'The right to adequate food, like any other human right, imposes three types or levels of obligations on States parties: the obligations to respect, to protect and to fulfil. In turn, the obligation to fulfil incorporates both an obligation to facilitate and an obligation to provide. The obligation to respect existing access to adequate food requires States parties not to take any measures that result in preventing such access. The obligation to protect requires measures by the State to ensure that enterprises or individuals do not deprive individuals of their access to adequate food. The obligation to fulfil (facilitate) means the State must proactively engage in activities intended to strengthen people's access to and utilization of resources and means to ensure their livelihood, including food security. Finally, whenever an individual or group is unable, for reasons beyond their control, to enjoy the right to adequate food by the means at their disposal, States have the obligation to fulfil (provide) that right directly. This obligation also applies for persons who are victims of natural or other disasters.'

7.3 Justiciability

This analysis of obligations can make the right to food instrumental to rights-based approaches to food security only if the individual or group protected by this right can hold the authorities to their obligations, that is to say if they lay a claim to it. For this reason it is an important question whether or not the right to food is 'justiciable'. In the view of the Special Rapporteurs, the right to food should be accepted – at the very least as far as its negative obligations are concerned – as justiciable. That is to say that individuals should have access to the (national) courts to defend their right to food in case national authorities unduly restrict it. Justiciability of the right to food, turns this right into an entitlement.

Once justiciability of negative obligations is accepted, it is accepted that the right to food in particular, and social, economic and cultural rights in general, are a matter for the courts. Once that foot is in the door, the question can be addressed to what extent exactly they are a matter for the courts.

The examples from the case law of the European Court mentioned above show that not only are strictly negative obligations suitable for litigation (Sunday Times case), but the obligation to protect as well (*Ärzte für das Leben*; *Öneryildiz*) and even positive obligations like the obligation to facilitate (Öneryildiz) and the obligation to provide (Benthem, the Netherlands).[251]

8. The human right to feed oneself

From these discussions emerges a human right to acquire access to food that is sufficient in quality to satisfy one's dietary needs, that is sufficient in safety not to cause adverse effects and that is culturally acceptable. This right can be exercised by applying within the prevailing legal environment all available economic and legal means.

It is the states' duty not to interfere unduly with the exercise of this right and to protect the enjoyment of this right from interference by others. In this sense this right can be labelled a 'freedom' as is done in the title of this chapter. In situations where people, due to circumstances beyond their control, cannot get sustainable access to adequate food, the state must pursue a policy to improve the situation and in case of emergency must lend a helping hand.

9. Catching up on doctrine

Although the Special Rapporteurs and several other authors present their paradigm as a matter of law – it is in law that negative obligations related to social, economic and cultural rights are justifiable – legal practice scarcely confirms this view. Courts seem to adhere to the generations paradigm described in Section 7 rather than to the holistic paradigm described in Section 7. In this sense the new paradigm reflects a legal-political ideal rather than fact.

Experience with the European Convention on Human Rights and Fundamental Freedoms shows that backing by the courts is seminal in giving human rights their place at the forefront of the legal system. The European Court of Human Rights has provided the national courts with the impetus and legitimisation to take on their national legal systems and policies. Where no supranational court exists, the development of the human right to food into a legal entitlement has to come from the national courts or the national constitutional legislatures. Unfortunately there are few signs that national courts have the courage to perform this task and also the legislatures show little inclination to include the right to food in national constitution.

[251] On positive obligations in the context of the European Convention on Human Rights, see Harris *et al.* (1995: 19-22).

This observation is not without exceptions. India seems to apply the ICESCR including the right to food.[252] South-Africa has included the right to food in its constitution (Khoza, 2004). A court in Belgium applied the right to food.[253] An application of the right to food by the supreme court in Switzerland heralded a change of the constitution to explicitly include it (Vlemminx, 2002: 31-38). There may be one or two other exceptions but the rest is silence.

On 10 December 2008 the UN General Assembly adopted an optional protocol to the ICESCR opening the possibility in one form or another for individuals to 'communicate' alleged violation of provisions in the Covenant to an 'expert body'. Although it does not seem very likely that at the global level where the UN is operating agreement can be reached on a system comparable to the European Court of Human Rights, any system that empowers concerned parties to force their state to discussion on the merits of a case in the light of the ICESCR will in all probability propel the development of the rights set out in the ICESCR far beyond the clichés 'too vague, too general to apply as yardstick for dispute resolution' (Vlemminx, 2003) and therewith beyond the traditional doctrine of the generations paradigm.

10. Towards a paradigm of 'real' human rights?

The holistic paradigm of human rights still has a long way to go before it will acquire full acceptance. Its strength is that it focuses attention on obligations that complement human rights. A shortcoming may be that it presents human rights as 'relative' rights, that is to say as rights defining the *relationship* between people and their national authorities. This relativity is valuable in a context where the national state holds decisive power over its inhabitants. Too much emphasis on state obligations, however, runs the risk of providing other actors with an excuse not to be beholden to respect human rights. Increasingly, decisive power over people is held by other actors than the state alone. This is not only true in failing or weak states. In a rapidly globalising world power relations increasingly stretch beyond national borders. The role of commercial enterprises and international organisations needs to be taken into account in the development of human rights theory and practice.

In private law a distinction is made between 'relative rights' or 'personal rights' and 'absolute rights' or 'real rights'. Relative or personal rights are confined to a specific relationship – usually of a contractual nature – between specific parties. Absolute or real rights can be invoked against anybody. Property rights are the key

[252] Ahluwalia (2004); Human Rights Law Network (2009); see also Supreme Court of India, 2000 SOL Case no. 673 mentioned by Ziegler (2003).

[253] Arrêt no. 36/98 of 1 April 1998 of the Belgian Court of Arbitration; arrêt no. 51/94 of 29 June 1994 of the Belgian Court of Arbitration.

example of absolute rights. The owner can upholds his right (to a thing – res in Latin, therefore 'real') against anybody regardless of a specific legal relationship.[254]

If we accept that human rights are rights that everybody at all times holds for the sole fact of being human and are not just rights the citizen holds because of his relationship to a certain state, then the option is open for other parties beside the state to have obligations as well.

The idea in itself is not new. The issue of so-called horizontal effect of human rights has been discussed off and on at least since the 1980s. Horizontal effect is the applicability of human rights (notions) in relationships between citizens. Discussions on corporate social responsibility often have a human rights dimension. More recent are discussions on extra-territorial obligations.

It was the right to food that inspired the development of the holistic paradigm of human rights law. It was again the right to food around which discussion on extra territorial obligations emerged. Article 11 ICESCR explicitly calls for international co-operation. In 2001 three NGOs, Brot für die Welt, EED (Evangelischer Entwicklungsdienst) and FIAN (FoodFirst Information and Action Network), presented the first civil society report on extraterritorial obligations to the UN Committee on Economic, Social and Cultural Rights. A second report, 'Germany's extraterritorial human rights obligations. Introduction and six case studies' followed in October 2006. Around the same time George Kent[255] posted a draft book on the Internet for comments. Extraterritorial obligations refer to the human rights obligations of a state towards people in other countries. The international protection of human rights should not be limited by exclusively emphasising the obligations of states towards people living within their territories. States have always acted beyond their borders. Their acts and omissions directly impact on the enjoyment of human rights in other countries (Brot, 2006). The tripartite classification of obligations to respect, protect and fulfil human rights not only applies to obligations on the national level, but also to extraterritorial obligations. Extraterritorial obligations require states to act unilaterally, as well as in a bi- and multilateral context. Extraterritorial obligations emanating from international human rights represent a much needed instrument for designing and implementing policies which provide a response to the challenges of globalisation (Brot, 2006).

Human rights need a legal infrastructure to be exercised as 'real' rights. This infrastructure can in part be derived directly from international treaties in the

[254] Interestingly, in an introduction to Dutch law we find a reference to human rights as example of absolute rights not pertaining to an object (Hage *et al.*, 2001: 151).

[255] Available at: http://www2.hawaii.edu/~kent/000 GORF FILE NAMES.doc, now published as *Global Obligations for the Right to Food* by George Kent (2008). Available at: http://www2.hawaii.edu/~kent/publications.html.

context of monistic legal orders,[256] but will need a legislative imbedding as well. This imbedding may be part of the state obligation to protect. A recognition as right will in many settings be sufficient to be applied in the context of tort law, that is to say to provide a legal basis for civil action. Protection through criminal law will always need specific legislation. The European Court of Justice emphasised in the Öneryildiz case the importance of criminal prosecution in case of violation of human rights.

Recognition of human rights as absolute rights that protect each human being in every social relationship is conceivable both within the context of the generations paradigm (with regard to civil and political rights) and within the context of the holistic paradigm. It need not replace either of them, but may serve to add another dimension.

11. Concluding remarks

The generations paradigm served to position civil and political rights as frontrunners in the development of human rights law. It thus provided a model showing how human rights can acquire their rightful place within the legal order. The holistic paradigm shows that further development is not served by a distinction of rights but of obligations, that in principle apply to all human rights. The obligations that make civil and political rights suitable for application as legal norms, apply to economic, social and cultural rights as well. The proposed real paradigm engages the discussion as to what extent and how the protection that human rights provide against the state apply in other social relations as well.

The generations paradigm still seems to be dominant in legal practice. The holistic paradigm based on the principle of indivisibility, interdependence and interrelatedness of all human rights as advocated by the UN is so far little more than theory.

This legal theory on the human right to feed oneself as recognised in Article 11 ICESCR and several other international documents is well developed and ready for application in practice. This development of theory on the right to food has greatly contributed to the understanding of economic, social and cultural rights in general. History shows that the final push, to turn convictions on human rights into positive law, must come from the judiciary. As long as an international body is missing it is up to the national courts to take on this responsibility. Unfortunately, in general so far they seem reluctant to do so.

[256] I.e. legal orders that recognise international law (under certain conditions) as a source of law without prior transposition into national law.

Civil society, and in particular legal scholars, must relentlessly point to this responsibility, show the possible ways and mobilise shame.

Some hope may be drawn from initiatives at UN level that have led to the creation of a procedure to address problems with regard to compliance with state obligations under the ICESCR. Such initiatives deserve our full support.

Bearing in mind that the better should not become the enemy of the good, it is worth addressing the question of whether further development of a paradigm of real human rights is useful for the further development of the understanding and application of human rights.

References

Ahluwalia, P., 2004. The implementation of the right to food at the national level: a critical examination of the Indian campaign on the right to food as an effective operationalization of Article 11 of ICESCR. Center For Human Rights And Global Justice Working Paper. Economic, Social And Cultural Rights Series Number 8.

Alkema, E.A., 1982. De internationale sociale rechten en het Nederlandse recht. In: Van der Ven J.J.M. (ed.), Het Europees sociaal handvest. Ars Aequi Libri, Nijmegen, the Netherlands.

Alston, P. and Eide, A., 1983. Advancing the Right to Food in International Law. In: Eide, A., Eide, W.B., Goonatilake, S. and Gussow J. (eds.). Food as a Human Right. The United Nations University, Tokyo, Japan.

Alston, P. and Tomasevski, K. (eds.), 1984. The Right to Food. SIM Publisher, the Hague, the Netherlands.

Arambulo, K., 1999. Strengthening the Supervision of the International Covenant on Economic, Social and Cultural Rights. Theoretical and Procedural Aspects. Intersentia, Antwerp, Belgium.

Bossuyt, M.J., 1975. La distinction juridique entre les droits civils et politiques et les droits economiques, sociaux et culturels. Human Rights Law Journal 8: 783-813.

Craven, M., 1995. The International Covenant on Economic, Social and Cultural Rights. A Perspective on its Development. Clarendon Press, Oxford, UK.

Eide, A., Eide, W.B., Goonatilake, S. and Gussow J., 1983. Food as a Human Right, United Nations University. The United Nations University, Tokyo, Japan.

Eide, A., 1987. Human Rights Study Series No. 1, United Nations publication.

Hage, J.C. Schlössels, R.J.N. and Wolleswinkel, R., 2001. Recht, vaardig en zeker. Een inleiding in het recht. Boom Publishing, the Hague, the Netherlands.

Halász, J.(ed.), 1966. Socialist Concept of Human Rights. Budapest, Hungary.

Harris, D.J., O'Boyle, M. and Warbrick, C., 1995. Law of the European Convention on Human Rights. Oxford Publishers, London, UK.

Human Rights Law network, 2009. Right to Food. 4th edition Human Rights Network, New Delhi, India

Ishay, M.R., 2004. The History of Human Rights; from ancient times to the globalization era. University California Press, Berkely, CA, USA.

Kearns, A.P., 1998. The Right To Food Exists Via Customary International Law. Suffolk Transnational Law Review 223: 255-256.

Kennedy, D., 2005. Freedom from fear. The American People in Depression and War, 1920-1945. Oxford History of the United States, Volume IX, Oxford, UK.

Kent, G., 2008. Global Obligations for the Right to Food (Another World Is Necessary: Human Rights, Environmental Justice, and Popular Democracy). Rowman & Littlefield, New York, NY, USA.

Koekkoek, A.K. and Konijnenbelt, W., 1982. Het raam van hoofdstuk 1 van de herziene Grondwet. In: Koekkoek, A.K., Konijnenbelt, W. and Krijns, F.C.L.M. (eds.) Grondrechten; commentaar op hoofdstuk 1 van de herziene Grondwet aangeboden aan mr. H.J.M. Jeukens. Ars Aequi Libri, Nijmegen, pp: 21-31.

Khoza, S., 2004. Realising The Right To Food In South Africa: Not By Policy Alone - A Need For Framework Legislation. South African Journal On Human Rights 20: 664-683.

Lewis, J.E., 2003. A Documentary History of Human rights. A Record of the Events, Documents and Speeches that Shaped our World. Caroll and Graff, New York, NY, USA.

Luhmann, N., 1974. Grundrechte als Institution. Ein Beitrag zur politischen Soziologie. Dunker & Humbolt, Berlin, Germany.

Van der Meulen, B.M.J., 1993. Ordehandhaving. Actoren, instrumenten en waarborgen. Kluwer, Deventer, the Netherlands.

Moskowitz, M., 1958. Human Rights and World Order. The Struggle for Human Rights in the United Nations. Oceana Publication, New York, NY, USA.

Novak, M., 2003. Introduction to the International Human Rights Regime. Roul Wallenberg Institute, Lund, Sweden.

Vlemminx, F.M.C., 2002. De autonome rechtstreekse werking van het EVRM. De Belgische en Nederlandse rechtspraak over verzekeringsplichten ingevolge het EVRM. Deventer, the Netherlands.

Ziegler, J. 2001. The right to food. Report to the UN General Assembly, distributed 23 July 2001, A/56/210.

Ziegler, J., 2002. The right to food. Report to the UN General Assembly, distributed 27 August 2002, A/57/356.

Ziegler, J., 2003. The right to food. Report submitted by the Special Rapporteur on the right to food, 10 January 2003, E/CN.4/2003/54.

Chapter 4
State obligations for human rights: the case of the right to food

Asbjørn Eide[257]

1. The right to adequate food is a human right

'Freedom from want is at the core of the human rights project.'
(Roosevelt, 1941)

One of the most influential inspirations in modern times for the adoption of a universal system of human rights was the famous 'Four Freedoms' speech of US president, Franklin D. Roosevelt, in January 1941. His vision for the future world order, to be initiated when World War II was brought to an end, included four freedoms: freedom of speech, freedom of faith, freedom from want and freedom from fear. While his immediate target audience on that occasion was the United States Congress, his vision reflected sentiments felt by groups in many parts of the world who were dismayed at the crisis of civilisation caused by extreme nationalism, totalitarianism and militarism, economic chaos in the wake of the Great Depression, and the legacy of slavery and colonialism with its racist underpinnings and consequences.

The world was in need of a new foundation. On that basis the initiatives were taken to establish a world-wide United Nations, which after four years of preparation was created in 1945. One of the main purposes was:

> '...to achieve international co-operation in solving international problems of an economic, social, cultural, or humanitarian character, and in promoting and encouraging respect for human rights and for fundamental freedoms for all without distinction as to race, sex, language, or religion'

In 1948 the Universal Declaration of Human Rights (UDHR) was adopted, setting out the list of human rights and freedoms to which the Charter referred. The right to food was further elaborated in the International Covenant on Economic, Social and Cultural Rights (ICESCR) in 1966. Full realisation of human rights requires that the right to adequate food and housing is ultimately universally enjoyed, as part of an adequate standard of living. The United Nations Committee on Economic,

[257] This chapter draws on a longer chapter by the author in Eide, W.B. and Uwe K. (eds.), 2007. Food and Human Rights in Development. Volume II, Intersentia, Antwerp-Oxford.

Social and Cultural Rights (CESCR), established to monitor that Covenant, stated, in its General Comment No 12 on the right to food, that:

> 'The right to adequate food is realized when every man, woman and child, alone or in community with others, has physical and economic access at all times to adequate food or means for its procurement[258].'

Even though considerable progress has been made since 1948, today hundreds of millions are undernourished and many more millions are malnourished due to the inadequate composition of their daily diet, with widespread and tragic consequences of illness and premature death. The commitments made by states individually and jointly have to be transformed into effective obligations and be seriously implemented in practice. This does not happen by itself: deliberate actions are required at many levels. Many steps have already been taken in that direction; much more has to be done in the future.

The following is the message of United Nations High Commissioner for Human Rights Louise Arbour on the occasion of Human Rights Day, 10 December 2006:

> 'The awareness of the stranglehold of poverty on billions of men, women and children around the world, and of how this state of deprivation and misery compromises our common future, has never been higher. Yet, despite an increasingly sophisticated understanding of the complex makeup of poverty, ranging from exclusion and discrimination to a skewed international trade system, approaches to poverty reduction are still often tinged with appeals to charity or altruism.
>
> On this Human Rights Day, we reaffirm that freedom from want is a right, not merely a matter of compassion. Fighting poverty is a duty that binds those who govern as surely as their obligation to ensure that all people are able to speak freely, choose their leaders and worship as their conscience guides them.
>
> All countries, independent of national wealth, can take immediate measures to fight poverty based on human rights. Ending discrimination, for example, will in many cases remove barriers to decent work and give women and minorities access to essential services. Better distribution of collective resources and good governance, exemplified by tackling corruption and ensuring the rule of law, are within the reach of every state.

[258] E/C.12/1999/5, 12 May 1999, Committee on Econonomic, Social and Cultural Rights, General Comment 12. paragraph 6

> But as much as States bear the primary responsibility for their own development, the international community must also meet the commitments it has made to support the efforts of developing countries. Many rich countries have yet to meet development assistance targets they have accepted, yet they continue to spend ten times more on military budgets. They also spend nearly four times their development assistance budget – an amount almost equal to the total gross national product of African countries – to subsidize their own domestic agricultural producers. Indifference and a narrow calculus of national interests by wealthy countries hamper human rights and development just as damagingly as discrimination at the local level.
>
> At the 2005 World Summit, global leaders recognized that development, peace and security and human rights are mutually reinforcing. In a world where one in every seven people continues to live in chronic hunger, and where inequalities between and within countries are growing, our ability to reach the goals the Summit reaffirmed in order to 'make poverty history' will remain in serious doubt if we do not tackle poverty as a matter of justice and human rights.'

With the High Commissioner's appeal, let me turn to the theme of this book. I will review the processes by which the vaguely formulated state obligations for the right to food have been given more specific and meaningful content through progressive interpretation and elaboration, and with a view to arriving at a common framework of how to operationalise human rights obligations also more generally. The main purpose is to revisit the analytical framework developed and widely used for exploring state obligations for human rights and which must be understood in this broader perspective.

2. On the right to food in human rights law, and on human rights as 'law'

2.1 Human rights as such are not initially subjective rights under positive law

Human rights emerge as broadly consensual conceptions of ideal rights. We shall not here enter into the historical evolution of human rights conceptions in parts of the West from the 17th century onwards, but focus on the developments starting with the adoption of the United Nations Charter in 1945 and the Universal Declaration of Human Rights (UDHR) in 1948, which is the foundation of the subsequent developments of human rights.

On 10 December 1948 when the General Assembly adopted the UDHR, it spelled out the list of rights which were intended but not made explicit in the UN Charter. The rights contained in the UDHR were not, however, universal in any possible meaning of the word. They were not universally recognised and far from being universally enforced or enjoyed.

The General Assembly therefore proclaimed the Universal Declaration of Human Rights as:

> '...a common standard of achievement for all peoples and all nations, to the end that every individual and every organ of society, keeping this Declaration constantly in mind, shall strive by teaching and education to promote respect for these rights and freedoms and by progressive measures, national and international, to secure their universal and effective recognition and observance, both among the peoples of Member States themselves and among the peoples of territories under their jurisdiction.'

The list of rights contained in the UDHR represented a future-oriented *project*. It was not a statement of existing facts or existing legal rights. Making human rights universal, including the right to food, was intended to be a forward-looking process. It could not be expected to move forward automatically, nor did that happen: it had to be pursued by deliberate action to follow up on the concerns, at the local, national and international levels. Committed actors had to take up the challenge, to push and to persuade where they could, in line with what the General Assembly said in the preamble to the Universal Declaration: '...by teaching and education to promote respect for these rights and freedoms, and by progressive measures to secure their universal recognition and observance'.

This required efforts to expand the moral and social commitments towards these rights, to set legally binding standards, to promote their acceptance, develop the institutions of supervision and monitoring, and to form alliances to pressure and encourage governments to live up to their commitments.

2.2 When and how do human rights become 'law' and subjective rights?

Human rights were met with scepticism by legal doctrine and large parts of the legal profession during the first decades of the first post-Second World War period. When the UDHR was adopted, legal doctrine had difficulties considering human rights listed therein to be 'law'. The dominant position had for some time been that only enforceable law is law. Both legal positivism and legal realism therefore had considerable difficulty in embracing human rights as part of the legal discipline. Subjective rights required enforceable duties, but human rights were seen almost as a throwback to natural law which had been chased out of legal doctrine during the 19th century through the works of Jeremy Bentham, John Austin and numerous others.

Whether law was conceived as 'commands from the sovereign backed by sanctions', typical of legal positivism from the time of John Austin (1832/1995) onwards, or 'prophesies of what the court will do'[259](1897) associated with Justice Holmes, the 'father' of American legal realism (which inspired Scandinavian legal realism), human rights law could not be law unless and to the extent that the rights contained therein had been absorbed into positive, national law and/or made justiciable in the practice of the courts. Positive law is however changeable law; thus in a domestic legal system, torture could be illegal yesterday, legal today, and illegal tomorrow. The variable content of law under legal realism depends heavily on the orientation of the judges; hence the selection of the judges of the Supreme Court of the United States is a highly political issue.

The purpose of human rights is to set limits on and give direction to the exercise of power and choice, and to establish guarantees that cannot be neglected or set aside by legislators, administrators or judges. Decision-making within any of the branches of governments must be within the framework set by these rights. This is a very ambitious purpose; so how can it have an impact on powerful figures who make the real decisions in real life?

Elements of a binding social contract limiting and directing exercise of public power had emerged in some constitutions, written or unwritten, during preceding centuries. Some legal orders had recognised and applied the principle of non-retroactivity of laws and other aspects of legality. Freedom from torture, freedom of religion, and elements of the freedom of expression and freedom of association had been recognised as components of 'fundamental rights' and cornerstones of the social contract. But even these limited guarantees were swept aside by European authoritarian states in the 1930s, abetted by legal positivism. Even the guarantees that continued to be respected were only limited fragments of the more comprehensive human rights system established after World War II.

Some states had included significant elements of economic and social rights in their domestic legal order during the first half of the 20th century, but many others had not. Political rights were lacking as positive, enforceable subjective rights in a great number of states.

The adoption of the UDHR was therefore the beginning of a development of human rights as law not only *constraining* but also giving *direction* to domestic political or legal options, a function which was greatly strengthened with the adoption of human rights conventions containing legally binding obligations for states under

[259] Full quote: 'The prophecies of what the courts will do in fact, and nothing more pretentious, are what I mean by the law', Available at: http://www.gutenberg.org/catalog/world/readfile?fk_files=223748&pageno=4.

international law. Two questions nevertheless remained. The conventions were of course part of international law and binding as such on States Parties, but were they binding under national law? This was the question of domestic applicability of human rights conventions. The CESCR has approached this question in the light of two principles of international law.

The first principle, as reflected in Article 27 of the Vienna Convention on the Law of Treaties of 1969, is that:

> '[A] party may not invoke the provisions of its internal law as justification for its failure to perform a treaty'.

In other words, States should modify the domestic legal order as necessary in order to give effect to their treaty obligations. The second principle is reflected in Article 8 of the UDHR, according to which:

> 'Everyone has the right to an effective remedy by the competent national tribunals for acts violating the fundamental rights granted him by the constitution or by law.'

The second was slightly different from the first. In order to be treated as subjective rights, the human rights in question had to be justiciable, that is, referring to matters which are appropriately resolved by the courts and norms which are self-executing (capable of being applied by courts without further elaboration). But there are no common legal standards determining when a right or a legal provision is justiciable or self-executing; there are considerable differences in the general approach of different legal systems in this respect. Many components of economic, social and cultural rights are capable of immediate implementation.

Still, it must be recognised that some of the rights in the ICESCR for quite some time in the future will be considered by many of the States Parties to be non-justiciable and/or non-self-executing. When this is so, are the human rights contained therein not proper rights?

The answer to this is that they are indeed proper human rights, and that to the greatest extent possible they should be transformed into subjective, legal rights, which has to take place through a process of legislation and/or through development of customary law through consistent administrative practice. Before that is done, human rights can be understood as the right to demand policies and legislation which genuinely pursue the objective to realise the right to the fullest extent possible. This is the essence of ICESCR Article 2(1), whereby states are required to take measures for the progressive realisation of the rights 'by all appropriate means, including particularly the adoption of legislative measures'.

2.3 The scope of state obligations

The purpose of international human rights conventions is to create legally binding obligations for states to implement the human rights contained in the conventions. States become legally bound by ratifying or acceding to the conventions. The most important conventions for that purpose were the two International Covenants adopted in 1966: the one on Civil and Political Rights (ICCPR) and the ICESCR. The latter had as per 1 January 2010, 160 States Parties, more than three fourth of the total number of existing independent states.

By Article 11 of the ICESCR, States Parties have recognised the right of everyone to food as part of an adequate standard of living, while the Convention on the Rights of the Child of 1989 (CRC) contains important obligations concerning the right of every child to 'nutritious food' in its Article 24, and to an adequate standard of living in Article 27.

'Under ICESCR Article 2(1), each State Party undertakes:

> ...to take steps...to the maximum of its available resources, with a view to achieving progressively the full realization of the rights recognized in the present Covenant'.

The CESCR has in its general Comment No. 12 on the right to food stated that:

> '[T]he right to adequate food, like any other human right, imposes three types or levels of obligations on States parties: the obligations to respect, to protect and to fulfil. In turn, the obligation to fulfil incorporates both an obligation to facilitate and an obligation to provide'

This was the first time in the practice of the CESCR that it made use of what has since been referred to as the *tripartite typology of state obligations* under international human rights law. It has since been extensively relied on in the practice of the Committee and widely used in the literature, though it is not entirely uncontroversial (see further below).

As recognised by the Committee, the typology was first introduced in the study entitled *Right to adequate food as a human right* which I prepared in my capacity as Special Rapporteur for the United Nations Sub-Commission on the Prevention of Discrimination and Protection of Minorities.

3. The tripartite typology of state obligations: origin and purpose

3.1 Overcoming the ideological schism

The two main human rights covenants were adopted in 1966 after lengthy debates deeply affected by the ideological polarisation during the Cold War. The Universal Declaration on Human Rights had been adopted in 1948, encompassing the comprehensive package of human rights that had evolved over the three preceding centuries, mainly in the Western countries. While the Communist countries of Eastern Europe abstained from voting on the adoption of the UDHR in 1948, all Western countries voted in its favour, though internally there were some differences of opinion concerning the inclusion of economic and social rights.

In simple terms, two conflicting conceptions of the functions of human rights emerged: one, most strongly articulated in US policies, that the role of human rights was to ensure freedom of the state; the other, articulated by the Socialist countries of the time, that their function was to ensure freedom through the active role of the state, providing basic security and satisfaction for all basic needs.

In the polarised and shallow debate resulting from ideological rhetoric, the notion gained some currency that civil – and to some extent political – rights were fundamentally different from economic and social rights. Linked to this was the notion that civil rights imposed only duties of abstention by the State from interference in the fundamental freedoms of the individual (often called 'passive duties'), while economic and social rights were considered mainly to consist of duties for the State to use its resources to provide for the needs of people. Labouring under this conception, it was understandable that the liberal mind in many Western circles was hesitant or negative to economic and social rights, while such a conception of those rights fitted well with the dominant thinking in the Socialist countries of Eastern Europe and the Soviet Union. This rather simplistic conception then also contributed to the decision to split the human rights listed in the Universal Declaration into two separate covenants in 1966.

As part of my examination of the rights contained in the two covenants when I was a member of the United Nations Sub-Commission on Human Rights, I became increasingly convinced that both sides were wrong in their conceptions of state obligations under both sets of rights. Through an analysis of the terms of the relevant provisions I came to the conclusion that all human rights start from the premise of the freedom of the individual. Hence, a major function of human rights is indeed to ensure freedom *from* the state, as had been traditionally held in Western liberal thinking. But freedom from the state is not enough, since it would leave open the insecurity and fear arising from violence, exploitation, corruption and fraud by private parties against other private parties. The liberal states had indeed emerged because of the need felt also to have a degree of *protection* through the state against such harmful acts by private parties.

Nor could such protection take care of all justified needs in modern societies. The state had to take care of some common goods which could not be dealt with solely by private action. Provision of education for all, at least primary school education, was obviously one of these; provision of social security or at least a social safety net and an extensive health care system had generally emerged as necessities in the complex industrial and post-industrial societies.

The conclusion was therefore that, rather than a simple division between 'passive' duties associated with civil and political rights and 'active' duties associated with economic, social and cultural rights, it was much more appropriate to refer to three different types of obligations which to varying degrees could be applicable to both categories of rights. The full package of human rights requires a combination of all and a proper balance between them. This tripartite approach to state obligations was introduced into the international discourse in the middle of the 1980s through my above-mentioned study for the United Nations on the right to adequate food as a human right. It was subsequently adopted through the Maastricht Guidelines (Van Boven *et al.*, 1998) and later made regular use of by the CESCR and an increasingly broad range of academic works, to the extent that today, nearly everyone working with the normative character of economic, social and cultural rights on a scientific basis or in human rights advocacy is familiar with the tripartite typology, to respect, protect and fulfil.

3.2 Foundation of the typology in the texts of international instruments

In interpreting the texts of the treaties it became possible to show that the different dimensions of state obligations in terms of respect, protect and fulfil were cross-cutting both civil and political rights and economic, social and cultural rights. The assumption was therefore untenable that civil and political rights were only passive and cost-free, and that economic, social and cultural rights obligations nearly always required costly provisions from state resources. The functioning of an adequate system of courts and law enforcement institutions, including legal aid where necessary, required active measures and expenses for the state; so did the operation of political systems based on elections and an appropriate administrative system. On the other hand, the right of everyone to work which was freely chosen or accepted, as set out in ICESCR Article 6, or the right to form trade unions set out in ICESCR Article 8, or the right to establish private educational institutions as set out in ICESCR Article 13, implied that the relevant state obligations were to respect these rights.

3.3 Categories of state obligations

The typology has been taken to imply that it reflects three different levels of burden for the state: to 'respect' is a low burden, involving no or very few resources; to 'protect' involves a higher burden, but still rather limited, while the obligation to provide is understood to imply a much higher level of burden for the state, which is also one of the reasons why the obligation to provide is met with greater hesitance or resistance.

It is quite understandable that the tripartite typology may have been understood in this way, and in regard to financial outlays by the state it is probably true that in most cases the obligation to provide costs more. In a somewhat broader conception of economic costs, however, this is not at all so clear-cut. If we introduce the economic concept of *opportunity costs* the picture looks rather different. As an example, think about a plan to build a huge hydroelectric power plant. It involves the building of dams, the scale of which can be monumental involving construction of hundreds of dams and dikes, and creating enormous reservoirs of water. The land surface flooded by such reservoirs is sometimes huge. The economic benefit of building such a dam can be enormous.

In dealing with such a prospect, the authorities are faced with several human rights problems. Should they respect the rights of the local population to maintain their land, and not to be displaced? This would cost nothing in direct outlay by the state, but would have an enormous opportunity cost. The other alternative is to decide to oust the local population, but provide them with certain benefits, in the short (compensation) and long term (some form of social security to avoid impoverishment). This would require a much higher outlay, but this might be much lower than the opportunity cost of *not* letting the dam be built.

These questions are normally dealt with from the perspective of expropriation and adequate compensation, but from a human rights perspective it is a question of a choice between the obligation to respect (avoiding harm) and fulfil (provide economic and social benefits to those who are harmed). In advance of the initiation of the project, the cost calculus would show that the respect for the land of the local people and their way of life would require very little in terms of financial outlays, but very much in terms of economic opportunities lost.

This is a very stylised and simple example, where one can postulate a direct relationship between the non-respect and the ensuing consequence of the need to provide some social benefits for those negatively affected. In real life, situations are much more complex, but elements of the same dilemmas can very often be found. Economic development of a society can improve the conditions for many, but others can as a consequence be impoverished, and it should be the responsibility of the state to ensure that those who are negatively affected are provided with some assistance in the form of a social safety net. The double rationale would be the necessity to take care of their immediate needs which they no longer can manage on their own because their entitlements have been lost, and also to provide them with opportunities to develop other self-sustaining entitlements.

As a general rule it would be more in line with human rights to respect and protect the functioning entitlements of groups who would otherwise become vulnerable. There are very serious other consequences, quite apart from the economic, when people's traditional way of life is suddenly undermined; even if pecuniary compensation is provided. De-culturation and loss of meaning and orientation is a very common accompaniment to displacement, which can lead to apathy, alcoholism, suicide or criminal behaviour: cases are numerous in practice.

The purpose of these examples is to underline that in broad economic terms, the obligations to respect or to protect may have stronger economic consequences than the obligation to fulfil, whether by facilitation or by direct provision. It is therefore seriously misleading, in dealing with economic and social rights, to focus only – or mainly – on the obligation to fulfil.

3.4 The category of respect for economic, social and cultural rights: its wide significance

The obligation to respect requires states to refrain from interfering with the enjoyment of economic, social and cultural rights. It calls for non-interference by the state in cases where individuals or groups can take care of their own needs without weakening the possibility for others of doing the same. It might imply the recognition of customary land rights, e.g. of indigenous peoples and generally of peasants in rural communities where their land tenure is unrecognised. The obligation to respect is also inherent in the right of persons to seek an income by which they can satisfy their food and other needs through their own free choice of work. Furthermore, the obligation to respect can also require abstention from projects which would undermine the ecological conditions required for the continued production of food in a given area through pollution or man-made desertification.

3.5 The category of 'protection': its wide range

The most important type of state obligation in regard to economic and social rights is the duty to protect, which has a wide range of applications. In general terms, the duty to protect requires the state to prevent private parties (individuals or non-state actors such as corporations) from interfering with the enjoyment of economic, social and cultural rights. Some degree of protection is required against more assertive or aggressive subjects, (e.g. corporations which act in a ruthless way) or protection against fraud, against unethical behaviour in trade and contractual relations, against the marketing and dumping of hazardous or dangerous products, and so on.

Regarding the right to health, the CESCR Committee has stated that obligations to protect include, *inter alia*, the duties of States to adopt legislation or to take other measures ensuring equal access to health care and health-related services provided by third parties; to ensure that privatisation of the health sector does not constitute a threat to the availability, accessibility, acceptability and quality of health facilities, goods and services; to control the marketing of medical equipment and medicines by third parties; and to ensure that medical practitioners and other health professionals meet appropriate standards of education, skills and ethical codes of conduct[260].

State actions to protect are widely used and their legitimacy is widely recognised in most societies, though to various degrees. Protection to safeguard economic and social rights can sometimes be challenged as impediments to the free functioning of the market. But in many contexts it is recognised that the functioning of the market itself needs some degree of protective measures.

There is much controversy within society about the scope of protection versus the scope of freedom of action; particularly in regard to the functioning of the market, but also in social and family affairs. One example is the question of privacy and respect for family life versus the protection of women or children against harm, maltreatment or neglect in the family. What should be the scope of state protection of the child when it comes into conflict with parents' assertion of privacy and respect for family life? This is an area where the Convention on the Rights of the Child requires that the interest of the child shall be the primary concern, which can imply a protective role by the state which goes far beyond what many traditional societies have allowed for.

3.6 The division of 'fulfil' into 'facilitate' and 'provide'

The right to food, like other human rights, can be fulfilled in several ways. What has often been the focus of attention and a source of controversy is to equate 'fulfil' with direct provision: to give poor people food, or money with which to buy the food.

In some regards this is a necessary response, particularly when sudden and unexpected events occur which destroy their pre-existing sources of food security. Natural and manmade disasters fall into this category. It is also necessary to directly provide foods or means for their procurement to persons who for reasons

[260] E/C.12/2000/4, 11 August 2000, Committee on Economic, Social and Cultural Rights, General Comment 14, paragraph 35.

beyond their control cannot take care of their own needs, due to sickness or disability, or for children whose parents die or disappear or otherwise no longer take care of them.

To a large extent, however, the obligation to fulfil can be met by adequate *facilitation*: the organisation of social security systems to which persons at least to some extent contribute themselves, and which caters for situations of subsequent illness, disability or old age. The provision of free educational opportunities is both a direct provision and a step in the facilitation of the ability of the person to care for her or his own situation in future life.

While the typology was formulated as a tripartite set of categories (respect, protect and fulfil) it was clear from a very early stage that further precision would be required, and that there would be a continuum from straightforward and well-known models of facilitation that might require modest amounts of direct financial outlays, to direct provision, which might or might not cost more in relation to the individual recipient. The better the facilitation, the less need there might be for direct provision.

4. Benefits, criticisms and responses

The tripartite typology has made it possible to lay to rest the simplistic distinction between civil and political rights on the one hand and economic, social and cultural rights on the other. Admittedly, implementation of economic and social rights may incur greater costs than civil and political, but these costs do not necessarily consist of financial outlays by the state. They can consist of opportunity costs for the state when it has to respect the rights of private parties, or opportunity costs for corporations who are barred from marketing a potentially profitable pharmaceutical product because of prohibitions by the state made necessary to protect the right to health.

What is more important is that the typology has made it possible to more meaningfully address the composite nature of obligations under economic and social rights. At first sight, the language used to describe the obligations under ICESCR Article 2(1) appears to focus almost exclusively on financial outlays: the States Parties shall take steps 'to the maximum of their available resources' to realise these rights. The awareness that the realisation of economic, social and cultural rights far from always requires a direct outlay of financial resources, has made it possible to give more attention to the other types of state obligations, to respect and to protect.

This is also what has happened in the practice of the CESCR Committee. In the General Comments, as well as its concluding observations addressed to states upon examination of their state reports, much more detail can now be found on the obligations to respect and protect.

5. A note on the main users of the tripartite typology

5.1 Its usefulness for the monitoring bodies

After states ratify the human rights conventions, particularly those containing economic, social and cultural rights, most of them tend to continue their policies more or less as before, while asserting that they do the best they can to reach the aims set out in the particular conventions. They consider the obligations contained in such provisions as ICESCR Articles 2(1) and Article 11 to be so vague and general that it leaves them with a very wide discretion to choose their own policies in these areas.

It is for this reason that international monitoring bodies are essential for the realisation of these rights, and one of the most important tasks of these bodies has been to give more detailed content to the vague, general obligations.

It is for this purpose that the typology has come to be a useful instrument. It has made it possible to elaborate a very detailed set of requirements to states. This elaboration can be found in almost every General Comment adopted since the pioneering GC12 on the right to food in 1999. It drew on and supplemented the other analytical frameworks elaborated before, such as the distinctions between availability, affordability and accessibility, starting with General Comment 4 on the right to housing.

It is of course possible for reluctant governments to point out that a monitoring body, here the CESCR Committee, is not a court and that its General Comments or even the concluding observations from a monitoring exercise are not binding on the states. But this reflects an unwillingness to take part in the international efforts to promote the realisation of human rights, and such unwillingness may have some political costs domestically and internationally.

Much will however depend on the role of the civil society, particularly the non-governmental organisations and community-based organisations. Can they make use of the clarification of state obligations made by the monitoring body?

5.2 Its use by NGOS and civil society: the case of FIAN

It is widely agreed that human rights would not have had as much impact as they have without the active role of non-governmental organisations. In the area of civil and political rights, impressive achievements have been made by organisations such as Amnesty International and Human Rights Watch, among others. In the area of economic, social and cultural rights there is also a growing activism among non-governmental organisations.

Specifically for the right to food, the most instrumental role has been played by FIAN (FoodFirst Information and Action Network), headquartered in Heidelberg, Germany, with local groups active on all continents of the world. Michael Windfuhr, FIAN Secretary-General until 2006, has started to analyse the experiences of FIAN in case-related right to food work. FIAN has found the tripartite typology extremely useful in its work (personal communication, 2006) and the case work experience has made it possible to categorise in considerable detail the different activities of governments under the three main headings respect, protect and fulfil, the latter through facilitation or provision.

One lesson stands out as being very important: the case work of FIAN shows that, if complaint procedures become available, judgements related to availability of resources or the other vague terms of economic and social rights will be possible without major problems. Windfuhr argues that the human rights analyses are normally much less complicated than often argued, when it comes to concrete cases.

5.3 The usefulness for governments: the typology in the new guidelines on the implementation of the right to food

In 1996, at the World Food Summit, Heads of State and Government reaffirmed the right of everyone to have access to safe and nutritious food, consistent with the right to adequate food and the fundamental right of everyone to be free from hunger. It took a long time to follow it up, but in 2003 interested governments did join an open Intergovernmental Working Group established by the FAO Council, to elaborate a set of Voluntary Guidelines to support the progressive realisation of the right to adequate food in the context of national security.[261]

The main elements of the tripartite typology are maintained in these guidelines. The obligation to respect and to protect are explicitly mentioned. The obligation to fulfil is not expressly included but its elements are there: promotion, facilitation, and the establishment of safety nets or other assistance for those who are unable to provide for themselves.' The relevant wording is as follows:

> 'States Parties to the International Covenant on Economic, Social and Cultural Rights (ICESCR) have the obligation to respect, promote and protect and to take appropriate steps to achieve progressively the full realization of

[261] The voluntary guidelines to support the progressive realization of the right to adequate food in the context of national food security, adopted by the 127th session of the FAO council, November 2004. Available at: http://www.fao.org/docrep/meeting/009/y9825e/y9825e00.htm.

> the right to adequate food. States Parties should respect existing access to adequate food by not taking any measures that result in preventing such access, and should protect the right of everyone to adequate food by taking steps so that enterprises and individuals do not deprive individuals of their access to adequate food. States Parties should promote policies intended to contribute to the progressive realization of people's right to adequate food by proactively engaging in activities intended to strengthen people's access to and utilization of resources and means to ensure their livelihood, including food security. States Parties should, to the extent that resources permit, establish and maintain safety nets or other assistance to protect those who are unable to provide for themselves.'[262]

6. Conclusion

The introduction of the tripartite typology of human rights has turned out to be very useful. It has eliminated the sharp distinction which in the past was often made between civil and political and economic, social and cultural rights. It has facilitated great advances in the study of economic, social and cultural rights by highlighting the obligations to respect and protect as being of equal importance to the obligations to facilitate and provide.

It may still be appropriate to consider some human rights, or some aspects of the rights, to be conceived initially as meta-rights, giving the rights-holder a right to demand certain policies, the exact content of which still remains within the discretion of the duty-bearer. Through the practice of the monitoring bodies in their dialogue with the state parties and through their general jurisprudence, that discretion can and should be narrowed considerably, moving the human right from a meta-right to a full-fledged subjective right either through national legislation and coherent administrative practice, or through a direct incorporation of international human rights law as interpreted by the treaty bodies. The clarifications of state obligations have been important steps towards the creation or consolidation of subjective rights.

Human rights are unlikely ever to be completely transformed into positive law, nor should they be. Positive law, or real operative law, is always an imperfect mix or compromise between the ideals as expressed in human rights and the standards required by the most influential or powerful groups or sections of society. Human rights must nevertheless forever remain a source of pressure on authorities and of inspiration for those who want a better social and international order, and thereby humanise social and international relations in line with Article 28 of the Universal Declaration of Human Rights: 'Everyone is entitled to a social and

[262] Voluntary Guidelines, Introduction, paragraph 17.

international order in which the rights and freedoms set forth in this Declaration can be fully realised.'

References

Austin, J., 1832/1995. The Province of Jurisprudence Determined. Cambridge University Press, Cambridge, UK.

Van Boven, T.C. Flinterman, C. and Westendorp, I. (eds.), 1998. The Maastricht Guidelines on Violations of Economic, Social and Cultural Rights. Human Rights Quarterly 20: 691-705.

Holmes, J.O.W. jr., 1897. The Path of Law. Harvard Law Review 10: 457.

Chapter 5
The Netherlands and the right to food: a short history of poor legal cuisine

Frank Vlemminx

1. Introduction

The right to adequate food is a human right. In our modern era human rights are held to be the hallmarks of civilisation. Human rights take the place of different old standards such as monotheism, clothing, technological skill or the three R's (reading, writing and arithmetic) which for centuries figured large in our notion of a civilised nation. While the old standards brought about dissension, racism, war and colonialism, the new ones are aimed at uniting all humans and all nations. The Universal Declaration of Human Rights that was adopted by the General Assembly of United Nations in 1948 makes very clear what it is all about. Article 1 reads:

> 'All human beings are born free and equal in dignity and rights. They are endowed with reason and conscience and should act towards one another in a spirit of brotherhood.'

It is beyond dispute that every single human being on earth is endowed with human rights. No one needs to show s/he actually deserves these rights. Simply belonging to the biological species Homo Sapiens suffices.

The Preamble to the Universal Declaration states that recognition of the inherent dignity and of the equal and inalienable rights of all members of the human family is the foundation of freedom, justice and peace in the world. Disregard and contempt for human rights have resulted in barbarous acts which have outraged the conscience of mankind. Of course, the old standards are not played out but in 1993 the General Assembly firmly fixed their position:[263]

> 'While the significance of national and regional particularities and various historical, cultural and religious backgrounds must be borne in mind, it is the duty of States, regardless of their political, economic and cultural systems, to promote and protect all human rights and fundamental freedoms.'

In 1966 these universal human rights were laid down in two Covenants. Civil and political rights such as the right to life, the right to freedom of expression, the right to freedom of religion or the right to vote and to be elected, were enshrined

[263] U.N. Doc. A/Conf. 157/23, part 1, par. 5.

in the International Covenant on Civil and Political Rights (ICCPR). Economic, social and cultural rights such as the right to work, the right to strike, the right to social security, the right to education or the right to health were enshrined in the International Covenant on Economic, Social and Cultural Rights (ICESCR; hereafter 'the Covenant'). The right to adequate food is part of the economic, social and cultural rights and is to be found in Article 11 of the Covenant.[264]

In most countries human rights are also secured at the national level. This holds for civil and political rights and economic, social and cultural rights. History shows that economic, social and cultural rights tend to be proclaimed in the wake of some kind of major social or political upheavals. In Germany, for instance, codification of these rights took place after World War I, in Italy after World War II, in Portugal after the revolution of 1975 and in Spain 3 years after Franco died. In spite of the notorious 'hungry winter' during the German occupation, the right to adequate food was not laid down in the Dutch Constitution after World War II. Economic, social and cultural rights were not acknowledged until the re-codification of the Dutch Constitution in 1983. The right to food as such was not set out. The government did not think it necessary to mention this right separately as it was considered to be implicit in the right to means of subsistence laid down in Article 20. This Article reads:

> '1. It shall be the concern of the authorities to secure the means of subsistence of the population and to achieve the distribution of wealth.
> 2. Rules concerning entitlement to social security shall be laid down by Act of Parliament.
> 3. Dutch nationals resident in the Netherlands who are unable to provide for themselves shall have a right, to be regulated by Act of Parliament, to aid from the authorities.'

On the other hand, the Netherlands is a state party to the ICESCR. The Covenant took effect in the Netherlands on 11 March 1979, almost three years after it entered into force on 3 January 1976 and more than twelve years after it was concluded.

Unfortunately, history also shows that after the codification of economic, social and cultural rights people may easily feel complacent. The rights set forth fall short of expectations, however beautiful they are. The authorities are inclined to disregard them and preferably the rights are not considered justiciable. As stated by Leckie (1998: 83):

> 'For reasons too well-known and extensive to outline here, most would recoil in horror at the deprivation of freedom of life when violence is involved, but display considerably more tolerance when human suffering or death stem

[264] For a literal quote of the text of this provision, see Chapter 3 by Van der Meulen (in this volume).

> from preventable denials of the basic necessities of life such as food, health care or a secure place to live.'

This fact raises the question of how the Netherlands deals with economic, social and cultural rights in general and the right to adequate food in particular. The Netherlands considers itself a champion of human rights. So is there any judicial monitoring of the enforcement of the Covenant in the Netherlands? Is there any domestic application of the right to adequate food? Before these questions are dealt with, the Dutch legal system that sets their stage is outlined briefly.

2. The Dutch legal system

This section will refer to Dutch case law and parliamentary proceedings, in an approach that follows on from the Dutch system of law. For a better understanding of this chapter some background information concerning this system of law is needed.

In comparative law the Netherlands[265] is included in the civil law family of continental Europe.[266] Among other things, this means that legislation is the most important source of law.[267] The Netherlands has a written constitution[268] with which lower legislation must be in conformity. The courts can annul legislation that departs from the constitution. There is one exception to this rule. The highest level of legislation, called 'formal legislation' or sometimes translated as: 'Acts of Parliament', is legislation forged in co-operation between the national government and parliament, and may not be annulled for unconstitutionality. The idea is that the formal legislator with its democratic basis has the power of final interpretation of the Constitution, not the courts. In all other cases the courts have the power of 'authentic interpretation'. That is to say, the courts have the final say on what legislation means, if it is valid, and more generally what the law is. This gives case law a status almost equal to legislation as a source of law. Although case law exercises considerable influence, there is no doctrine of binding precedent. In theory, the courts are free to depart from earlier decisions given at the same or

[265] Adjective: Dutch. After its seafaring provinces on the western coast the Netherlands is also known as 'Holland'.
[266] For an introduction to Dutch law in English see: Chorus (1999); Taekema (2004); Besselink (2004). See also Jansen and Middeldorp (2001), Researching Dutch Law, available at: http://www.llrx.com/features/dutch.htm#top. A sociological portrait of Dutch law is available on the Internet, F.J. Bruinsma (2003), Dutch Law in Action, available at: http://www.uu.nl/content/dutchlawinaction2003.pdf.
[267] But it is not necessarily the highest. In Dutch doctrine usually four sources of law are distinguished: international treaties (including secondary international law), written legislation, case law, and unwritten law (e.g. general principles of law, general principles of proper administration and customary law).
[268] An English translation is available at: http://www.minbzk.nl/contents/pages/6156/grondwet_UK_6-02.pdf.

even higher levels. In practice, however, they will be reluctant to do so in cases of established case law. For this reason case law is a key factor in legal analysis.

There are four courts that are highest in their respective field(s) of law. The *Hoge Raad* ('Supreme Court') is the 'Court de cassation' in civil, criminal and fiscal cases. It judges on matters of law, not of fact. The three other highest courts are all administrative courts that pass judgements both on the law and on the facts. They are the *College van Beroep voor het bedrijfsleven* ('Industrial Appeals Board'[269]) specialising in economic administrative law. It will not be mentioned again in this chapter; the *Centrale Raad van Beroep* ('Central Court of Appeal') specialising in civil servants law and social security law; and the *Afdeling bestuursrechtspraak van de Raad van State* ('Administrative Judicial Review Division of the Council of State') for all other administrative law cases.[270]

The courts use different instruments for the interpretation of legislation. Important among these is the so-called 'legal-historic' method. When they apply this method the courts rely on the intentions of the legislators as are apparent from the published proceedings of parliament.

The Dutch Constitution was re-codified in 1983. The new Constitution opens with a chapter on fundamental rights ('grondrechten'). This chapter contains civil and political rights as well as economic and social rights.

3. The nature of economic, social and cultural rights in the eyes of the government

In the Netherlands, the lot of all rights in the ICESCR, including the right to food contained in Article 11, has been intertwined from the outset with that of the equivalent rights in the Dutch Constitution. Under Article 91 of the Constitution a treaty should be approved by Act of Parliament. When the Covenant was submitted for approval to Parliament in the mid-Seventies, the government in the explanatory memorandum on the bill – addressing Article 2 of the Covenant – described the nature of the Covenant as follows:[271]

> 'The first paragraph clearly shows that the drafters of the Covenant have considered that the economic, social and cultural rights in terms of nature and contents because of their generality do not offer a fixed standard for the rate and degree of realisation of those rights.'

---

[269] Known from ECHR 19 April 1994, case no. 16034/90 (Van der Hurk case; the Commission decided on 10 December 1992, no. 16034/90), and in EC law from many preliminary questions.
[270] Before 1994 it was called the *Afdeling rechtspraak van de Raad van State*: 'Judicial Devision of the Council of State'.
[271] Parliamentary Documents II 1975-1976, 13 932 (R 1037), no. 3, p. 45.

In the government's opinion the Covenant solely contained open standards whose contents should be defined by politics and policy. With regard to Article 11 of the Covenant the government pointed out:[272]

> 'that in particular in developing countries the living conditions of large population groups are far below an acceptable level and that in those countries the realisation of ever-improving living conditions is an urgent necessity for many years to come.'

During virtually the same period Parliament was considering the inclusion of economic, social and cultural rights in the Dutch Constitution and the government provided those rights with similar comments. From the edit of the constitutional articles in question:[273]

> 'an ample policy margin has been left for the bodies entrusted with regulations or care. The policy margin means primarily that the rate and speed of realisation of the interests phrased in the provisions are left to the government body concerned.'

The government further took the position that the implementation of the economic, social and cultural rights in the Constitution had been almost realised.[274] For this reason the government dismissed judicial control of enforcement. Only in very exceptional cases would the Court be able to establish a violation of the rights. The government considered the possibility that such a case would arise to be virtually academic.[275] In brief, the government lumped together the rights contained in the Covenant and in the Constitution. That the wording of the rights contained in the Constitution was extremely vague where the Covenant was rather detailed on many points escaped notice or was considered irrelevant.

4. The rights contained in the constitution and in the covenant in legal practice

The government's position about the justiciability of the economic, social and cultural rights contained in the Constitution has had a huge effect on legal practice. Since the rights were embedded in the Constitution in 1983 courts have established just two violations.[276] In addition, the National Ombudsman has held once that

[272] Parliamentary Documents II 1975-1976, 13 932 (R 1037), no. 8, p. 26.
[273] Overall constitutional revision, Deel Ia Grondrechten, Den Haag 1979, p. 258.
[274] Overall constitutional revision, Deel Ia Grondrechten, Den Haag 1979, p. 255.
[275] Overall constitutional revision, Deel Ia Grondrechten, Den Haag 1979, p. 258.
[276] Council of State, 10 May 1989, *AB* 1989, 481 on the right to education in Article 23 Constitution and District Court of Utrecht, 18 June 1991, *NJ* 1992, 370 on the right to housing in Article 22.2 of the Constitution.

a right was being violated.[277] This does not mean that the government carefully honours the rights, for in almost all cases courts dismiss evaluation against the rights with a reference to the government. As will be shown below, the rights contained in the Covenant do not fare any better.

Of course the significance of the economic, social and cultural rights contained in the Constitution does not depend entirely on the possibility of judicial control. It is conceivable that the rights are incorporated into legislation and administration entirely voluntarily. An extensive evaluation study has been performed, which showed that the economic, social and cultural rights contained in the Constitution play absolutely no role in the legislation process and administrative actions (Gerbranda and Kroes, 1993: 334). More recent, but less extensive studies point in the same direction (Vlemminx, 2002: 20-22). Although the significance of the Covenant for legislation and administration was not studied separately, it is safe to assume that this gloomy picture applies just as much to the rights contained in the Covenant.

5. Direct applicability of the covenant

The answer to the issue of judicial control of the enforcement of the Covenant should first of all consider the manner of reception of the Covenant's provisions. The Netherlands applies a monistic system. This was decided by the Supreme Court as early as 1919.[278] This means that within the Dutch judicial system treaty provisions can be invoked if they have direct applicability. The Netherlands, however, applies a special term in connection with this direct applicability. This is the concept 'binding on all persons' that was included in Article 94 of the Constitution in 1956. It is a rather obscure phrase, and besides the question whether the provisions of the Covenant award discretionary powers to the legislature and/or administration, other factors play a role as well (Vlemminx, 2002). These other factors aim at evading direct applicability.

First of all, especially when the Covenant is at stake, the courts place great value on the position taken by the government in its explanatory memorandum on the approval act about direct applicability. In view of earlier remarks it will come as no surprise that the government dismisses direct applicability of the Covenant:[279] 'in general the provisions contained in this Covenant will not have direct applicability.' When the Covenant is invoked courts often refer to this comment.

Secondly, because of the phrase 'binding on all persons' direct applicability may require that the treaty provision addresses citizens directly. The word 'may' already

[277] No 18 July 1995, report no. 95/271 on the right to health of Article 22.1 Constitution.
[278] Supreme Court, 3 March 1919, *NJ* 1919, p. 371. See: Vlemminx and Boekhorst (2000: 455-465).
[279] Parliamentary Documents II 1975-1976, 13 932 (R 1037), no. 3, p. 13.

shows that this requirement is certainly not always set. However, partly for this reason direct applicability of the Covenant is often dismissed.

Especially with regard to the Covenant a third characteristic of the phrase 'binding on all persons' is that direct applicability is assessed in an extremely abstract manner. The case to be resolved by a court does not play any role whatsoever in the considerations. For instance, direct applicability of the right to adequate food contained in Article 11 of the Covenant may be denied because in the court's view the term 'adequate food' is too vague, while in the case at issue *no food at all* is supplied and Article 11 is undeniably violated.[280]

This specific approach to direct applicability of the Covenant implies that the General Comments of the U.N. Committee on Economic, Social and Cultural Rights on the assessment of direct applicability appear to play no role at all. No matter how much the Committee emphasises that the policy freedom is restricted or even absent, this does not make a difference for Dutch courts with regard to direct applicability. It is not the issue of policy freedom but the other factors mentioned above that are decisive. Even the discussion of direct applicability in General Comment 3 on the nature of States parties obligations (paragraph 3) and General Comment 9 on the domestic application of the Covenant do not change this. The effect seems to be non-existent. The same applies to the typology of obligations, whose relevance has been stressed for years by Dutch scholars (Vlemminx, 2002). Because of this approach to direct applicability of the Covenant the U.N. Committee in response to the third periodic report that was filed by the Netherlands requested the Netherlands:[281]

> 'to reassess the extent to which the provisions of the Covenant (IVESCR) might be considered to be directly applicable. It urges the State Party to ensure that the provisions of the Covenant are given effect by its domestic courts as defined in the Committee's General Comment 3, and that it promotes the use of the Covenant as a domestic source of law.'

6. The case law in broad outline

This section does not discuss all court rulings denying direct applicability of the Covenant. A dry and depressing enumeration of that sort is something most readers can do without. Instead this section singles out the courts most likely to be faced with Article 11 of the Covenant, that is to say, the civil courts including the Supreme Court on the one hand and the Central Court of Appeal which is

[280] District Court of The Hague, 6 September 2000, *RAwb* 2001, 55.

[281] UN Doc. E/C.12/NLD/CO/3, 24 november 2006. UN Committee on Economic, Social and Cultural Rights, Concluding Observations on the Netherlands, paragraph 19.

the highest court with regard to social security on the other. Furthermore, only the highlights will be dealt with.

The civil courts have never been in favour of the Covenant's direct applicability. For instance, in 1983 the right to strike contained in Article 8 of the Covenant was denied direct applicability by the Supreme Court because the Article, given the words 'The States Parties to the present Covenant undertake to ensure', does not address citizens directly, but the state.[282] The civil courts also strongly stress the importance of the government's position in the explanatory memorandum on the act adopting the Covenant. Partly because of this explanatory memorandum the Supreme Court in 1989 did not award direct applicability to Article 13 of the Covenant,[283] explicitly stating a year later that some meaning should definitely be attached to the explanatory memorandum.[284]

The Central Court of Appeal pursues a slightly different course. It seems to appreciate that in cases of reliance on the Covenant important human rights are at stake. It appears to weigh the pros and cons of direct applicability by delivering rulings that are quite promising, but keeps opting out in the end.

In 1986 the Central Court of Appeal opened the possibility of direct applicability by ruling that it need not be excluded in advance that the right to equal remuneration (Article 7 of the Covenant) by its nature would be suitable for direct applicability.[285] However, in 1989 the Central Court of Appeal held that the direct applicability of any provision of the Covenant constituted a total exception to the Covenant's general character.[286] In other words, the main rule is formed by the absence of direct applicability. With this ruling this Court quashed a verdict which established a violation of Article 7 of the Covenant.[287] In 1991 the Central Court of Appeal dismissed direct applicability of the right to equal remuneration altogether.[288] It did so with reference to a similar negative ruling of the highest civil court, the Supreme Court.[289]

In 1995 the Central Court of Appeal took a new route, primarily with regard to Article 11 of the Covenant and the right to subsistence. The Court did not state whether or not Article 11 had direct applicability and established that the right compelled the government to guarantee an adequate minimum. The Court then

[282] Supreme Court, 6 December 1983, *NJ* 1984, 557.
[283] Supreme Court, 14 April 1989, *AB* 1989, 207.
[284] Supreme Court, 20 April 1990, *NJ* 1992, 636.
[285] Central Court of Appeal, 3 July 1986, *AB* 1987, 299.
[286] Central Court of Appeal, 16 February 1989, *AB* 1989, 164.
[287] Civil Service Tribunal Amsterdam, 12 March 1984, *NJCM-Bulletin* 1984, p. 245 et seq. This verdict is the only ruling that establishes a violation of the Covenant during the 25 years that the Covenant has been valid for the Netherlands.
[288] Central Court of Appeal, 17 December 1991, *RSV* 1992, 164.
[289] Supreme Court, 20 April 1990, *AB* 1990, 338.

assessed whether the cutback measure at issue had an adverse effect on that minimum. Because no such adverse effect occurred and Article 11 had therefore not been violated, the direct applicability issue need not be considered.[290] Various later rulings, concerning not only issues regarding the Covenant[291] but also issues regarding the European Social Charter and ILO Conventions, confirmed this new approach. This again opened up the possibility of the Covenant's direct applicability.

But the case law of the civil courts apparently made the Central Court of Appeal back off for a second time. On 6 September 2000 the District Court of The Hague held that Article 11 of the Covenant was worded so generally that no minimum could be distilled and that the Article therefore did not have direct applicability.[292] On 25 May 2004[293] the Central Court of Appeal argued, further to reliance on Article 11 of the Covenant, among other things, that given the wording and the purport this Article contains generally formulated objectives rather than a right that can be invoked by citizens. With reference to the explanatory memorandum on the approval act of the Covenant direct applicability was denied. The same happened on 18 June 2004 with regard to Article 9 of the Covenant.[294] At present therefore the Central Court of Appeal is back to square one as far as the direct applicability of the Covenant is concerned.

7. The attitude of the Dutch government

With regard to the reporting duty imposed on the Netherlands by Articles 16 and 17 of the Covenant, it should be noted first of all that this duty does not appear to be taken seriously. Reports should be filed every five years and the Dutch government constantly fails to file them in time. The 1996 report discusses the right to food and the government's comment is very brief:[295]

> 'Adequate food
>
> 235. Food production in the Netherlands greatly exceeds the population's requirements. During recent decades food production and agricultural production in general have increased rapidly and the production and import/export statistics show that the Netherlands is one of the countries with a large net export of food. There are, however, substantial imports of food and cattle feed from various industrial and developing countries.
>
> 236. Product quality (e.g. in terms of hygiene, residue levels, contaminants and nutritional quality) is high and meets legal as well as other consumer requirements.

[290] Central Court of Appeal, 31 March 1995, *JB* 1995, 161.
[291] Central Court of Appeal, 22 April 1997, *JB* 1997, 158.
[292] District Court of The Hague 6 September 2000, *RAwb* 2001, 55.
[293] Central Court of Appeal, 25 May 2004, LJN: AP0561.
[294] Central Court of Appeal, 8 June 2004, LJN: AP4680.
[295] Second periodic report, 05/08/96, E/1990/6/Add. 11.

237. The prices of common foodstuffs are relatively low. Low-income groups are able to buy adequate amounts of food. In periods of severe recession extra attention is paid to the low-income groups. Specific measures may be taken. At the moment there are no programmes to secure the food supply of special groups. An important consideration in this respect is the percentage of income that is spent on food. This decreased from 37% in 1960 to 14.9% in 1993. There are no regional differences which would be relevant here.
238. The foregoing shows that food and nutrition policy in the Netherlands focuses on food quality and the promotion of healthy eating habits by the various population groups, rather than on the supply of food as such. General policy is described in the Food and Nutrition Policy report presented to Parliament in 1984, which was followed by progress reports in 1989 and 1993.
239. Nutrition surveys covering 6,000 people were made in 1987 and 1992. The next one will most probably take place in 1997. The figures show an adequate supply of micro and most macronutrients. However, from the health point of view, fat consumption and to a lesser degree energy intake are considered to be too high.
240. In 1986 a long-term campaign aimed at reducing fat consumption was started. A positive downward trend (fat represented 40% of total calorie consumption in 1987 and 38% in 1992) has been noted. The campaign will continue in the next few years, and will include both a direct and an indirect approach to lowering fat consumption till the optimum of 35% is reached.
241. Nutritional education is an important instrument of food and nutrition policy in the Netherlands. The leading organization in this field is the Food and Nutrition Information Office. This organization is mainly financed by the Government, that is to say the Ministries of Health, Welfare and Sport and the Ministry of Agriculture. The Office directs its activities at the public, educational and information institutions and industry.
242. Environmental considerations are of growing importance in food production and consumer choice. Important topics are the use of pesticides and a more balanced use of fertilizers. Attention is particularly focused on organic farming, both nationally and by the EU. The area under cultivation and the number of organic farmers are slowly increasing. High prices, due to lower production levels and distribution problems, prevent organic farming from expanding any faster.'

When the report was considered in 1998[296] Mr. Ahmed of the Committee on Economic, Social and Cultural Rights remarked:

[296] E/C.12/1998/S.R. 14.

> 'that, according to the Netherlands section of the International Commission of Jurists, 240,000 households, or almost 1 million persons, were living on an income below the social minimum and some 250,000 children belonging to poor families participated very rarely in recreational and cultural activities. He was surprised that a country as wealthy as the Netherlands was unable to solve those problems.'

To this Mr Potman of the Dutch delegation replied: 'There seemed to be no basis for the allegation that 250,000 children were unable to exercise the rights provided for in the Covenant.'

That this figure of 250,000 children was taken from the second annual report on poverty issues, which had been presented to the Minister of Social Affairs and Employment in 1997, was not mentioned. It should further be noted that three years later the official study by the Social and Cultural Planning Office and the Central Bureau for Statistics revealed that one out of seven households in the Netherlands were living below the poverty line.[297] To all appearances the Dutch government, or the delegation representing the government, has deliberately provided the Committee on Economic, Social and Cultural Rights with incorrect information. The same seems to apply to the direct applicability issue. When the report was considered in 1998 members of the Committee:[298]

> 'said they knew Dutch courts incorporated international law references in court decisions, in particular in the field of human rights, and that was admirable. Had there been any recent court decisions in which the Covenant had been mentioned?'

The Dutch delegation replied:

> 'The direct application of the covenant was not needed in the Netherlands as such because whoever had problems with one of the rights enshrined therein could refer to the other instruments the Government had ratified, including International Labour Organisation conventions and the European human rights convention.'

Where the ILO conventions are concerned, this answer is also misleading. Dutch courts almost always deny direct applicability, as with the Covenant. But

[297] Social and Cultural Planning Office and Central Bureau for Statistics, *Armoedemonitor 2000*, Den Haag 2000, pp. 16-17. Every other year the Dutch Central Bureau of Statistics and the Socio-Economic Planning Bureau publish a joint detailed report by the name of 'Armoedemonitor' (Poverty Monitor) which reflects the state of affairs in cold statistics. In the even years they publish a slightly less detailed *Armoedebericht* (Poverty Bulletin). See the official website for more details: www.armoedemonitor.nl.
[298] DPI-Press Releases, Committee on Economic, Social and Cultural Rights starts consideration of report of the Netherlands, 5 May 1998.

there have been a few exceptions. In 1996 the Central Court of Appeal awarded direct applicability to a provision of the ILO Convention 102 and established that a violation had occurred.[299] The government then set out to terminate a similar European convention on social security (ILO Convention 102 could not be terminated because the applicable term had not yet arrived).[300] Fortunately parliament refused to approve the termination.[301] This affair occurred shortly before the delegation gave the above response. The delegation knew or should have known that the government did not appreciate and would not accept the establishment of a violation of an ILO Convention, but did not mention this. On 14 March 2003 the Central Court of Appeal further awarded direct applicability to a provision contained in ILO Convention 118 and again a violation was established.[302] Again the government wanted to terminate the convention and a bill to approve the termination was submitted to parliament. This time Parliament conceded. The termination of ILO Convention 118 was effected retroactively as from 14 March 2003.[303]

The manner of reporting, like the case law of the courts on the Covenant, clearly shows that the Dutch government makes every effort to minimise the significance of this convention for the Netherlands.

8. Conclusion

All human rights, including economic, social and cultural rights such as the right to adequate food, are a laudable ambition, a beautiful dream, and easily incite authorities to express the loftiest thoughts. When the ICESCR was submitted to the Dutch Parliament, the government remarked that the Netherlands, being a member of United Nations, should undertake to participate in the promotion of respect for human rights in the world community. Moreover, the government thought it improper to deny Dutch citizens and other people living in the Netherlands the safeguards created by United Nations.[304] Lofty words indeed. The government was right, though. As the saying goes, if you want to change the world, start with yourself. Sad to say, as far as the domestic enforceability of the safeguards is concerned, the Dutch government has not made this dream come true and this especially holds for the right to adequate food.

---

[299] Central Court of Appeal, 29 May 1996, *RSV* 1997, 9.
[300] Parliamentary Documents II 1996-1997, 25 524.
[301] Acts II 1997-1998, p. 4381.
[302] Central Court of Appeal, 14 March 2003, *RSV* 2003, 114.
[303] Act of 9 December 2004, Staatsblad 715.
[304] Parliamentary documents II, 1975-1976, 13 932 (R 1037), no. 3, p. 11.

References

Besselink, L.F.M., 2004. Constitutional Law in the Netherlands. Ars Aequi Libri, Nijmegen, the Netherlands.

Chorus, J. (c.s. eds.), 1999. Introduction to Dutch Law. 3rd ed. Kluwer Law International, Deventer, the Netherlands.

Gerbranda T.J. and Kroes, M., 1993. Grondrechten evaluatieonderzoek. Final Report: Stichting NJCM Boekerij, Leiden, the Netherlands.

Leckie, S., 1998. Another Step Towards Indivisibility: Identifying the Key Features of Violations of Economic, Social and Cultural Rights. Human Rights Quarterly 20: 81-124.

Taekema, S. (ed.), 2004. Understanding Dutch Law. Boom Uitgevers, the Hague, the Netherlands.

Vlemminx, F.M.C., 2002. Een nieuw profiel van de grondrechten. Boom Uitgevers, the Hague, the Netherlands.

Vlemminx F.M.C. and Boekhorst, M.G., 2000. Article 93. In: A.K. Koekkoek (ed.), De Grondwet, Een systematisch en artikelsgewijs commentaar. W.E.J. Tjeenk Willink, Deventer, pp. 455-465.

Chapter 6
Declared, not acquired: claiming hunger as a violation of the right to food, with a case study from Indonesia

Irene Hadiprayitno[305]

1. Introduction

Human rights may be defined as the protection of human dignity by means of law. In theory, the transformation of political commitment to human rights into (inter) national law is an important step towards implementation. The key assumption is that provisions *declared* in international or national laws and regulations have coercive powers and will guarantee access to justice. The official signing and ratification of human rights by the State confirm its acceptance of the role as the duty-bearer, which in concrete situations has the obligation to enforce human rights (Raz, 1984: 5). The transfer of mandates and monitoring by States to UN bodies is based on the expectation that this will help to create more than lip-service to human rights at the national level (Habermas, 2001: 113).

Whether this is a valid assumption depends largely on how the officially declared rights can be *acquired* in domestic politics (De Gaay Fortman, 2006: 263). Here one talks about a practice of enforcement; particularly on the issue of access to justice, entailing rules, regulations, and, if necessary, penalties that can assure compliant behaviours of the State or other third parties to non-violations of human rights. Important to guarantee compliance are a well-functioning legal system and socio-political culture, in which rights are honoured. Achieving such a situation may be easier if the right in question has been part of a historical struggle. An example is the declaration of freedom of speech that is based on a long practice of free expression in some countries.

This chapter is about the reverse of that rather unproblematic reality. In this case, the focus is on the right that has been signed, ratified and/or declared by the State, but is neither embedded in socio-political culture, nor based on institutionalised practices. In these situations, the question is to what extent and how the declaration of human rights by the State does or does not provide legal means for right-holders to establish violations of human rights and to hold the State accountable.

---

[305] The author wishes to thank Prof. B. van der Meulen, Dr. D. Roth and Dr. O. Hospes for their helpful comments on the earlier draft, all errors remain the author's own.

The particular focus is on hunger and the right to food in Indonesia. The chapter will depict the complexities of addressing hunger and acquiring the right to food at the national and local level. For this purpose, the chapter will first provide the framework of thinking on the right to food by presenting the substantive and procedural contents of the right to food from international human rights laws. Thereafter, the judiciary multi-level context of the right to food in Indonesia is analysed in the second and third parts of the chapter. The second part of this chapter provides a comparison and analysis of existing rules and regulations at the national level, assessing to what extent these rules and regulations could be used to address the entitlement[306] to food as a human right in Indonesia. Special attention will be given to the Indonesian human rights law, food law and disaster management law (in which mitigating measures for food emergency such as hunger are regulated). Then the case of hunger in Yahukimo, Papua, will be examined to discuss whether hunger can be judged as a violation of the right to food at the local level in the given judiciary contexts. In the final part of the chapter some remarks are put together on how the right to food can be acquired.

2. The right to food and its contents as part of international human rights law

This section briefly discusses the substantive and procedural contents of the right to food.[307]

Article 11 of the International Covenant on Economic, Social and Cultural Rights (ICESCR) encompasses two connotations. The first derives from paragraph 1 that states 'the right of everyone to an adequate standard of living for himself and his family, including adequate food'; and hence can be concluded as the right to adequate food. The second, proclaimed in paragraph 2, is 'the fundamental right of everyone to be free from hunger'.

Academic commentaries on these articles suggest that the first provision (Article 11(1)) may be included as 'relative' standard (Narula, 2006: 706) and it is substantially much broader than the second (Alston, 1984: 32). In the General Comment No. 12, the Committee on Economic, Social and Cultural Rights has concluded that the 'core content' of the right to adequate food implies ensuring:[308]

> 'The availability of food in a quantity and quality sufficient to satisfy the dietary needs of individuals, free from adverse substances, and acceptable within

[306] Sen (1984: 497) defines entitlement as 'the set of alternative commodity bundles that a person can command in a society using the totality of rights and opportunities that he or she faces'.
[307] For more comprehensive discussions on the right to food as a human right, see the contributions of Van der Meulen, Eide and Wernaart in this volume.
[308] E/C.12/1999/5, 12 May 1999, General Comments 12, The Right to Adequate Food, paragraph 8

> a given culture; The accessibility of such food in ways that are sustainable and that do not interfere with the enjoyment of other human rights'.

In contrast the right to be free from hunger, originating from paragraph 2 of Article 11, is 'absolute'. Even stronger, it is the only right to be qualified as 'fundamental' in the whole Covenant. Hunger as the minimum core of the right to food is being reiterated in General Comment No. 12. In Paragraph 17, it is stated that 'violations of the Covenant occur when a State fails to ensure the satisfaction of, at the very last, the minimum essential required to be free from hunger'.

Another legal basis is found in Article 1(2) of the International Covenant on Civil and Political Rights, which says that 'in no case may a people be deprived of its own means of subsistence', and in addition Article 6 which states that 'every human being has the inherent right to life'. These clearly imply the fundamental right to the necessities to sustain life, which including adequate food, and particularly those required to be free from hunger.

Comparing the right to food with food security as another transnational concept, Kerstin Mechlem noted the similarities of both in emphasising food availability, accessibility, safety, and cultural acceptability (Mechlem, 2004: 643). Yet, the attributes of their substances are rather distinct. Here, Mechlem explained that the aim of food security can be based on a number of grounds, ranging from moral grounds to more market-oriented motivations. Human rights, on the other hand, are exclusively based on the very idea of human dignity. In a similar vein, Philip Alston argues that human rights require *a priori* values. Thus from a human rights perspective, all other considerations would be secondary in nature (Alston, 1998: 104). Furthermore the acknowledgement and international acceptance of the right to food, which are demonstrated by States ratifications, may conclude that as a human right, the right to food has a precise content. It can be violated and the violation can be the subject of judicial or quasi-judicial remedies (Mechlem, 2004: 644). Here one touches on the procedural issue of the right to food.

As with any other human right, the State is the duty-bearer of the right to food, with the obligation to respect, protect and fulfil. There are two ways in which State Parties should approach their obligation. First, State Parties must act *immediately* 'to mitigate and alleviate hunger even in times of natural or other disasters'.[309] Second, State Parties to the ICESCR are required to take steps to *progressively* achieve the right to adequate food, better known as the principle of progressive realisation. The principle compels the State Parties to move 'as expeditiously as possible' toward this goal.[310] Article 2(1) of the ICESCR further asserts that the State Parties should:

[309] E/C.12/1999/5, 12 May 1999, General Comments 12, The Right to Adequate Food, paragraph 6

[310] E/C.12/1999/5, 12 May 1999, General Comments 12, The Right to Adequate Food, paragraph 14.

> '...take steps, individually and through international assistance and co-operation, especially economic and technical, the maximum of its available resources, with a view to achieving progressively the full realisation of the rights recognised in the present Covenant by all appropriate means, including particularly the adoption of legislative measures'.

The principle of progressive realisation may be interpreted as a balance between the activation of the economic, social and cultural rights, in accordance with the extent of resources available for appropriation in the country in question. Consequently, States would always have the margin of discretion to choose the necessary measures[311], including budget appropriation regarding food policies. State compliances are assessed on whether they are unable or unwilling to meet their obligations as regards the right to food. Claims of inability of the State would not be accepted, unless the State in question can prove that it has unsuccessfully sought to obtain international support to ensure the availability and accessibility of the necessary food.[312]

As food becomes a concern of international human rights law, theoretically, the State should at least consider the adoption of a framework of law as a major instrument for realisation[313] and apply corrective legislation and administrative measures when violations occur.[314] Based on such legal modalities, any individual who is the victim of a violation of the right to adequate food would have access to effective judicial or other appropriate remedies at both the national and international level in order to acquire restitution, compensation, satisfaction or guarantees of non-repetition.[315]

The adoption of legislative measures is only the first step in guaranteeing the right to food. The Limburg Principles on the Implementation of the International Covenant on Economic, Social and Cultural Rights assert that legislation alone is not sufficient to fulfil the obligations of the Covenant.[316] States should also adopt administrative, economic, social and educational measures to achieve progressive realisation of the right to food. In this regard, one touches on the operational methods behind good practices in designing and executing policies, programmes and projects on economic, social and cultural matters. The method is popularly known as the right-based approach; a conceptual framework for the process of human development that is normatively based on international human rights standards and operationally directed at promoting and protecting

[311] E/C.12/1999/5, 12 May 1999, General Comments 12, The Right to Adequate Food, paragraph 21.
[312] E/C.12/1999/5, 12 May 1999, General Comments 12, The Right to Adequate Food, paragraph 17.
[313] E/C.12/1999/5, 12 May 1999, General Comments 12, The Right to Adequate Food, paragraph 29.
[314] E/C.12/1999/5, 12 May 1999, General Comments 12, The Right to Adequate Food, paragraph 31.
[315] E/C.12/1999/5, 12 May 1999, General Comments 12, The Right to Adequate Food, paragraph 32.
[316] UN doc. E/CN.4/1987/17, Limburg Principles on the Implementation of the International Covenant on Economic, Social and Cultural Rights, paragraph 17-18.

human rights.[317] Specifically, the General Comment No. 12 further explains this method by mentioning that the formulation and implementation of the right to food strategies or policies require full compliance with principles of accountability, transparency, people's participation, decentralisation, legislative capacity and the independence of the judiciary.[318] Notably, there is a limit to the methods of designing and executing policies, programmes or projects based on the right to food, particularly in comparison with those made based on food security concerns.

Notably, the principle of progressive realisation, asserted and elaborated in many international documents, leaves a lot of room for interpretation. It stipulates a series of normative points for State's positive and negative conduct. These normative frameworks conclude that to enable the right-holders to acquire their right to food, both legislative measures as well as good practices in designing and executing policies are required. How to enforce compliance with these normative frameworks is omitted, albeit unintentionally. This may be of advantage since food-related struggles not only fall under the jurisdiction of the international and/ or national human rights laws. In many countries, access to food and the means to feed oneself are regulated in different types of laws and regulations, namely general food laws, food security laws, land laws or agrarian laws. In acquiring the fulfilment of the right to be free from hunger one should therefore not only deal with the competing substance of the right national laws and regulations but also consider the procedural overlap or deficits that affect its acquisition processes.

3. The right to food in Indonesia

This section presents the substantive and procedural contents of the right to food in Indonesia. Attention is paid to the right to food in Indonesian Human Rights Law as well as the legal basis for food security and food emergency measures.

3.1 The right to food as a human right

Since the ICESCR is an international human rights treaty, it creates legally binding international obligations on, currently, 160 States that have agreed to be bound by the standards contained in it. In October 2005, Indonesia adopted and ratified the ICESCR by Law No. 12 of 2005. Based on this ratification the Indonesian government officially assumes the obligation to respect, protect and fulfil the substantial contents of the rights according to the normative content of the ICESCR and declares its commitment to take steps by all appropriate means to achieve their full realisation.

[317] The a rights-based approach includes the elements of: (1) express linkage to rights, (2) accountability, (3) empowerment, (4) participation, and (5) non-discrimination and attention to vulnerable groups. UN Doc. A/51/950, Renewing the United Nations: A Programme for Reform.

[318] E/C.12/1999/5, 12 May 1999, General Comments 12, The Right to Adequate Food, paragraph 23.

The legal basis for human rights in Indonesia can be found in the fourth amendment of the 1945 Constitution.[319] Article 28 of the amended Constitution is devoted solely to human rights principles. This article guarantees a list of the universally accepted human rights, namely the right to freedom of assembly, the right to life, the right to establish a family, the right to personal development, the right to be treated equally before the law, the right to work and employment, freedom of religion and freedom to express opinion, the right to information, freedom from torture and inhuman and degrading treatment, the right to a healthy environment, and the right to be free from discriminative treatment. Equality before the law, the protection to the freedom of religion and the right to education remain in the similar texts in Article 27, Article 29 and Article 31 respectively.

The Human Rights Law No. 39 of 1999 defines human rights as follows:[320]

> '... a set of rights bestowed by God Almighty in the essence and being of humans as creations of God which must be respected, held in the highest esteem and protected by the state, law, Government, and all people in order to protect human dignity and worth.'

The Human Rights Law adopts the principle of equality and non-discrimination, the principle of indigenous rights, but not the principle of self-determination. This law protects the right to be treated equally, right to life, right to justice, freedom of the individual from slavery, religious rights political beliefs, and freedom of speech. With regards to economic, social and cultural rights, a wide range of internationally recognised rights are also guaranteed by this law here, such as the right to property and ownership, the right to work, and the right to education.

The Human Rights Law only implicitly addressed the right to food. Article 9 stipulates that 'everyone has the right to life, to sustain life, and to improve his or her standard of living'. The term 'to sustain life' and 'to improve his or her standard of living' would entail 'the adequate standard of living', which thus includes, amongst others, the entitlement to food.

A distinctive feature of legal protection over human rights in Indonesia is the recognition of human obligations. Paragraph J of Article 28 of the Indonesian Constitution stipulates that:

> 'in exercising her/his rights and freedoms, every person shall have the obligation to accept the restrictions established by law for the sole purposes of guaranteeing the recognition and respect of the rights and freedoms of

[319] The 1945 Constitution has been amended four times: October 1999, August 2000, November 2001, and August 2002.

[320] Article 1(1), Law No. 39 of 1999.

> others and of satisfying just demands based upon considerations of morality, religious values, security and public order in a democratic society'.[321]

The concept of human obligations is addressed in the preamble[322] and Chapter 4 of the Human Rights Law No. 39 of 1999.[323]

Additionally, the recognition of the concept of human obligations is reflected in both the National Plan of Action on Human Rights of 1998-2003[324] and the National Plan of Action on Human Rights of 2004-2009.

> 'Efforts to respect, promote, fulfil and protect and fulfil human rights in Indonesia should be based on the principle of unity and equilibrium and the realisation of the national condition. The principle of unity implies that the civil and political rights, the economic, social and cultural rights, and the right to development are indivisible in their implementation, monitoring as well as assessment. The principle of equilibrium signifies there should be balance and harmony between individual and collective rights and obligations towards his community and nations. The matter is in accordance with the natural attribute of human as an individual and social being. Balance and harmony between freedom and responsibility are the crucial factor in respecting, promoting, fulfilling and protecting human rights.'[325]

Moreover, the latest Plan of Action explains that the promotion and protection of human rights should be inspired by Indonesian values, customs, cultures and traditions.

> 'It is commonly acknowledged that human rights are universal and the international community has also recognised and concurred that their implementation is the duty and responsibility of States, taking fully into consideration the various value systems, history, culture, political systems, level of social and economic development, and other relevant factors.'[326]

Notably, human obligations as well as culture and social and economic development are being asserted to provide a balanced way of looking at human rights. This is not uncommon in the context of Indonesia or South-East Asia. Emphasis on

---

[321] Article 28 J(2) of the Fourth Amendment 1945 Constitution.
[322] Considerant 2 of Law No. 39 of 1999 says 'whereas besides basic rights, humans also have basic obligations to one another and to society as a whole, with regard to society, nation and state'.
[323] Chapter 4 on Human Obligations, of Law No. 39 of 1999.
[324] Decree of The People's Consultative Assembly, No. 11/MPR/1998 on the National Plan of Action on Human Rights 1998-2003, Paragraph 2.
[325] Presidential Decree No. 40 of 2004 on the National Plan of Action on Human Rights 2004-2009, paragraph 3.
[326] Presidential Decree No. 40 of 2004 on the National Plan of Action on Human Rights 2004-2009, paragraph 4.

differences in culture, value systems, history and/or political systems has been the prevailing way of framing human rights (Twinning, 2009: 156). Compared with the aforementioned principle of progressive realisation, which is also argued to offer a similar purpose, stipulating the concept of human obligations signifies that the balance is not only on resources at the country's disposal, but also on how the right-holders may acquire their entitlement in Indonesia. This may be accurate when one consider the procedural enforcement of human rights in Indonesia.

Human Rights Law No. 39 of 1999 regulates that the system of claims and remedies of human rights is the subject of the National Human Rights Court. Article 4 of Law No. 26 of 2000 addresses the jurisdiction of the Human Rights Court. This law only confirms the authority of the Human Rights Court to hear and rule on cases of gross violations of human rights, which include the crime of genocide and crimes against humanity.[327] The Human Rights Court, however, does not have jurisdiction over violations to other rights guaranteed in the Indonesian Constitution or the Human Rights Law No. 39 of 1999, such as the right to work, the right to health, and freedom of speech. Consequently, failures in respecting, protecting or fulfilling these entitlement could not be regarded as human rights violations in the legal system.

In terms of judicial procedures, Article 10 asserts that cases of gross violations of human rights will be conducted in accordance with provisions in the existing Code of Criminal Procedure (*Kitab Undang-Undang Hukum Pidana* or KUHAP). Under Article 11, the Attorney General is given power of arrest,[328] and is authorised as the investigator and public prosecutor.[329] Hitherto, the activities of the Human Rights Court are mostly centred on the prosecution of previous human rights perpetrators in several *ad hoc* tribunals.[330]

3.2 The right to food and food security

At present, the General Food Law No. 7 of 1996 is arguably the only document that regulates the issue of food extensively. The law includes provisions on food safety, food quality, labelling, and food security. Article 1(17) of Law No. 7 of 1996 defines food security as 'a condition where food necessity is fulfilled at the household level, manifested in its availability, amount and quality, safety, equally

[327] Article 4 and 7, Law No. 26 of 2000.
[328] Article 11 (1), Law No. 26 of 2000.
[329] The Attorney General is also authorised as investigator and public prosecutor '...to undertake the detention or extend the detention of a suspect for the purposes of investigation and prosecution'. Article 11 (2), Law No. 26 of 2000.
[330] Article 43 (1), Law No. 26 of 2000 states that gross violations of human rights occurring prior to the coming into force of this Law shall be heard and ruled on by an ad hoc Human Rights Court.

distributed and accessible'.[331] For a specific directive on food security one can refer to a newer regulation: the Government Regulation No. 68 of 2002.

The government is responsible for 'the regulation, management, control and supervision of food availability both in terms of quantity as well as quality; its safety, diversity and economic accessibility'.[332] In carrying out these tasks, the government should aspire to achieve sufficiency in food stock and reserve; availability, adequacy and distribution of basic food supplies; and diversification of food production.[333] The government is also responsible for taking measures to prevent and/or combat indications about food inadequacy, food emergency; and/or the speculation and manipulation in food supply and distribution.[334]

In order to achieve food availability, the Government Regulation No. 68 of 2002 mentions four methods. These are (1) development of a food production system based on local resources, institutions and culture; (2) development of a food efficiency system; (3) development of means for food production; (4) maintenance and development of productive natural resources.[335] Furthermore, food supply can be obtained from foreign inputs, domestic production, and national food reserve.

The General Food Law allocates specific provisions on food reserve as one of the important conditions to realise food security. The source of national food reserve is from the government as well as the general public. The first implies food reserve that is managed and controlled by the government. Accordingly, the general situation of food reserve is decided by the government periodically; by means of calculating actual food demands and supplies, with the purpose to anticipate food inadequacy and food emergency.[336] The role of the general public is manifested in the location of food reserve, which is to be found at the level of village, region, province and state.[337] At these different levels, food reserve is managed and controlled by the people themselves, including peasants, credit unions, traders, and household industries.[338]

To a certain extent the substantive content of the right to food is found in the General Food Law and the Government Regulation No. 68 of 2002. The issue of food safety, nutrition or dietary needs and economic and physical accessibility (but not cultural acceptability) are explicitly regulated. People participation is also acknowledged as a method for inputs or suggestions to achieve better food

[331] Article 1(17), Law No. 7 of 1996.
[332] Article 45 (1), Law No. 7 of 1996.
[333] Article 46, Law No. 7 of 1996.
[334] Article 46 (d), Law No. 7 of 1996.
[335] Article 2 (2), Government Regulation No. 68 of 2002.
[336] Article 47 (2), Law No. 7 of 1996.
[337] Article 5 (2), Government Regulation No. 68 of 2002.
[338] Explanatory Notes Article 47(1), Law No. 7 of 1996.

policies or to solve food insecurity problems.[339] Participation, however, is the only resemblance of the normative content of the right to food in the two laws. The principle of transparency, accountability, or non-discrimination are not included as requirements for designing and implementing actions necessary for food security. Furthermore, as stipulated in Law No. 7 of 1996, the recourse mechanism is regulated based on Indonesian Criminal Law and is possible for complaints limited to food sanitation, food safety and food distribution.

In practice, the issue of food security in Indonesia is often associated with the country's rice policy. The production of rice to feed the whole nation is considered as one of the most important development policies, which therefore requires full state control and is based entirely on government decisions. To many, this approach would imply that the government would keep the issue of food availability and food accessibility under control. However, the situation of the food market is oligarchic, meaning that it is controlled by both the government and private companies. Hence, competition instead of control is frequently the contributing consideration to food policies (Timmer, 2004: 12-19). The agent responsible for the food production chain from rice and other basic commodities is the Logistic Agency, known as Bulog (*Badan Urusan Logistic*) and in practice the Bulog system of purchase and distribution is carried out with the help of several private companies.

Well aware of the potential implication of rising food prices, the government always aims to keep the price of rice relatively stable. Data gathered from 1969 to 2003 show that the official price of rice was maintained between 750 IDR and 1,100 IDR per kilogram unit, with some fluctuations spotted in 1973-1975 and 1997-2002 (Warr, 2005: 430). If one compares the prices of rice with the political stability in the country during these periods, it is evident that two major incidences of social unrest occurred on two occasions: firstly, the Malari Riots (*Malapetaka Lima Belas Januari*) on 15 January 1974 and the May Riots of 14-15 May 1998. Without presuming a direct and linear connection between these events and the price of rice, one could inevitably argue that the rising price of rice was one of the relevant factors behind the looting and stealing of food and other valuable goods from burning shops. Against this background, the strategy of the Indonesian government of using food policy as a tool to prevent domestic unrest makes sense. Yet, the perception of connecting food policy with national stability has the effect of concentrating on merely the accumulation of food, as opposed to the fulfilment of the entitlement to food.

3.3 The right to food in food emergency situations

The Government Regulation No. 68 of 2002 on Food Security defines an emergency situation on food: ' a volatile critical condition threatening the social livelihood

[339] Article 51, Law No. 7 of 1996.

of the society, which therefore requires exceptionally immediate and precise measures'. Although the definition does not mention the word hunger explicitly, it reflects common characteristics of hunger. Additionally, the Explanatory Notes of the General Food Law No. 7 of 1996 refer to a food emergency as a situation resulting from natural disasters; harvest failures or conditions outside the control of human beings.[340]

A specific document regulating the mitigating measures against emergency is Law No 24 of 2007 on Disaster Management. The Law was issued as a response to the Tsunami in Aceh and North Sumatera, 26 December 2004. It aims to regulate comprehensive emergency measures in all chains of disaster management from prevention, domestication, promptness, and control, to rehabilitation and restoration, in order to reduce its extremely negative impacts.

The law provides for three categories of disasters: natural, non-natural and social disasters. Natural disasters are events caused by natural activities, such as earthquakes, tsunami, volcano eruption, flood, drought, typhoon or storm, and landslides.[341] Non-natural disasters are events causes by a series of non-natural activities, such as failure of technology, failure of modernisation, epidemic or diseases.[342] Social disasters are events caused by a series of human activities, including social and communal conflict or terror.[343] In the following text, the law does not differentiate between the measures adopted for these types of disasters.

The Law stipulates the principles of humanity, justice, equality in law, balance and harmony, order and the rule of law, togetherness, protection of the environment and the use of science and technology in implementing measures to mitigate both natural or non-natural disaster.[344] Measures applied in emergency events should consider the principles of promptness and precision; priority, co-ordination and integrity; efficiency and effectiveness; transparency and accountability; partnership; empowerment; non-discrimination and non-proletition.[345]

The emergency status of a disastrous event will be declared by the Government.[346] This will be done by looking at the number of victims, the loss of property, the level of damage to facilities and infrastructures, the geographical scope, and the social and economic impacts.[347] In doing so, the government will receive

[340] Explanatory Notes Article 47 (2), Government Regulation No. 68 of 2002.
[341] Article 1(2), Law No. 24 of 2007.
[342] Article 1(3), Law No. 24 of 2007.
[343] Article 1(4), Law No. 24 of 2007.
[344] Article 3(1), Law No. 24 of 2007.
[345] Non-proletition is used by Indonesian Disaster Management Law and indicates that it is forbidden to preach the religion and faith while donating emergency aids. Explanatory Notes Article 3(2i), Law No. 24 of 2007.
[346] Article 1(19), Law No. 24 of 2007.
[347] Article 7 (2), Law No. 24 of 2007.

recommendations from the National Authority for Disaster Management.[348] This body has the mandate, amongst others, to recommend, standardise and inform the Government about the measures necessary for mitigation. The Authority is also authorised to design and decide policies related to emergency measures and co-ordinate their implementation.[349] In the field, mitigating emergency events might involve the engagement of corporations and international organisations.[350]

Article 26(2) stipulates that 'everyone in disaster struck areas has the right to receive supports for fulfilment of basic needs'[351], which includes clean water and sanitation, food, clothing, health care, psychosocial services, and refugee settlement.[352] Victims also have the right to 'receive compensation for construction failures' that take place in disaster areas.[353] Compensation is available to victims who have partly or entirely lost their property because of the emergency status of the area.[354] In contradiction to the rather general clause on the entitlement to receive basic supports, the law specifically asserts that victims are entitled to receive compensation for the loss of property.

As mentioned before, Law No. 24 of 2007 on Disaster Management was adopted against the traumatic background of Tsunami 2006. It therefore emphasises measures relevant in particular to mitigate unavoidable natural disasters. In this light, the principle of promptness is observably important in this law, as is also the case in the General Food Law. Notably important in Law No. 24 of 2007 is the integration of the right-based approach reflected by the assertion of the principles of non-discrimination, participation and transparency. However, the Law does not specify the subject of food availability and accessibility for victims and limits compensation to property.

3.4 A cornucopia of law

The foregoing sections reveal that there is an abundance of legal provisions dealing with food security and food availability in Indonesia. Notably the right to food remains implicit, but based on these stocks, one can already assess the possibility of acquiring the right to food in Indonesia within the legal system.

The ratification of the ICESR by the Indonesian government as stipulated in Law No. 11 of 2005 means that the right to food has a legal standing. The incorporation of the language and the principles of human rights in the legal system regulating

---

[348] Article 10, Law No. 24 of 2007.
[349] Article 13, Law No. 24 of 2007.
[350] Article 28-30, Law No. 24 of 2007.
[351] Article 26 (2), Law No. 24 of 2007.
[352] Article 53, 55(1), Law No. 24 of 2007.
[353] Article 26 (3), Law No. 24 of 2007.
[354] Article 32 (1 and 2), Law No. 24 of 2007.

food policies serves a similar purpose. The Indonesian government would become accountable to the obligation for progressive achievement to the right to adequate food and for adopting immediate responses to the fundamental right to be free from hunger. Yet assessing their actual impact on the right-holders' position to acquire the right to food through the judiciary system is not a straightforward matter.

Law No. 39 of 1999 Human Rights stipulates the protection of a wide range of entitlements, including those contained in civil and political rights, economic, social and cultural rights, and implicitly on the right to food. Law No. 26 of 2000 on Human Rights Courts, however, merely asserts the jurisdiction of the *ad hoc* Human Rights Court to hear cases of gross violation of human rights, i.e. genocide and crimes against humanity. This discrepancy creates an area of uncertainty in the promotion and protection of the right to food, particularly in the Indonesian judiciary. People with food insecurity would have to employ other channels, legal or non-legal, to claim against the non-fulfilment of food availability and food accessibility.

According to General Food Law No. 7 of 1996, individuals have the opportunity to bring proceedings regarding food policies as regulated under Indonesian Criminal Law. The complaints however are limited to food sanitation, food safety and food distribution, and do not extend to cases on food insecurity, or specifically hunger. Failure in securing access to and available food for everyone to be able to sustain their daily life is obviously outside the jurisdiction of the General Food Law No. 7 of 1996. Notably, being a product of the New Order regime and issued before Indonesia incorporated human rights principles into its legal system, this legal document is considered as outdated and cannot be expected to protect the substantive content of the right to food. The contained provisions are not sufficient to provide victims with recourse mechanisms based on human rights standards nor to promote accountability, transparency or non-discrimination as the principles in regulating food policies. Yet, the condition is rather similar in the newer legislations too. Government Regulation No. 68 of 2002 indeed asserts some provisions of good practices containing the spirit of the right to food, for example Article 8(1) which stipulates the entitlement to participate in designing policies regarding food security, but it still does not regulate the possibility to legally claim against non-fulfilment of food security.

In the case of hunger, although the adversity may be connected to what is defined as a food emergency by Government Regulation No. 68 of 2002, failures on implementing necessary mitigating measures are not being regulated either. Law No. 24 of 2007 on Disaster Management can only act as the legal basis for adopting measures and policies in the course of food emergency. Nevertheless, the law asserts some limit to the methods in adopting measures to combat disasters in general, including non-discrimination, rule of law, transparency and accountability. The most important contribution of this Law is that it asserts the principles of promptness and precision, which are consistent with Paragraph 16 of the General Comment No. 12 that compels States to act immediately in case of hunger even in times of natural disasters.

Notably, one can easily find the language of human rights in the legal documents adopted after the collapse of the New Order, as the discourse of non-discrimination, participation and accountability becomes part of the daily vocabulary in Indonesian politics. Yet it is observed that the incorporation of the principle of the right to food neither grants the right-holders the possibility to seek recourse in the court of law and nor empowers the court to actually treat cases concerning food policies.

A consequence of the limitation of a procedural deficit to actually acquire the human right to food in the court of law is the wide discretion that the government may have in executing the necessary measures. Litigation in the light of the human right to food is not viable when the government fails, for example, to immediately address hunger strikes or prevent the reoccurrence of hunger. Access to justice is a useful strategy for ending past human rights violations. Court cases often act as a deterrent, reducing the likelihood that similar things will happen in the future.

Without prescribing the possibilities for victims to actually acquire justice in cases of human rights violations related to economic, social, and cultural rights, the right to food would have to be activated in the form of policies, programmes or projects. In this light, incorporation of the right to food in the Indonesian legislation may serve two functions: first, as an ideal objective for policies and, second as a prescription for government's good practices in the area of food availability and food accessibility. These are not of limited importance; they are still relevant in the context of achieving progressive realisation of the right to food, which understandably cannot happen overnight.

4. Hunger in Papua: a violation of the right to food?

4.1 The case of Yahukimo

The Province of Papua compromises the majority of the Western part of the island of Papua New Guinea. The province originally covered the entire western half of the island, but in 2003, along with the establishment of the Special Autonomy Status of Papua, the eastern portion of this province was declared as a separate province (now West Papua).

In spite of its abundant national resources, statistics show that Papua as the largest province of Indonesia ($\pm$ 421,981 km^2) has grown slowly. The report by the Central Bureau of Statistics (BPS) shows that the Human Development Index (HDI) of Papua in 2007 only reached 63.4, having the lowest position in Indonesia. Similarly, in 2008 the Human Poverty Index (HPI) of Papua reached a score of 37.08, which was much higher than the average score of the Indonesian HPI, which was 15.42 (BPS, 2009). In domestic politics, various impediments (in terms of transportation, communication, education, health, and so on) of Papua are identified by the many Indonesian bureaucrats as the contributing factors in Papuans' backwardness and isolation.

Yahukimo is a regency established in 11 December 2003 by Law No. 26 of 2000. Its formation is a result of the creation of sub-national administration or *pemekaran*. In the case of Papua *pemekaran* has delivered 36 regions while in 1996 there were only 13 regions in this island. The name of Yahukimo is derived from the names of four indigenous tribes living in the area: Yali, Hubla, Kimyal, dan Momuna. The area is being governed from Wamena, the capital of the Regency Jayawijaya.

Hunger is not new in Yahukimo regency. In 2005, the national media reported a food emergency situation in the area. In response to this, the central government built 12 storage facilities in 2006 to stock food in anticipation of any future harvest failure. In September 2009, it was again reported by the media that since January 2009 hundreds of villagers in Yahukimo had died of hunger associated with diseases. The storage facilities built previously to ensure food availability in the area were reported to be empty and unused.

As with the hunger strike in 2005, the exact number of victims was not known in 2009. Some said that it concerned seven districts including Suntamon, Langda, Bomela, Walma, Seredala, Proggoli and Heryekpine, and a total of 26 subdistricts. Others reported 113 deaths in the whole of Yahukimo district, others stated that the total was 222 in two districts alone. In Walma and Bomela districts, it is also mentioned that as many as 60 and 160 people, respectively, have died of hunger. Some also declared that food was still available, but people needed to reduce their daily intake (Somba, 2009a).[355] Both medias and the government mentioned that the adversity increased because the people affected were generally those living on mountain slopes or areas of higher elevation which could only be reached by foot or small plane. The emergency situation led to a spike in diseases, such as malaria, diarrhoea and gastric illnesses, as an indirect result of which many people lost their life.

Although the central government responded quite promptly by sending food aid of 100 tonnes of rice, sweet potatoes and other foodstuffs including noodles to the affected area, they also repeatedly refused to use the word 'hunger' to explain the deaths in the area. The Press Release from the Ministry of Social Welfare explains that the widespread famine was attributed to heavy rains and harvest failures

[355] 'Korban Kelaparan di Yahukimo Mencapai Ratusan', Jawa Pos, 8 September 2009.

leading to starvation.[356] The head of Yahukimo disaster response team reported that staple crops, such as tubers, were rotten due to incessant rain. Similarly, the Chief of Papua's Yahukimo Regency, Ones Pahabol, announced the failures of the sweet potato harvest (Somba, 2009b). It was explained that because of the lack of proper access to the highland areas of Yahukimo, daily food sources came from home-grown produce. With such unsustainability, when villagers fail to harvest enough produce to support their families, they immediately face starvation. The lack of infrastructure, such as roads and public transportation in Yahukimo is also identified as a contributing factor that aggravates the food insecurity status of the area. Furthermore, the government has admitted that villagers suffered from various diseases, and medical aid could not easily be sent to the affected areas. The Secretary of Yahukimo administration, Robby Langkutoy, as well as the Ministry of Social Welfare, have stated that diseases and failed harvest were the major causes of death among victims (Somba, 2009c).[357]

Switching the concern of hunger to disease is not a winning case in Papua. According to the survey conducted by *Sekretariat Keadilan dan Perdamaian* (SKP) and the Institute for Social Transformation (INSIST), there are two problems regarding the quality of health care in Papua. The first is the lack of availability of proper health care services, and the second is the inadequate budget allocated for health care. The research revealed that 44.9% of people living in Jayawijaya do not have access to health care. The data show that one *Puskesmas* (local health service) has to serve 9,842 people. This is considered to be an unreasonable coverage, especially with the problems of distance and concentration of population in the area. Regarding the second problem, the research also disclosed that in 2005 health care accounted for just 2.75% of the total Local Development Budget (APBD) (HamPapua, 2006).

The above description demonstrates that hunger in Yahukimo is embedded in the context of harvest failures, the social and economic situations in Papua Province, the extent of extreme poverty as well as the lack of facilities and resources. As a result, one cannot expect much more than the immediate measures from the government, which are inevitably charity-oriented and are not followed by anticipation programmes aimed at advancing people's entitlement to feed themselves. Further examination in this regard is provided in the following paragraphs.

---

[356] Press Release, Ministry of Social Welfare of 11 September 2009, available at: http://www.menkokesra.go.id/content/view/12586/39/. Accessed 8 February 2010.

[357] Nethy Dharma Somba, 'Starvation no, food shortage yes: Local official', *the Jakarta Post,* 5 September 2009.

4.2 Concluding observations from the case study

Three main observations were made from the case study of Yahukimo, Papua. The first is that hunger is responded to in a prompt manner, in the sense that government immediately adopted measures to deploy emergency food supplies. The second is that the denunciation of the occurrence of hunger, in which the government preferred to identify the adversity in terms of harvest failures or drought and not according to the nature of the adversity itself. The third is that the very poor economic and social infrastructures of Papua are used as legitimate reasons for the persistence of hunger.

With respect to the first issue, at a glance, one can associate the immediate measures taken by the central government with the procedural requirement stipulated in Article 6 of General Comment No. 12 on how State Parties should approach their obligation regarding hunger. The action is also consistent with the definition of food emergency provided in Government Regulation No. 68 of 2002, in which it is stipulated as a requirement of exceptionally immediate and precise measures. Similar provision can also be found in Law No. 24 of 2007 on Disaster Management that states combating food emergency should do so according to the promptness principle.

Comparing this positive assessment with the second issue on government denunciation of the occurrence of hunger, one might need to consider the broader impact of the word 'hunger'. In general, words like hunger and starvation have a strong emotional impact and are therefore rarely used as technical terms (Kent, 2005: 21). In this light, deaths caused by hunger would inevitably increase the negative impact of the situation in Papua. Politically, this would not be strategic, given the delicate relationship this province has with the central government (Chauvel and Nusa Bhakti, 2004). In Yahukimo, the denunciation of hunger means that the adversities are caused more by unavoidable circumstances, i.e. drought and harvest failures, which the government would be unable to anticipate.

The denunciation has the effect of decreasing the gravity of the situation. In such a context, the adopted immediate response would be considered as a sufficient measure. Arguably, focusing on food supply as the approach to eradication would also result in such complacency. This does not mean disregarding its function in solving the problem in the short term, which is undoubtedly necessary to ease the suffering resulted from hunger. Yet, Yahukimo teaches us that opting for immediate measures is an inadequate way of improving people's ability to feed themselves. It merely focuses on the result of the problem not on the problem itself.

Hunger in Yahukimo is obviously not a rare occurrence. The case study of food emergency presented in this chapter could have been prevented, since a similar situation had taken place before. In a perfect environment for human rights entailing a well-functioning legal system, the right-holders could count on the compliant behaviour of the State to the obligation to respect, protect and fulfil in order to prevent similar events from happening again. Exactly at this crossroads, the occurrence of hunger ought to be eradicated beyond immediate measures, i.e. as a violation of the right to food.

Normatively, Article 2(1) of the ICESCR stipulates that the State obligation to the progressive realisation of economic, social and cultural rights in this regard is of possible relevance. The principle is a recognition that States Parties may not be able to ensure instant realisation of the rights contained in the Covenant, including the right to food. State Parties are, however, required to adopt measures at a national level that might better shield vulnerable segments of the populations. Further, States are also required to incorporate human rights principles to define the objective and formulation of the policies and corresponding benchmarks.[358]

In an effort to operationalise the principle of progressive realisation, the Special Rapporteur on the Right to Food, Olivier de Schutter advises the establishment of appropriate institutional mechanisms in order to:[359]

- identify, at the earliest stage possible, emerging threats to the right to adequate food, by adequate monitoring systems;
- assess the impact of new legislative initiatives or policies on the right to adequate food;
- improve co-ordination between relevant ministries and between the national and sub-national levels of government, taking into account the impact on the right to adequate food, in its nutritional dimensions, of measures taken in the areas of health, education, access to water and sanitation, and information;
- improve accountability, with a clear allocation of responsibilities, and the setting of precise time frames for the realisation of the dimensions of the right to food that require progressive implementation; and
- ensure the adequate participation, particularly of the most food-insecure segments of the population.

Notably, the proposal made by the Special Rapporteur puts an emphasis on the implementation of the right to food outside the judicial mechanism. This is the point where progressive realisation on the right to food should take precedence to aim at eliminating the persistence of hunger. In this regard, progressive realisation means that the effective instigation of a human rights culture in designing and executing right-based food policies, programmes and projects requires a wide-ranging effort undertaken on as many fronts as possible and that it could not be achieved solely or even primarily through the traditional approaches involving mechanisms and procedures that are largely reactive or curative in nature.

[358] E/C.12/1999/5, 12 May 1999, General Comments 12, The Right to Adequate Food, paragraph 21.
[359] Report of the Special Rapporteur on the right to food submitted in accordance with resolution S-7/1 of the Human Rights Council (Olivier de Schutter), A/HRC/9/23, paragraph 14.

The requirement to adopt a wide-ranging effort seems to be crucial given the current socio-economic situation in Yahukimo, Papua, which is the third observable fact derived from the case study. The above statement made by the Special Rapporteur on the right to food mentions that national strategies should include measures taken in the areas of health, education, access to water and sanitation, and information as well. The poor availability of basic infrastructures, such as roads, creates many isolated areas, including Yahukimo. These too would fall under the principle of progressive realisation. Thus, the logical question now would be the extent of this principle, particularly whether the government is willing to allocate the maximum available resources; including, first of all, making painful compromises about methods of food production that can actually improve the ability of people to feed themselves, and second, implementing development policies that imply progress in the sense of a structural improvement in people's possibilities to sustain their daily livelihoods.

Nevertheless, one should avoid succumbing to similar caveats of normative bullet points of the principle of progressive realisations suggested by the Special Rapporteur. Advising on efforts to maximise allocation of available resources could also be misinterpreted as rigorous intervention, centralisation and concentration, massive appropriation of natural resources and forced change of livelihoods. In this light, the normative bullet points listed by the Special Rapporteur are unlikely to be realistic in the context of Yahukimo, Papua. Once appropriated, these suggestions could serve as new 'political technologies' (Lemke, 2002) and modes of governance by the Indonesian government, or at least it could be interpreted as such.

The statement made by Amartya Sen almost three decades ago is still proving to be significant here. Based on studies of famines in Africa and South Asia, Sen radically influenced the evolution of famine theory by arguing the main focus on eradicating hunger should not be on food supply failure. He mentioned that food supply statements say things about a commodity (or group of commodities) only, while starvation statements are about the relationship of persons to the commodity (Sen, 1981). Connecting Sen's thesis with hunger strikes in Yahukimo, Papua, suggests that mitigating measures for hunger should indeed move from a 'food-supply' approach to taking human suffering as a result of hunger seriously.

Against the above, categorising hunger as a violation of the right to food would carry no significant weight if one only scrutinises it from the perspective of human rights as legal trajectories. As argued before, an accountable recourse mechanism to address the food emergency may be able to prevent the likelihood of similar events, but to set it as a benchmark for the effort to eradicate hunger would be too unrealistic, given the complex and overlapping mechanisms on food in Indonesia. A focus on State obligations to the right to food serves merely as a diplomatic embellishment without considering the values and attributes of food and the positioning of the right to food itself in the Indonesian legal system.

In this light, there is not much to expect from the State of Indonesia, except the immediate efforts and responses.

As a further attempt, inevitably, one is required to analyse hunger in combination with other orientations that are also right-based, such as the sociological and political economy focus of hunger, or other private and administrative arrangements of food availability. In these complex arenas, entitlement to food or people's actual command of food is governed according to the prevailing struggles and relationships of people, context and power.

5. Conclusion

Everyone would agree that hunger is a fundamental breach of one's dignity, particularly as it is not an unavoidable disaster.[360] Its persistence becomes a question of life or death. It is understandable therefore that States pledge their position against hunger in international human rights laws.

The article is written within the framework that States' declarations to commit on human rights norms are distinct from the rules, policies, and practices that signify the actual capabilities of the right-holders to acquire their entitlements. The discrepancy presents itself because struggles, priorities and norms at the domestic level are negotiated differently from those at which States declare their commitments. In this regard, for the right to food to actually improve the relationship between the individual and the food commodity requires more actions than merely incorporating human rights norms and values in the regulations, rules and policies.

The body of international law on access to food as a human right suggests two ways of promotion. The first is through an immediate measure and the second is through a progressive realisation. The array of laws and regulations coming from the international level and the national level of Indonesia suggest that the immediate measure is applicable and particularly crucial as a response to hunger. Yet the mass of laws are flawed in asserting the possibility of legal claims based on the very idea of human rights, which is human dignity. Given this shortcoming, it is argued that in the context of Indonesia it is crucial to improve individual access to justice, entailing the ability and the independence of the court of law to accept and to treat hunger as a violation of the right to food. From an optimistic point of view, considering hunger as a breach of one's legal rights would eventually reduce the likelihood of its reoccurrence.

[360] Jean Ziegler, the former Special Rapporteur on the Right to Food once argued that 'persistent hunger in today's world is neither inevitable nor acceptable. Hunger is not a question of fate; it is manmade. It is the result either of inaction or of negative actions that violate the right to food. It is therefore time to take action.' E/CN.4/2003/54, p. 51, paragraph 58. Report of the Special Rapporteur on the Right to Food.

On the other hand, enforcing the right to food in policy measures would require a broader understanding of the principle contained in the right. The case of Yahukimo reveals that although the immediate measures taken are in line with the human rights way of thinking, it would be logically legitimate to move beyond this approach in combating the adversity. Notably, the awareness to adopt broad measures in improving people's access to food and feed themselves has been adopted in many normative guidelines on the right to food. It has been repeatedly proposed that combating violations of the rights contained in ICESCR should be done through measures that adhere to the principle of progressive realisation. It is important here to regard hunger as a symptom of a bigger problem, encompassing all aspects related to people's ability to sustain their daily livelihoods. Accordingly, regarding hunger as the violation of the right to food means not only taking the measures based on the assessment of the symptoms but also adopting policies, programmes and projects necessary to tackle the root of the problem.

Yet, this might not always be realistic in many social, political or cultural contexts. One can expect little when normative points are being used as the standards for improvement. In this regard, the question is not so much whether the State as the duty-bearer complies with the obligation to respect, protect, or fulfil; but rather how to facilitate the right-holders to advance their ability to feed themselves, how much regulation is needed to protect the poor and vulnerable, as well as how to facilitate good arrangements that take human suffering seriously. Of course, the concept of the human right to food is important, as is human dignity, but it is also very flexible to interpretation, especially where it is related to specific locations such as Yahukimo, Papua. Furthermore, for a country such as Indonesia, a judiciary mechanism may only be one of the many channels, if there are any at all, that govern people's capability to acquire access to food.

References

Alston, P. and Tomasevski, K. (eds.), 1984. The Right to Food. Martinus Nijhoff Publishers, Dordrecht, the Netherlands.

Alston, P., 1998. What's in a name: Does it Really matter if Development Policies Refer to Goals, Ideas or Human Rights?. SIM Special 22. SIM Publishers, Utrecht, the Netherlands.

Biro Pusat Statistikk (BPS), 2009. Perkembangan Beberapa Indikator Utama Sosial-Ekonomi Indonesia. BPS Press, Jakarta, Indonesia.

Chauvel, R. and Nusa Bhakti, I., 2004. The Papuan Conflict: Jakarta's Perceptions and Policies. East West Centre, Washington, D.C., USA.

De Gaay Fortman, B., 2006. Human Rights. Elgar Companion to Development Studies. Edward Elgar Publishing, Cheltenham Glos, UK.

HamPapua, 2006. Sehat Itu Sa Pu Hak, News-Report, 11 December.

Kent, G., 2005. Freedom From Want. Georgetown University Press. Washington D.C., USA.

Lemke, T., 2002. Foucault, Governmentality, and Critique. Rethinking Marxism 14, 3: 1-17.

Mechlem, K., 2004. Food Security and the Right to Food in the Discourse of the United Nations. European Law Journal 10: 641-648.

Narula, S., 2006. The Right to Food: Holding Global Actors Accountable under International Law. Columbia Journal of Transnational Law 44: 691.

Habermas, J., 2001. Remarks on Legitimation through Human Rights. The Postnational Constellation, Political Essays. MIT Press, Cambridge, MA, USA.

Raz, J., 1984. Legal Rights. Oxford Journal Legal Study 4: 1-21.

Sen, A., 1984. Resources, Values and Development. Blackwell, Oxford, UK.

Somba, N.D., 2009a. Dozens die from starvation in Yahukimo once again. The Jakarta Post, 3 September.

Somba, N.D., 2009b. No famine but food shortage in Yahukimo: Government. The Jakarta Post, 4 September.

Somba, N.D., 2009c. Starvation no, food shortage yes: Local official. The Jakarta Post, 5 September.

Timmer, C.P., 2004. Food Security in Indonesia: Current Challenges and the Long-run Outlook. Working Paper No. 48. Centre for Global Development.

Twinning, W. (ed.), 2009. Human Rights: The Southern Voices. Cambridge University Press, Cambridge, UK.

Sen, A., 1981. Poverty and Famines: An Essay on Entitlements and Deprivation. Clarendon Press, Oxford, UK.

Warr, P., 2005. Food Policy and Poverty in Indonesia: A General Equilibrium Analysis. The Australian Journal of Agricultural and Resource Economics 49: 429-451.

UN Documents

A/51/950, 14 July 1997, Renewing the United Nations: A Programme for Reform.

A/HRC/9/23, 8 September 2008, Report of the Special Rapporteur on the right to food submitted in accordance with resolution S-7/1 of the Human Rights Council (Mr. Olivier de Schutter).

E/C.12/1999/5, 12 May 1999, General Comments 12, The Right to Adequate Food.

E/CN.4/1987/17, 8 January 1987, Limburg Principles on the Implementation of the International Covenant on Economic, Social and Cultural Rights.

E/CN.4/2003/54, 13 January 2003. Report of the Special Rapporteur on the Right to Food.

Indonesian Laws/regulations

Indonesian Constitution, Fourth Amendment, 11 August 2002.

Decree of The People's Consultative Assembly, No. 11/MPR/1998, 25 June 1998, on the National Plan of Action on Human Rights 1998-2003.

Government Regulation No. 68 of 2002, 30 December 2002, on Food Security.

Law No. 7 of 1996, 4 November 1996, on Food.

Law No. 39 of 1999, 13 September 1999, on Human Rights.

Law No. 26 of 2000, 23 November 2000, on Human Rights Court.

Law No. 24 of 2007, 26 April 2007, on Disaster Management.

Presidential Decree No. 40 of 2004, 11 May 2004, on the National Plan of Action on Human Rights 2004-2004.

Part 2

Law, science and politics in securing food safety

Chapter 7
Food politics: science and democracy in the Dutch and EU food polity

Henri Goverde

> 'People who want to make the world a better place cannot do so
> by shifting their shopping habits: transforming the planet requires duller disciplines, like politics'
> The Economist on 'Good Food', December, the 9th, 2006.

1. Introduction

In this chapter the institution building in European and Dutch food affairs will be discussed from the perspective of 'food politics'. Food politics is a rather new and therefore not exhaustively developed field in political science. Marion Nestle (2002), one of the very few authors in this new field, has linked food, nutrition and health to politics. She wrote as a central thesis for her book *Food politics* that '...diet is a political issue. Because dietary advice affects food sales, and because companies demand a favourable regulatory environment for their products, dietary practices raise political questions that cut right to the heart of democratic institutions.' There are many '...struggles over who decides what people should eat and whether a given food is 'healthy'. As a result, they inevitably involve struggles over the way government balances corporate against public interests.' (Nestle, 2002: 28).

In short, the driving force behind 'food politics' concerns the following key issues:
- uncertainty about the availability and access to food as a basic need for all human beings;
- doubts about the quality of food (public health and public safety);
- doubts about the appropriateness of the organisation that produces and distributes the available quantity and quality of food.

This chapter presents key issues of food politics in the Netherlands and EU from a historical perspective. The first aim is to assess whether or not there is a dominant or even hegemonic discourse which has sustained the power relations in Dutch and EU food polities (as configurations of stakeholders and, formal and informal, institutions involved in food politics). For this purpose the role of governments and science in addressing uncertainties and doubts on food and in defining public space for meaningful debate on this, will be highlighted. The second aim is to identify and develop a perspective or approach to enhance democratic food politics.

While the renaissance of Dutch food institutions has started at the end of 19th century, the Dutch and the EU polities concerning food affairs only became parallel in the last quarter of the 20th century. This parallel development provides an interesting background for a historical and comparative review. On the one hand, a path-dependency between both institutional projects is imaginable. On the other hand, it cannot be excluded that the role of public organisations regarding access of food and food safety has been (slightly) changed in the transition from a pure national Dutch food regime into an EU food polity and, reciprocally, that the recently established EU food regime has had an impact on the further development of the Dutch food institutions particularly in the last two decades.

From the perspective of social and political theory, the historical development of institutions can be analyzed as a process that results from the intermingling of a normative stance concerning the political role of the state in society, the existence of a dominant or even hegemonic discourse in the field under consideration, and the checks and balances in the power relations between the relevant stakeholders in this specific context.

Two contrasting lines of argument can be distinguished in the role of the state in food politics (Korthals, 2003: 51-73). The first argument takes a normative-liberal stance. This argument holds that consumers should make decisions about food affairs mainly by themselves. State institutions have a limited role, i.e. setting conditions for standards of safety and health. The second argument is based on a communitarian philosophy. This argument claims that food is of such high importance for a society that citizens should have the right to make (collective) decisions about it: then food is understood as an exceptionally political commodity (McInerney, 1983: 162, 172). On the basis of a brief review of these two arguments, a deliberative perspective or hybrid approach will be introduced in the concluding part of the chapter.

To address key issues of food politics in the Dutch and the EU polity and to explain its institutions, the chapter is organised as follows. The second section offers a theoretical reflection on concepts like (counter-) hegemony, power and institution-building especially in the field of food politics. In Section 3 a short history of Dutch food polity will be presented. The main features of it will be 'modernity' and 'consensus-seeking behaviour'. The next section will analyse the formal institutions in the EU General Food Law, with emphasis on the role of scientists. Section 5 will focus on what is recognised as the 'truth' in the multi-level governance game between the EU and the national food authorities, and how this truth is recognised. Section 6 employs some normative guidelines for more democratic food politics in order to find openings for public debate about new food-styles within the contemporary institutional framework. In Section 7 the different dimensions of a hegemonic discourse in the food system and how these sustain the power relations in Dutch and EU food politics are discussed.

In addition some suggestions for enhancing opportunities for democratic food politics are launched. This implies a more transparent role of science in the food (safety) system and a better facility for alternative food- and life-styles.

2. Hegemony, power and institution building in the food system

According to Gramsci (1971) hegemony refers to the ability of a dominant class to exercise power by winning the consent of those it subjugates, as an alternative to the use of coercion. In line with this author's terminology (Gramsci, 1971: 56-61) hegemony in the Dutch and EU food system is not based on the ideology of the bourgeoisie as a comprehensive social class. Rather it should be perceived as the result of discursive activities (texts, story-lines, discourses) in a community of practices among 'intellectuals', the thinking and organising element of a particular relevant denomination (Gramsci, 1971: 3) i.e. Dutch and EU managerial civil servants and recruited (Research and Development [R + D])-scientists. In addition, hegemony should be sustained by a political and economic grand narrative. For instance, the neo-liberal or modern liberal approach to economy combined with epistemic rules of the game in the emergence of positivist science will often function as a belief system and worldview. It is suggested here that this belief system of the denomination of 'intellectuals' is probably echoed in the regulative, disciplinary and control system of the European General Food Law (GFL). More precisely, this food system is supported by the rules of the game concerning scientifically based risk analysis and risk management. Furthermore, if the hegemony of neo-liberalism supported by modernist science is reflected in the formal institutions, these can probably authoritatively speak 'truth' not only to governments but also to profit and non-profit organisations in the food system.

As theorised elsewhere, hegemony can also be analysed as a specific result of power dynamics, perceived as a permanent process in a three-layered configuration of power relations among mutually dependent political actors (Goverde, 2006; Haugaard and Lentner, 2006). These layers are called: 'episodical power' (power over), 'dispositional power' (power to), and 'systemic or facilitative power' (Clegg, 1989; Goverde and Van Tatenhove, 2000). The power relations in each layer and between these layers contribute to a final position of the power balance in a political situation that, though always in labile equilibrium, can be very robust: i.e. hegemonial. It is plausible that this corresponds to the current political situation in the EU food system.

On the other hand, all regulations in a tripartite (government, civil society, market) political system can only be perceived as a temporal social compromise. In theory, social change is possible as a result of transitions in the episodical, dispositional or systemic layers of the power balance. However, it is also obvious that for any changes in a hegemonial configuration the driving force of a counter-hegemonic act needs to be extremely comprehensive in its contesting of the dominant discourse.

In theory, this brings us to the question if there is enough 'public space' (in terms of Habermas, 1992) to demarcate the public dimension of the food polity. In other words, whether or not the *res publica* in food politics can be debated in a meaningful way. Though this chapter starts from the assumption that a translucent public dialogue is still possible, a critical issue concerning 'modernity' can not be neglected, i.e. that differentiation, individualisation and rationalisation of society have almost ruled out the opportunities to debate 'what for' questions (Blokland, 2001: 13). Therefore, public debates seem to focus mostly on instrumental, functional-rational questions, in other words on 'how much'- and 'how to do' matters.

To some extent, it is imaginable that this fundamental conflict is echoed in the institutions of the EU food system. Suppose, a specific discourse concerning 'what is good food' no longer has hegemonic power, then the question is raised if the institutions of the food system are still consistent with the on-going political debate about access to and criteria for 'healthy food'. The question then is, how to build capacity and institutions that fit appropriately with, on the one hand, the emergence of differentiation in systems of governmental food control and, on the other hand, self-government of businesses enhanced to limit governmental interference within the ideological boundaries of neo-liberalism and new public management.

As expected, the instrumental and pragmatic notion is reflected especially in the capacity of new institutions concerning food safety in the EU and the Netherlands. Because the food safety regime is established on the connectivity of public and private scientific networks holding a common discourse embedded in a belief system that claims the objectivity and political neutrality of scientific knowledge, the result is a rather hybrid polity. In practice, the 'hybrid' character of both EU and Dutch food polities seems to be sustained by the hegemonic control of technocratic governmental civil servants supported by (independent but business-oriented) committees of experts.

3. Phases and features of the Dutch food polity: modernity, consensus, fragmentation

A review of the history of food institutions in the Netherlands reveals a great many argumentative political struggles. In line with Nestlé (2002), Dutch food politics conceives debates about the definition of 'healthy food', about the norms, about who is allowed to set the norms as well as how these norms are monitored and controlled. As will be explained below, to a certain extent the (formal and informal) configuration of the main stakeholders and the food institutions, i.e. the 'food polity', has been changed during the attempts to tackle these political struggles. For instance, in a book about nutrition a master-chef complains about significant institutional developments:

> 'In 1997 Hazard Analysis Critical Control Points (HACCP) was implemented. The aim of this rule is to control the processes of all goods through the firm, i.e. how it enters the firm, how the ingredients are stocked, and how these goods are manufactured. I'm responsible for checking all critical moments in the cooking process.' (Van Otterloo, 2000: 371).

However, in *Nutrition re-debated* (Voeding opnieuw ter discussie) (2000: 369-374), Van Otterloo claims that debates about food quality and -safety also show great continuity. This author concludes:

> 'While all attention was focused on caloric value of food as well as on hygienic production, manufacturing and distribution of food at the end of the nineteenth century, just one century later the different qualities of food are *hot issues* [emphasis in original] again'.

3.1 Dutch food politics: towards scientific food management

The first feature in the history of Dutch food politics seems to be an enormous increase in the number of actors involved. At the end of the nineteenth century actors in the food chain such as producers, distribution and trade companies were confronted with newcomers such as medical doctors, scientific experts in physics and chemistry and also some governments. In addition, mostly during the twentieth century, new actors were free associations of citizens (civil society), such as the Association of Vegetarians (1894), supporters of bio-dynamic agriculture and/or reformist social movements (during the twenties and the thirties), unions for better education for house-keeping and farming (1890-1940), and the Dutch Association of Female Housekeepers (NVH), and the Consumers Platform (1954).

Some of these organisations, for example the NVH, became 'mediators' between the producing firms and the consumers. The public relations of these mediators included a dominant discourse promoting the rationalisation of housekeeping and the preparation of meals in accordance with scientific hygienic norms. The household that was perceived as a firm full of efficient equipment, became the place for all types of innovations. Here, ergonomic architectural interior design of the kitchen is just one of the modernist examples. In addition to advertisements, marketing and selling venues this discourse of modernisation also impeded mechanisms and institutions of education, research, public relations and control mostly initiated by the government.

The Consumers Products Safety Act[361] and the Meat Inspection Act[362] were enforced in 1921. From an international perspective (France 1851, England 1875, Germany 1879), however, the Netherlands introduced a national food law rather late in the day. Since the French under Napoleon had abolished the rules of the guilds, the Netherlands had no system for food control besides some local rules. According to Van Otterloo (2000: 258) this was the result of '...the dominance of economic liberalism on the one hand and the permanent troubles between interest groups, scholars and civil servants about finances, standards and methods' on the other, as well as the total lack of 'political will'.

Throughout the twentieth century time and again food and its ingredients have caused '...debates, controversies and contests between ever growing social and political interests'. In the beginning there were scientists and governments that produced the norms for 'good food' and which ingredients could be perceived as 'healthy'. A strong belief in the capacity of science as a potential arbiter in struggles over security, safety and sanity of food seems to be dominant. As a consequence, the history of Dutch food polity shows clearly that chairs in the home of science were mainly offered to those who promoted utility and urgency of scientific research in their professional life. So, it is mainly practices in the food community that have stimulated the urgency for scientific knowledge in food affairs. Therefore, in the food polity, 'speaking truth to power' (Wildavsky, 2005) means above all the mediation of expert knowledge, mostly produced according to the Cartesian research design in the natural sciences, to food practices monitored by the national government.

Later, as a result of seepage during the neo-corporatist era (see below), these norms and rules became part of the belief system of the public in general. On the other hand, the diversity of consumers has simultaneously created different images of good and bad food. These images often neglected scientific perceptions of the food quality (Van Otterloo, 2000: 308, 309).

In summary, the Dutch food polity was rather dynamic in the period 1890-1940. The power relations among the stakeholders were not yet fully established. For sure, there was an institutional trend in the public domain to set the food norms at the national level and to get legitimacy for the norms by referring to statements of the scientific elite. In addition to governmental regulation in food affairs, rationalisation in food production, food manufacturing and food distribution as well as in consumption was also based on scientific and technological assessments in the private sphere (firms, mediators, house-keepers). So, generally speaking there was a positive attitude in the public and private domain towards 'scientific food management', based on rationalisation and mechanisation of the food system.

[361] Also known as Commodities Act. Act of 28 December 1919, Statute Book (Staatsblad) 1935, 793.
[362] Act of 25 July 1919, Statute Book 524.

3.2 Organising availability of food: state action in a neo-corporatist polity

Shortly before the Second World War there was a serious political debate about the urgency of a government organisation for food security. In particular Dr. Ir. S.L. Louwes, civil servant responsible for Agriculture and Livestock in the Ministry of Agriculture, enhanced initiatives for public policy in agriculture and food security since 1934. He did not agree with *ad hoc* measures in these fields and recommended a comprehensive policy approach. The first result was the foundation of a national office for monitoring food security in wartime in 1937. During the Second World War this central governmental office organised practically the whole food service, including transport and distribution throughout the country. Furthermore, a significant aspect of its activities was the foundation of a complex of new institutions, for example the Agricultural Economic Institute (1940) – a think-tank and research centre –, a National Advisory Board for Food (1940), an organisation for the distribution of food ration-cards, the Food Inspectors Agency as well as the regulation for incorporating retailers into the food system. Louwes (1980: 228) explains how the governmental food polity became a rather autonomous system after the war. It was self-referential, mainly because many officials were either directly involved in agriculture or had an open mind about the goals of the farmers' organisations. In addition, he wrote – apparently with some regret –: '... that in the democratic political context, featured by its mechanisms for regular elections and behaviour to attract voters, attempts to solve problems were often congruent with the short term perspective of political parties and interest groups. For this reason the agricultural sector has had a great influence on the selection of policy instruments used for policy implementation. The choice of policy tools was very much oriented to the short term too'. (Louwes, 1980: 228-9)

Notwithstanding the short-term focus under democratic conditions, the institutional autonomy of the agricultural and food sector became manifest in the re-introduction of a separate ministry (June, the 23rd 1945) embedded in a corporatist structure based on a constitutionally recognised public institution that compromises 'capital' and 'labour' in an organisational and policy perspective (*publiekrechtelijke bedrijfsorganisatie*). This reflects a neo-corporatist polity, coined by Frouws (1994: 47) as a: *'social-political structure of articulation of interests and policy formation, grounded in co-operation of functional interest organisations – who have a monopoly in representation and a political-economic consensus among its leaders – and governmental institutions, in which the organisations were supposed to produce discipline among their member farmers in exchange of influence on policy-making and the way policies are implemented.'* (italics in original). Louwes (1980: 226) perceived the neo-corporatist approach of the political conflict in the agricultural sector as a 'societal compromise'. 'The idea, as such an 'ideal type' of economic order, was oriented neither to the unlimited liberal game of economic driving forces nor to an all-mighty government, but to an order of firm-fellows – employers and employees – who by the organisational tool of corporations were equipped with competences to set economic as well as social rules.' Indeed, this institutional form to manage political and economic struggles was based on a temporal societal compromise which nevertheless lasted for about fifty years. However, with the exception of the meat and fishery sectors, it became obvious that some driving forces (mainly in different sectors of agriculture and floriculture) succeeded in getting rid of this societal compromise in the nineties of the last century.

3.3 Dutch food institutions in the 20th century: modernism and consensus: an interpretation

Historically, modernity reflects '...the institutions and modes of behavior established first of all in post-feudal Europe, but which in the twentieth century increasingly have become world-historical in their impact' (Giddens, 1991:14-15). In contrast the chapters about nutrition (*Voeding*) in 'Technology in the Netherlands during the Twentieth Century' (*Techniek in Nederland in de 20ste eeuw, deel III*) characterise institution-building in food affairs in terms of modernity only in relation to the last two or three decades. Though rather late, the Netherlands developed as an industrialised country in the last two decades of the nineteenth century. In addition to a capitalist industrial development, modernity in the Dutch nutrition and food industry also depends upon the specific notion, which had already emerged in the seventeenth century, that of 'Enlightenment'[363]. This notion assumes that the key to human progress and social order is objective knowledge of the world through scientific exploration and rational thinking. It also assumes that the social and natural worlds follow laws that allow food affairs be measured, calculated and therefore predicted.... The science of probability and statistics was developed as a means of calculating the norm and identifying deviations from the norm, thus embodying the belief that rationalised counting and ordering would bring disorder under control.

So, the institutionalisation of food affairs in the Netherlands fully mirrors the characteristics of modernity, especially up to the 1980s. In addition, since the end of the Second World War the Dutch food polity has also been an example of the consensus-seeking model of policy-making, coined as the *Rheinland* model. Political decisions are made when consensus has been reached after a deliberation among the socio-economic and governmental partners[364]. The result was an almost

[363] For an informative review of the principles and history of the Enlightenment in Britain as well as on the Continent, see Jonathan Israel, Enlightenment Contested (2006) and Radical Enlightenment (2001). For an intellectual history of the 20th century and how modernity has penetrated it, see Peter Watson, The Modern Mind (2001).

[364] It should be noted, however, that this consensus-seeking social network of interdependent (public and private) actors in the agricultural sector, has also often been labelled as an example of the discourse of 'modernisation' (Van der Ploeg, 1999: 292).

autonomous polity of the combined activities concerning education, scientific research, and public relations.

3.4 Autonomy agro-food polity contested

In the last decades of the 20th century, however, both modernity and consensus-seeking as features of this Dutch autonomous agriculture and food polity were contested. In academia modernism was often accused of creating high (social and environmental) risks, and even a risk society (Beck, 1982; Lupton, 1999: 5-6). In the agro-food praxis, ideas and interests to tackle the recognised risks and to improve the image of the sector became rather fragmented. In addition to a dominant capitalist, globally oriented and high-tech discourse in agriculture and food production, some proto-discourses, for instance organic farming and slow food that incorporated traditional and layman's knowledge, were seeking recognition. As a consequence, the authority of the agro-food polity could no longer be sustained and the political decisions became more dislocated from the functional *Rheinland*-model. Strategic policies were centralised not only at the national but also at the intergovernmental (EU) level, though the implementation continued to be territorially decentralised, mainly in the hands of regional authorities.

The fragmentation of ideas and interests, especially in the food polity, echoed to some extent the idea that consumers wanted to express themselves in different life-styles and that eating specific food symbolises a particular fashion and identity. This development fits in well with the concept of 'reflexive modernity' that interprets current modernity as a new situation in which individuals demand the right to make choices about their own identity and its meaning in the wider world (Taylor and Flint 2000: 374). This is mirrored to some extent in the (marginal) presence of organic farming in the Netherlands, and also – as will be explained in Section 3 – in the ideas, practices and rules aimed at openness and transparency in new institutions like the Dutch Food Authority (*Voedsel en Waren Autoriteit*) and the European Food Safety Authority (EFSA).

3.5 Organic farming: example of counter-hegemony and power dynamics

The case of the Dutch organic farming sector can shed light on this mechanism (Arts and Goverde, 2006: 85-87). Although the number of organic farms, their land area as well as their turnover, more or less doubled in the Netherlands in the period 1997-2003, the Dutch organic farming sector is still small compared to other countries in Europe: Netherlands 2.2%, Austria 13%, Switzerland 10%. The Action Plans on Organic Farming (1996 and 2000) aim at: (1) improving professionalism of the sector, (2) transparency in the production chain and (3) growth of production figures, while heralding more of a market philosophy. However, there is little chance that the final goal (10% organic farming in 2010) will be realised. In fact, the

Dutch policy arrangement concerning organic farming is dominated by mainstream agricultural organisations. They perceive organic farming as an interesting niche market for those farmers who cannot or will not survive in mainstream globally oriented agricultural business. In this respect, it is no surprise that 'conversion' and 'assimilation' are the main policy themes in this arrangement. The state also endorses a market philosophy, facilitating the sector at best, while refraining from direct intervention. On the other hand, organic farmers are supported by organisations that propagate a more eco-centric and holistic discourse and strategy, which in fact challenges mainstream agriculture. These organisations also promote active state intervention in order to overcome structural inequalities. Of course, these views clash with the attitudes of traditional farmers' organisations and of the state. Furthermore, embedding the Dutch organic farming policy arrangement in its wider institutional context, i.e. the EU regulations, is rather problematic as well. There is even an 'organic farming policy paradox': mainstream agricultural policy constrains the sector, while organic farming policy enables the mainstream sector. Manure policy is a good example here. EU regulations allow farmers to distribute manure to some extent on their own land and only during certain periods of the year. Though organic farms have organic manure that is not mixed with the residues of fertilisers and heavy metals, in accordance with the same EU regulative regime, they are not allowed to use it on their own land all year round.

In summary, organic farming is a sub-culture that is in the margins recognised by the Dutch as well as the EU food authorities. However, its discourse is fragmented (niche in the market versus challenge to mainstream agriculture) and the special policy style treated by the dominant agricultural authorities is oriented more to assimilation and encapsulation rather than to accepting a counter-hegemonial device. That is why the policy arrangement concerning the organic farming sector is in itself a good example of the hegemony in the actual food system. In fact, organic farming could develop some episodical power reflected in a specific policy arrangement. However, the hegemonial forces changed neither its dispositional power, nor its systemic power.

4. Formal institutions in the General Food Law: scientific 'truth' in multi-level governance

To better understand the above-mentioned trends in the food polity, this section will outline the formal rules in the European General Food Law (GFL) and more precisely the role of scientific experts. For this purpose the focus will be on the dependence of the food polity on expert knowledge as well as on how scientific knowledge is produced.

4.1 Formal institutions in the General Food Law

After the food safety scares of the 1990s, the European legislature undertook an extensive overhaul of food legislation. The general principles of the new system of food law have been laid down in Regulation 178/2002, nick-named the General Food Law (GFL)[365]. The GFL is based on a systemic approach. All stages of production, manufacturing, and distribution of food are included in this system of law. According to Van der Velde and Van der Meulen (2005: 203) this results from the experience that the origin of many food crises was found in the first links of the food chain, i.e. in agriculture, fishery, and other ways of acquiring food and even in animal feed. In order to maintain the highest possible level of protection against food diseases and food crises, risk analysis (= assessment, management and communication of risks) is the foundation of food law. In particular, the risk assessment is based on the belief that science can satisfy the need for an objectively functioning arbiter (Van der Velde and Van der Meulen, 2005: 208). Risk assessment is the main task of the independent European Food Safety Authority (EFSA). This new European institution produces advices, i.e. suggestions for policies and risk management, on its own initiative or at the request of the European Commission, the European Parliament, or the EU Member States.

EFSA bases its scientific assessments on an organisation that consists of a Scientific Committee and Panels of experts. There are eight panels[366] at the moment, but this number is flexible. EFSA has internal rules to select members for the Committee and the Panels. These members are nominated for a three-year term in office, which is renewable, and their names are published. The members of the Committee function as presidents of the Panels. The Panels are supplemented with six independent academic experts. But their recruitment is not exactly transparent. It should be noted that EFSA functions in a tradition and culture of 'expert committees'. Their members are recruited among national officials and experts from 'consultative committees', composed of representatives of sectional EU-level interest organisations appointed by the EU Commission (Christiansen and Larsson, 2007; Larsson, 2003:58-59, 125). So, 'independent' scientists are not easy to define. It is therefore not hard to imagine that recruited scientists are often part of a dominant discourse that helps to satisfy the main function of this specific institution, i.e. 'precooking' norms, guidelines, and certificates which bind public and private organisations and associations operating in the EU. It is not surprising that deliberation will often be based on principles such as scientific

[365] Regulation (EC) No 178/2002 of the European Parliament and of the Council of 28 January 2002 laying down the general principles and requirements of food law, establishing the European Food Safety Authority and laying down procedures in matters of food safety, OJ 2002, L 31.

[366] There are panels on food additives, animal feed, GMOs, dietetic products, pesticides, animal health and welfare, contaminants, and biohazards (see EFSA: http://www.efsa.europa.eu; Van der Meulen and van de Velde, 2004: 163, Diagram 20).

evidence by 'probability and statistics' together with a strong belief that science can bring 'disorder under control'.

Additionally, EFSA operates not only to eliminate the risk of illness and to deal with crises, but also to prevent new risks. This task requires a scientific infrastructure to provide technical and scientific support for new rules and policies, organisation and co-ordination of new scientific studies, development of relevant data-bases, search for new risks, designing an early warning system as well as effective communication. Therefore, EFSA has organised a network with member-state institutions to promote the exchange of information, know-how of experts, and best practices. If EFSA and another (national) institution do not agree in a specific case, they are obliged to exchange information and produce a common document that will be published.

4.2 Role of science in the General Food Law

In order to estimate the role of science in more detail, not only the official text of the GFL (1.2.2002) is relevant, but also its first amendments (22.7.2003). First of all, some definitions refer to the world of science. In particular 'risk assessment' as one of the components of risk analysis, means '... a scientifically based process consisting of four steps: hazard identification, hazard characterisation, exposure assessment and risk characterisation.' A 'hazard means a biological, chemical or physical agent in, or condition of, food or feed with the potential to cause an adverse health effect'.

The scientific reliability of the process of risk assessment is supported by specific rules of the game. According to GFL article 6(2) 'risk assessment shall be based on the available scientific evidence and undertaken in an independent, objective and transparent way.' In order to produce such assessments the Scientific Committee is '...responsible for the general co-ordination necessary to ensure the consistency of the scientific opinion procedure, in particular with regard to the adoption of working procedures and harmonisation of working methods' (Art 28).

In contrast to the scientific approach of risk assessment, risk management is a political-administrative process (GFL articles 30-40). In general, the risk management process concerns the weighing of policy alternatives in consultation with interested parties, considering risk assessment and other legitimate factors, and, if necessary, selecting appropriate prevention and control options. The political-administrative character of risk management requires guarantees that the scientists deliver a scientific account within a specified time limit. For this reason scientific opinions do not need unanimity but the Committee and Panels can act by majority rule. However, minority opinions will be recorded. In certain issues several panels are asked for advice. In case of mutual disagreement EFSA is responsible for identifying the divergence at the earliest stage and ensuring that all relevant information is shared. If the divergence of opinions is between an EFSA agent versus a food security body of a Member State, a joint document should be published, clarifying the contentious scientific issues and identifying the uncertainties in the data. Part of the EFSA mission is to provide scientific assistance when technical criteria should be applied, established or evaluated at the request of the EU Commission. In all cases the best independent scientific resources available should be used, but duplication with Member State or Community research programmes should be avoided. Because procedural rules concerning all phases in the risk management process are not only crucial but also multi-interpretable and even contestable, the EFSA institutional framework contains conditions concerning independence, transparency, confidentiality, access to documents, and communication. However, the EU (Commission and Council of Ministers) is not well known for open access to documents (see different cases initiated by whistle blowers). Some rules seem to confirm this reputation. For instance, art 39.2 reads: 'Members of the Management Board, the Executive Director, members of the Scientific Committee and Scientific Panels as well as external experts participating in their working groups, members of the Advisory Forum and members of the staff of the Authority, even after their duties have ceased, shall be subject to the requirements of confidentiality pursuant to Article 287 of the EU Treaty'. As a consequence the different functionaries involved are disciplined in a rigorous regime of confidentiality. In my view, the question arises as to whether this is consistent enough with the intentions of independence and transparency (GFL section 4).

On the other hand, there is also a rule that 'the conclusions of the scientific opinions delivered by the Authority relating to foreseeable health effects shall on no account be kept confidential'. The Authority shall communicate on its own initiative, ensure that the public and any interested parties are rapidly given objective, reliable and easily accessible information as well as guarantee wide access to the documents which it possesses. In contrast, however, among other issues the access to EFSA documents was amended (July, 22, 2003) to bring it in line with the general principles of access to EU documents provided for in Article 255 of the EU Treaty. Regarding this emphasis of EFSA as embedded in a wider, hierarchical EU framework, this amendment can be interpreted as a sign that its autonomy, though scientifically legitimated, is limited. Notwithstanding this, the EFSA decisions are open to appeal at the EU Ombudsman and the EU Court of Justice.

5. Food polity and science: negotiated 'truth' in multi-level governance game

So far, the rules of the game in the EU food safety institutions confirm that the EU food polity believes in science as a reliable advisor and arbiter. Though the rules are managed primarily 'internally', the experts recruited seem to have a rather good reputation in business research and development at least. Their names are not kept a secret but are made public. At the same time, the EU food authorities recognise their dependency for expertise of 'independent scientists' in R&D departments of the main industries, particularly when there is a knowledge shortage concerning specific (risky) production processes. Of course, the disadvantage of this dependency is that new ideas, new story lines, and new practices cannot easily enter the international panels in these circumstances, except when introduced by the industries themselves. That is why the EU GFL Scientific Committee and the Panels need to be complemented with independent experts from the academic world.

Both types of experts are legally obliged to co-operate in order to produce evidence-based risk assessments using the best available scientific resources. If the best resources are available at the national level, then these will be used (no duplication of expensive research). If the experts cannot conclude in unanimity in a specific case, all opinions are registered.

5.1 Negotiated 'truth'

It should be noted, however, that many ingredients in the process of scientific consultancy give room for 'negotiation'. Scientists will not get an infinite amount of time to search for unanimity. External deadlines can force them to come to a majority statement, though the minority opinion will be registered as well. The GFL regulation includes many 'rules of the game', concerning for example the scientific organisation, the scientific network between EU and the Member States, the co-ordination of research designs, the production of joint documents, the confidentiality to be respected, the rules concerning data collection, identification of risks, access to documents, openness, transparency. Because of these negotiable elements in specific cases, speaking the 'truth' to powerful decision-makers (i.e. EU institutions like the Commission, the Parliament and the Council) will not simply be grounded on evidence-based positivist scientific research agreed by all involved scientists only. On the contrary, the recommendation will often have the character of a social construction negotiated in an epistemic community of rather self-referential scientists. As such this knowledge will be partial and contingent. As a consequence, politicians in the field of food policy will probably question time and again whether they should rigidly follow the negotiated 'truth' recommended by the representatives of this epistemic community in an aspiring democracy (Wildavsky, 2005: 404).

5.2 Food safety and multi-level governance

Because scientists are obliged to avoid duplication of scientific research, the negotiation about best practices is often a multi-level governance game too. On the national scale (see Section 3), the governmental authorities have also accepted responsibility for the safety of food products as well as the guarantee of public health by risk assessment, risk management and risk communication. Risk assessment is strictly based on scientific principles and is upgraded permanently according to the most recent results in food safety research as well as state-of-the-art measurement methods and instruments (Steegman, 2005: 225). The Dutch Food Authority (DFA), for instance, is responsible for the national risk assessment, the monitoring as well as control. Its risk assessments are based on national and international scientific research. The starting point is the idea that independent scientific knowledge is crucial for effective enforcement of the DFA statements. That is why the effectiveness of the DFA depends largely on good collaboration and adequate steering of related research organisations like RIVM (National Institute for Public Health and the Environment), Rikilt (Institute for Food Safety, part of Wageningen UR), and CIDC-Lelystad (Central Institute for Animal Disease Control, part of Wageningen UR). These research institutes have a good record of being independent. Their research reports are usually grounds for implementing the DFA mandate. The assessments include norms for (half-) products, prohibition of particular ingredients and contaminants, demands for food handling, labelling products, advices for consumers, etc.

On the other hand, the risk assessments in the Netherlands are still open to debate. It is contested, for example, whether and to what extent the advanced assessment methods produce enforceable norms. For instance, the zero-tolerance for particular ingredients cannot be implemented on a global scale and would imply an economic disadvantage for (Dutch or European) firms in the global market. As a consequence, the Dutch food authorities claim that the co-operation of national inspectors in the EU should increase to prevent false competition in the market. Next, the DFA believes that the level of food safety cannot really be improved in advanced industrial countries. Instead a focus on healthy diet is suggested to be more helpful.

Of course, the health and safety dimension of food is not a simple universal phenomenon. Therefore, communication of risk assessments is a sensitive affair (Steegman, 2005: 229-232). For example, many people accept some risks if they believe it is communicated in an honest and transparent way. On the other hand, some people accept more risks than others, particularly when they experience personal or societal advantage. Of course, consumers have different demands for information and it is not yet clear who is best placed to cope with this diversity - private enterprises or public institutions. Another aspect of food health and safety is that many products have a traditional and well-known place of origin (for example, wine or cheese produced at particular 'chateaux'). Then it is not always easy to combine these specific production conditions with global or European 'scientific criteria' used for risk assessments (and risk management). This is also true for the communication of risk based on scientific risk assessment criteria in combination with different consumer concerns (e.g. organic farming, animal welfare) as well as with different lifestyles (e.g. obesity and its consequences). So, if one can recognise several problems that probably cause different perceptions of food safety in different countries as well as among different sub-cultures, a plea for a normative-liberal approach in which free associated individuals are allowed to set their own norms for safety and healthy food seems appropriate. On the other hand, safe and healthy food is so important for a society that a communitarian stance is no surprise either. In this respect national and EU food authorities are involved in bi-lateral and multi-lateral co-operation in food affairs. Obviously, the use of scientific results does not produce direct conformity between the national and EU administrations. Instead it is negotiated in all relevant institutions, horizontally as well as vertically.

5.3 *Rheinland* model in trans-national context

As a consequence of the mutual interdependence of the national and EU food authorities, the pursuit of the multi-level governance game does not claim that the introduction of food institutions at the EU level has eroded *per se* the Dutch *Rheinland* model of consensus-seeking public decision-making. On the contrary, it seems that the production of scientific accountable but contingent 'truth' concerning safe and healthy food has introduced unavoidably a rather similar institutional food polity at the EU level. This institutional congruency is arguably largely the consequence of a common hegemonic discourse concerning the role of positivist science as a trustful arbiter in food risk analysis and risk management.

Nonetheless, it cannot be denied that the GFL and its main formal institution, EFSA, are a strong expression of trans-national centralisation in the food system. According to Van der Meulen and Van der Velde (2004: 212) there are two dimensions in this trend of centralisation. First, the introduction of a law (General European Food Law) as such is an act of centralisation, because the food rules are no longer based on 'directives' but on a 'regulation'. So, the Member States have no policy discretion for implementing the EU rules into their juridical polity. And secondly, the network of EFSA connected to Member-State food authorities potentially has a strong capacity to become a supranational co-ordinated system of scientific evaluation of food risks. Consequently, in my opinion, such a formal inter-governmental institution is very capable of speaking authoritatively 'truth' to power. In particular then, EFSA is legitimated to bind by common EU law not only public organisations but also (multi-national) corporations. Moreover, EFSA can influence, directly and indirectly, new scientific programmes as well as the supply of food in regional and local markets.

Having said this, the practices of EFSA as a rather new modernist institution have probably not established patterns of top-down policy-making yet. Besides the guarantees for openness, transparency, internal and external validity of the published research results, the modernist practices of science concerning food risks are dispersed at two administrative levels, the national and the intergovernmental. In the culture of academia it is not plausible to expect one level to dominate the other permanently. In addition, the scientists engaged in these multi-level governance procedures form an epistemic community with a common belief system in the scope of science to control disorder which is fruitful ground for compromises, if necessary.

On the other hand, because there is much room for discussions and negotiations in this specific multi-level governance game, sometimes decisions by majority rule will be taken to force experts to deliver a scientific recommendation in good time. Particularly when there is a shortage in expertise, it is obvious that the functioning of EU rules and decisions is rather dependent on the integrity of the scientists involved, and how they interpret (European) 'economic interests'.

To sum up, the overall result is a far-reaching process of centralisation in European food affairs and the culture of decision-making is only clear for people directly involved in the process. That is why a well-organised food research, assessment and management system at the national level is still relevant for coping with food safety risks throughout the production chain. As the functioning of the existing trans-national dimensions of the food safety polity, including the role of science in this, seems to be complex and operate rather behind the scenes, a plea to create more transparency between stakeholders in the Dutch food polity and the food safety institutions at the EU level seems justifiable.

Concerning the three key issues distinguished in food politics, Sections 4 and 5 focused primarily on food safety institutions and on the institutional organisation related to make enough safe food available in the Netherlands and in the EU. We have not examined specific bottom-up political initiatives to make food accessible to the poor (*voedselbanken*), though the phenomenon is well known in almost all EU countries. Politics concerning what is healthy food is an extension of the main political debate that focuses on the safety of food whether or not this is related to animal diseases. The debate about healthy food requires a strong link with the medical dimension of healthcare and health cure. Sectors like the health service, agriculture, and the food industry are embedded not only in totally different institutions and specific epistemic and professional communities, but also have been internationalised at completely different speeds. This might explain why in the institutions of the food sector the priority seems to be on food safety rather than on healthy food.

6. Government and food: enhancing democratic food politics

The plea in the last section to create more transparency among stakeholders in the food polity requires more insight into normative positions in food politics. In political philosophy and ethics, lines of argument are available not only to bring 'the political back in' (i.e. to make antagonisms explicit) in a search for an image of 'the good life', but also to gain food for thought concerning a combined position for institution-building in a democratically oriented food system (Korthals, 2003: 54-55). After a short elaboration of the normative-liberal and the communitarian, two positions recognised already above, a 'deliberative perspective' will be added. Finally, in order to develop a more democratic approach to food politics and regarding some guidelines for institution-building in the food system, a 'combined or hybrid normative perspective' will be suggested. This will allow for a focus institutionally on social and cultural values so far underexposed. For instance, the former sections concerning the operative European food regime did not show much evidence of the fact that, though members of a community with a plurality of life-styles, individuals are allowed to develop independently a specific life-style in food affairs.

6.1 Three political philosophical perspectives and governance

From a *normative liberal perspective* the political dimension of food affairs requires a public administration that intervenes pre-emptively and serves the security and health of all individuals. According to John Locke[367], public intervention is only legitimate concerning neutral, non-philosophical civil commodities under the condition that science has proven its utility for health. So, the classic liberal perspective incorporates an appeal to independent, value-free science. Furthermore, (semi-)public institutions should organise the markets in order to realise a justifiable distribution of available food resources. This part of the liberal perspective is based on additions to the classical position made by Rawls (1993). In summary, all social primary resources – freedom and equity, income and wealth, foundations of self-respect – should be distributed equally, unless an unequal allocation of one of these commodities favours the less fortunate. However, a government is not allowed to act according to a perfectionism implying a certain ideal type of 'the good life'. A justifiable distribution of available (food) resources requires equal opportunities for any conception of the good life that respects the principle of fairness. Under these conditions, individuals (as consumers) can act according to their personal preferences on the market.

From *a communitarian perspective,* however, individuals should be allowed to *participate* in political decision-making and *interactive* policy-making. Furthermore, different *(sub-)cultures* should have opportunities to produce and consume the *food* that is relevant for their *identities*. In this regard individuals (as citizens) act in the political arena as members of a community that develops common preferences concerning political and ethical matters in food affairs. As a consequence of communitarian decision-making, governments can be given the task of preventing

[367] John Locke (1632-1704) in: A letter concerning toleration, 1689, p. 241 (cited in Korthals, 2003: 56).

risks in the food chain and to inform the citizens about the risks food implies for personal health. However, risks and safety are not perceived as neutral, universal phenomena produced by rational, value-free scientific work. On the contrary, they are specific translations of normative consumer orientations. Individuals (either as citizens or as consumers) will evaluate information about food risks and food safety first on the grounds of the information given by the experts, and second on how they perceive the trustfulness and integrity of the expert who communicates the information.

Furthermore, there is a plurality of food-oriented communities in social life. Citizens participate in different communities around food- and life-styles. Because governments are no longer the central spider in the web of societal driving forces, if ever, some conclude that political and ethical issues concerning food risk and food safety are first and foremost a matter for the community of actors in the markets (producers and consumers).

While both former perspectives assume a cleavage between politics and economy as well as between civic and non-civic commodities, the *deliberative perspective* starts from a tripartite view of society: government and law, market, and (public opinion in) civil society. Jürgen Habermas in particular has argued that there are three types of normative discourses – moral, ethical, and pragmatic or prudential – which though intermingled are institutionalised in these three domains of society. In a model of deliberative processes of democratic politics, Habermas (1992) suggests that – in addition to the three discourses – there are strategic negotiations focused on producing fair compromises which are simultaneously judged on juridical criteria.

In an ideal-typical process the model functions as follows: a specific issue is qualified in pragmatic debates by expertise and counter-expertise. Experts produce a written recommendation that is evaluated in the public arena in ethical and moral debates. Next, in strategic negotiations it is tested if the participating actors in society can tolerate a compromise at least for a certain period of time. These negotiations are guided in parallel by juridical debates in order to search for a congruence of the results with existing juridical laws and rules.

In the deliberative perspective it is recognised that acts of governments and juridical systems are not neutral, but reflect the history of specific organisations of social life (i.e. path-dependency) and as such form a temporal compromise. It is also recognised that this societal compromise can represent a dominant discourse that can become institutionalised in a robust public organisation.

6.2 Guidelines for democratic institution-building in the food system

In my view, the meaning of these three political and ethical perspectives can imply the following considerations for the public institutional dimension of the food system:

First, though by definition both liberal perspectives see a limited task for governments concerning health (and by deduction for food) of individuals, the normative liberal perspective gives science the role of a 'priest' between the political elite and the citizens. Scientists are supposed to be able to judge independently and authoritatively about what individuals should perceive as healthy food. This role is institutionalised by law in order to legitimise governments to make pre-emptive interventions in the market to protect the health and safety of the citizens. However, the modern liberal position adds that governments are not allowed to act from a perfectionist position, i.e. from a self-nominated ideal concept of the 'good life'. That is why individuals are ultimately free to define 'the good life' for themselves. Therefore, governments should create rules which enhance free access to markets for all individuals (producers and consumers) with different food-styles.

Secondly, promoters of the communitarian liberal perspective are not convinced that modern liberals, particularly if supported by scientists embedded in a hegemonic discourse, will create access to markets for all different food- and life-styles. Individuals as members of a (food- and life-style) community (i.e. all 'citizens') should legally get access to the arenas where decisions are made concerning what is good food and what food is good for your health.

Next, the deliberative perspective recognises that governments, though they receive objective 'truth' from the scientific community, are not neutral in decision-making about food and health. In contrast to the liberal dichotomy between government and market, the deliberative perspective conceives society as a tripartite entity of government and law, market, and civil society developed in a path-dependent historical process. The moral, ethical and pragmatic discourses are intermingled and by strategic negotiations institutionalised in this tripartite polity. Therefore, and in contrast to the ideal-typical model of deliberation, it is feasible that a certain mixture of these discourses acquires a hegemonic position in society and as such is able to keep new or alternative story-lines, narratives, and even proto-discourses (concerning healthy food) out of power. In particular, governmental parliamentary authorities should pay attention to how systems of expertise can be crucial in these processes of inclusion and exclusion.

As a result, these considerations shed light on a *combined or hybrid perspective*. This, rather radical democratic, perspective would probably plead for a food policy infrastructure that encompasses different attitudes to food-styles and food fashions in a transparent public dialogue. Governments and related public authorities, in particular, should behave impartially to every rather 'crystallised food-style' (Korthals, 2003: 72). Free access to the market should be given, if a specific food-style operates according to rules for safety, health and conditions of civilised production (environmental sustainability, animal welfare, fair trade). It is even imaginable that some underexposed food-styles are publicly supported to get easier entrance to the market, based on a right of access and on the political principle that equality among sub-cultures should be enhanced.

7. Discussion

As far as the first purpose of this chapter is concerned, to address whether a hegemonic discourse sustains power relations in Dutch and EU food politics, it has been explained that the public space to discuss the right issues concerning access to food and food safety are rather limited. In particular it has been shown that most issues concerning the food system are discussed behind closed doors in rather insular institutions. Partly this is the consequence of prudence and precaution in circles of R&D specialists in the (multi-national) firms as well as in academic epistemic communities. On the other hand, it cannot be denied that interests of business economics and commercial marketing are obstacles to adequate transparency in the political-administrative arena as well as to the general public. Though not researched in detail here, the recent case of Q-fever in the Netherlands (2008-2010) seems to be a clear example of the food safety institutional network operating as a closed shop. In this case the *res publica* and the health safety interests of the people are completely dependent on the integrity and the courage of a few professionals who are prepared to take on the role of whistle-blower.

In addition, because the political process concerning the transition from a pure Dutch food regime into an intergovernmental and/or supranational EU food polity reflects the existence of institutions comparable at both political and administrative levels, the claim is that this position is based in a particular dominant discourse. This is also founded on the idea that the congruency of Dutch and EU food institutions is grounded in collective decision-making (mostly communitarian, or rather mixed with a normative-liberal stance into a hybrid model) fed by (probably) a modernist instrumental policy style emphasising regulatory rules, and the major influence of methodology in the positivist sciences. Furthermore, there is evidence that alternative ideas and proto-discourses concerning food strategies at the margins of mainstream food policies cannot be actively debated in the public arenas either on the national or international level. Most politicians and administrators seem to orient their statements and voting almost completely on the (majority) recommendation of the formal advisory committees based on a hegemonic positivist scientific discourse.

Finally, the multi-level game between food management agencies shows some degree of path-dependency. When EU institutions have to negotiate with national food safety institutions, the procedures at both levels seems to be highly comparable.

The second purpose of this chapter, to identify a perspective to enhance democratic food politics, requires more debate about facilities to discuss alternative food- and lifestyles and to learn more about how to live with 'divergence and difference' in the existing plurality. There should also be room for marginal food affairs in which small business owners, local traders with a global mind-set and consumers with different ideas, story-lines, narratives and practices about food express their own life-styles. Because food and eating have a huge capacity to give people identity, tolerance for different cultures concerning food production and consumption seems to be a pre-condition for living (apart) together in complex societies. To begin with, it was explored what dominant paradigms and alternative views on governance of food affairs by the state could be relevant. As part of this exploration, an alternative political philosophy can be distinguished that contests the hegemonic modernist approach in the EU GFL. Then, it seems also possible to win some power democratically for 'the other' in farming and food manufacturing and to support the different life-styles. Here it is suggested to take a combined or hybrid political and ethical perspective to allow public dialogue concerning different food styles and food fashions, to give right of access to under-exposed food-styles, as well as to enhance equity among food sub-cultures.

In contrast, the advantages and benefits of the existing two-tier food regime should not be neglected. Actors in the market as well as in civil society are supposed to be disciplined and controlled by the episodical power of the European Food Safety Authority (EFSA) and the national food authorities. The actors are ready to accept the dispositional power claimed by these authorities, either because they have influenced the production of the rules of the game themselves or because they trust the science-based statements concerning food risks and food safety expressed by these authorities. In addition, the GFL has institutionalised and authorised the analysis, assessment and management of risks in the total food chain in the EU-25 since 2004. This institutional transnational legal framework contributes significantly to the predictability of the behaviour of all actors in the food system, particularly within the EU. Of course, this predictability of every actor's efforts is in itself a political catalyst for stabilising the balance of power. In fact, this systemic power mirrors the final step towards a hegemonial position of the balance of power in food politics.

The contemporary established multi-level governance food regime is a hybrid polity of rules as well as a collection of formal and informal institutions. 'Hybrid' because this polity is founded on the connectivity of public and private scientific networks and on a common discourse based on a belief system that maintains the objectivity and political neutrality of scientific knowledge. In practice, however, this polity seems to be mainly under the control of an administrative elite of technocratic governmental civil servants supported by (independent but business-oriented) committees of experts.

If a more radical political perspective aims to help open up the venues for more democratic food politics, a two-way approach ought to be recommended: more transparency concerning the intermediary role of science and a better institutional facility for proto-discourses of alternative or marginal food- and life-styles.

References

Arts, B. and Goverde, H., 2006. The Governance Capacity of (new) Policy Arrangements: A Reflexive Approach. In: Arts, B. and Leroy, P. (eds.). Institutional Dynamics in Environmental Governance. Springer, Dordrecht, the Netherlands, pp. 69-92.

Beck, U., 1982. Risk Society. Polity Press, Oxford, UK.

Blokland, H., 2001. De modernisering en haar gevolgen: Weber, Mannheim en Schumpeter. Boom Uitgevers, Amsterdam, the Netherlands.

Christiansen, T. and Larsson, T. (eds.), 2007. The Role of Committees in the Policy Process of the European Union. Legislation, Implementation and Deliberation. Edward Elgar, Cheltenham, UK.

Clegg, S.R., 1989. Frameworks of Power. Sage, London, UK.

Frouws, J., 1993. Mest en Macht. Een politiek-sociologische studie naar belangenbehartiging en beleidsvorming inzake de mestproblematiek in Nederland vanaf 1970. Landbouwuniversiteit Wageningen, Wageningen, the Netherlands.

Giddens, A., 1991. Modernity and Self-Identity. Polity Press, Oxford, UK.

Goverde, H., 2006. Mars and Venus in the Atlantic Community: Power Dynamics under Hegemony. In: Haugaard, M. and Lentner, H.H. (eds.). Hegemony and Power. Lexington Books, Rowman & Littlefield Publishers, Lanham, MD, USA, pp. 109-130.

Goverde, H. and Van Tatenhove, J., 2000. Power and Policy Networks. In: Goverde, H.J.M., Cerny, P.G., Haugaard, Mark and Lentner, H.H. (eds.). Power in Contemporary Politics. Theories, Practices, and Globalizations. Sage, London, UK, pp.96-111.

Gramsci, A., 1971. Selections of the Prison Notebooks. Lawrence and Wishart, London, UK.

Habermas, J., 1992. Faktizität und Geltung. Beiträge zur Diskurstheorie des Rechts und des demokratischen Rechtsstaats. Suhrkampf, Frankfurt, Germany.

Haugaard, M. and Lentner, H.H., (eds.), 2006. Hegemony and Power. Consensus and Coercion in Contemporary Power. Lexington Books, Rowman & Littlefield Publishers, Lanham, MD, USA.

Korthals, M., 2003. Voor het eten. Filosofie en ethiek van voeding. Boom Uitgevers, Amsterdam, the Netherlands.

Larsson, T., 2003. Precooking in the European Union. The World of Expert Groups. Ministry of Finance, Stockholm, Sweden.

Louwes, S.L., 1980. Het gouden tijdperk van het groene front; het landbouwbeleid in de na-oorlogse periode. In: Kooy, G.A., De Ru, J.H. and Scheffer, H.J., (eds.). Nederland na 1945. Beschouwingen over ontwikkeling en beleid. Van Loghum Slaterus, Deventer, the Netherlands, pp. 223-249.

Lupton, D., 1999, Risk. Routledge, London, UK.

McInerney, J., 1983. A Synoptic View of Policy-making for the Food Sector. In: Burns, J., McInerney, J. and Swinbank, A. (eds.). The Food Industry: Economies and Policies. Heineman, London, pp. 163-172.

Nestle, M., 2002. Food Politics. How the Food Industry influences Nutrition and Health. University of California Pres, Berkely, CA, USA.

Rawls, J., 1993. Political Liberalism. Columbia University Press, New York, NY, USA.

Steegmann, C., 2005. Voedselveiligheid en voedselkwaliteit. In: Silvis, H., Oskam, A. and Meester, G. (eds.). EU-beleid voor landbouw, voedsel en groen. Van politiek naar praktijk. Wageningen Academic Publishers, Wageningen, the Netherlands, pp. 215-233.

Taylor, P.J. and Flint, C., 2000. Political Geography. World-Economy, Nation-State & Locality. Prentice Hall, Harlow, UK.

Van der Meulen, B. and Van de Velde, M., 2004. Food Safety Law in the European Union. An introduction. Wageningen Academic Publishers, Wageningen, the Netherlands.

Van Otterloo, A.H., 2000. Voeding opnieuw ter discussie. In: Schot, J.W., Lintsen, H. and Rip, A. (eds.), Techniek in Nederland in de twintigste eeuw. Deel III Landbouw en Voeding. Walburg Pers, Zutphen, the Netherlands, pp. 369-374.

Van Otterloo, A.H., 2000. Nieuwe producten, schakels en regimes 1890-1920. In: Schot, J.W., Lintsen, H. and Rip, A. (eds.), Techniek in Nederland in de twintigste eeuw. Deel III Landbouw en Voeding. Walburg Pers, Zutphen, the Netherlands, pp. 249-262.

Van Otterloo, A.H., 2000. Voeding in verandering. In: Schot, J.W., Lintsen, H. and Rip, A. (eds.), Techniek in Nederland in de twintigste eeuw. Deel III Landbouw en Voeding. Walburg Pers, Zutphen, the Netherlands, pp. 237-248.

Van Otterloo, A.H., 2000. Prelude op de consumptiemaatschappij in voor- en tegenspoed. In: Schot, J.W., Lintsen, H. and Rip, A. (eds.), Techniek in Nederland in de twintigste eeuw. Deel III Landbouw en Voeding. Walburg Pers, Zutphen, the Netherlands, pp. 263-280.

Van Otterloo, A.H., 2000. Ingrediënten, toevoegingen en transformatie: heil en onheil. In: Schot, J.W., Lintsen, H. and Rip, A. (eds.), Techniek in Nederland in de twintigste eeuw. Deel III Landbouw en Voeding. Walburg Pers, Zutphen, the Netherlands, pp. 297-310.

Van der Velde, M. and Van der Meulen, B., 2005. Het Europees levensmiddelenrecht. In: Silvis, H., Oskam, A. and Meester, G. (eds.). EU-beleid voor landbouw, voedsel en groen. Van politiek naar praktijk. Wageningen Academic Publishers, Wageningen, the Netherlands, pp. 183-214.

Wildavsky, A., 2005. Speaking Truth to Power. The Art and Craft of Policy Analysis. Transaction Publishers, London, UK.

Chapter 8
From food security to food quality: spreading standards, eroding trust?

Gerard Breeman and Catrien Termeer

1. Introduction

After the Second World War most European agricultural policies were aimed at securing the national food supply. In that time, production chains were aimed at food security. This meant that bulk producers dominated the agricultural policy communities and the food production networks. Since the 1980's, however, the aims of public food policies and production chains have been shifting. Public policies are now aimed more at sustaining food quality than increasing food quantity and chain co-ordination is used as a tool to do that.

In this chapter we explore some potential and neglected negative consequences of the introduction of chain co-ordination, using theories from the policy- and political sciences. The literature covers the positive effects of chain co-ordination sufficiently but we believe it has overlooked the potential negative effects. Based on a theory driven analysis, we intend to give some suggestions to fill this gap in the literature.

We focus on the Netherlands, because it has organised one of the most effective bulk-producing agricultural sectors in Europe. In the first section we describe how Dutch agriculture came to embrace chain co-ordination. In the second section we briefly summarise the concept of policy community and introduce our research questions. The third section considers the potential effects of chain co-ordination on trust relations. The fourth section discusses the possible institutional lock-in effects of chain co-ordination, which could hamper the existent co-operative structures. And we finish in our fifth section with some conclusions and an agenda for further research.

1.1 From food security to food quality

The notion of co-ordinated chain co-ordination came into full swing when agricultural policies made a shift in focus from food security to food quality. This section tells how, in Europe and in particular in the Netherlands, this shift in policy focus came about.

The dominant agricultural policies in Europe were initially always aimed at securing the national food supply. The principal reason for this, so students of economics were told, was that agricultural economics is different from mainstream economics. Professors pointed out time and again that farmers deserve special treatment in public policies as well as in the study of the economics. Food should be considered as a strategic product and therefore governments should always protect their national food production and support farmers in bad times.

The need for this 'special treatment' of agriculture finds its origin in late 19th century Europe. At that time, cheap grain was coming in from America and Russia and distorted the European markets. Most European countries imposed various policy initiatives to stand up to this competition. Most of them imposed tariffs on agricultural products. The liberal governments of Britain, Denmark, Belgium, and the Netherlands, however, adhered firmly to the free trade system. From the beginning, these countries tried to improve the competitiveness of the primary sector, rather than to impose protective barriers (Tracy, 1989). At the same time the governments supported the building of self-help organisations, such as farmer's unions, co-operatives, and collectively financed knowledge organisations. The Dutch government tried, for instance, to improve agricultural production by supporting the sector with subsidised research, consultation, and education.[368] In 1935 it created a separate Ministry of Agriculture and Fisheries.

The focus on food security in agricultural policies grew even stronger after the Second World War. In the Netherlands, the war ended with a traumatic famine (1944) and in the rest of Europe with many undernourished people. All over Europe food security was at the top of the political agenda. The main goal of agricultural policies was to secure the production of sufficient food at affordable prices. The name of the Dutch ministry was changed to the Ministry of Agriculture, Fishery *and* Food Supply. Its policy was mainly oriented at ensuring a strong, viable, and competitive primary sector. Public services were set up to increase knowledge, encourage technological development, improve financial means, and promote land consolidation (Kooij *et al.*, 2000). The establishment of the Common Agricultural Policy at the EU level (CAP in 1958) further secured farmers' incomes. At that time, the central idea of most public policies was that food security could only be achieved and maintained with a strong primary sector.

Through these national and European policies, the Dutch achieved self-sufficiency in food around 1965, as well as stability in the agricultural markets, and a fair standard of living for farmers. The issue of food security disappeared from the political agenda and the name of the Ministry was changed back to Ministry of Agriculture and Fisheries. The many successful achievements of the agricultural policy were widely appreciated, but this did not last long.

---

[368] The so-called OVO-triangle (*onderzoek, voorlichting en onderwijs*).

General support from society started to diminish in the late 1970's when public raised concerns about issues of animal welfare, environment, European production surpluses, and the dumping of these surpluses in developing countries, causing the distortion of fragile local markets. Even more disturbing were the increasing reports about health and safety hazards. Especially in the 1990's when 'mad cow disease' (BSE: Bovine spongiform encephalopathy) shocked the public, consumers began to question the ability of the modern food system to produce safe food. This caused a major shift in policy focus (Grin *et al.*, 2004; Kalfagianni, 2006; Hendriks and Grin, 2007).

The government started to focus not only on the position of farmers, their incomes, and export positions but also on the sustainability of agricultural development, liberalisation of trade, and the improvement of food quality. In 2002, food quality had become politically important enough that the ministries of agriculture and health got into a fight about which department was going to get the inspection service for food quality. The Ministry of Agriculture won and in 2003 the name of the ministry changed to the Ministry of Agriculture, Nature, and *food quality*.

1.2 Chain co-ordination

Once the policy focus was put on food quality the entire policy community, consisting of the minister, civil servants, farmers, the farmer unions, and the processing and retail industry, aimed to improve and maintain food quality through so-called chain co-ordination. Two types of chain co-ordination can be distinguished: vertical and horizontal co-ordination (Kalfagianni, 2006; Lazzarini *et al.*, 2001).

A vertical chain is a set of relationships between the several stages in the food production process. This runs from primary production by farmers, to industrial processing, distribution, and the final consumption. A vertical chain is about all aspects from 'seed to cutlet' or from 'farm to table.' The EKO-certificate for organic food in the Netherlands is an example of such vertical chain co-ordination. A product receives the certificate only if all stages of the production process meet some predetermined criteria. Due to globalisation and liberalisation an exceedingly complex system of vertical chains has evolved. Production and distribution networks are spread across a multitude of geographical locations.

Horizontal chain co-ordination (sometimes referred to as 'netchains', Lazzarini *et al.*, 2001) is based on commonly agreed quality standards and covers a specific group of food producers on a specific level (Hofstede, 2003). It is a type of co-ordination that aims at agreements and co-operation between groups of actors of the same production stage. For instance, if a self-selected group of food-processing industries commits itself to a set of quality standards and signs some kind of agreement, then this is usually labelled as horizontal chain co-ordination. They usually engage in long-lasting partnerships. This type of chain co-ordination is geared to a select set of actors, which are considered to be excellent players in a specific area of the production stage. An example of horizontal chain co-ordination is the FAMI Quality System. This system was established by the feed additive industry and consists of a code of conduct covering feed additives and pre-mixtures. In theory, horizontal and vertical chain co-ordination can be distinguished; in practice, however, they are usually mixed, they overlap and sometimes challenge each other.

The objectives of chain co-ordination are transparency of trade relations, reduction of transaction costs, sustainability, stability, continuity, traceability of food products and the development of commonly agreed quality standards. The desire for transparency embodies the need to reduce the complexity of the chains. A high degree of traceability is thought to be important to ensure the accurate and rapid identification of products up and down the chain. Transparency, so it is believed, contributes to qualitatively better food products, because it becomes better to separate the good from the bad food and to pinpoint the weak links in the chains (Kalfagiani, 2006). Chain co-ordination provides consumers with information about the origins of food products, the product processing, and retail channels.

Chain co-ordination has become popular. The European Union (EU), for instance, has imposed concrete demands on processes of chain tracking and tracing and has introduced the HACCP system (Hazard Analysis of Critical Control Points). The purpose of this HACCP system is to establish a general control system and to regulate the different food production chains Europe-wide.

2. Policy communities

In order to discuss the potential negative effects of vertical and horizontal chain co-ordination, which we believe are neglected in the literature, we use the institutional actor approach from the political sciences. This approach (Scharpf, 1997) takes the individual positions (interests, stakes, orientations) and the interactions between the individual actors (their games, the rules) explicitly into consideration. It does so because,

> 'it is unlikely, if not impossible, that public policy of any significance could result from the choice of any single unified actor. Policy formation and policy implementation are inevitably the result of interactions among a plurality of separate public and private actors with separate interests, goals, and strategies' (Scharpf, 1978).

During these interactions actors negotiate about all kinds of different matters, ranging from norms, values, rules and resources, to relationships and formal structures. If these interactions and negotiations reach a certain degree of stability we speak of institutionalisation. This means that subsequent interactions tend to develop in the same direction as the preceding ones. Hence a fixed, institutionalised, pattern of interactions emerges that both restricts and enables action opportunities (Scharpf, 1997).

An institutionalised pattern of interactions between actors from a specific policy sector, politicians, and specialised policymakers can be coined as a policy community. A policy community is characterised by restrictive membership, frequent interaction between members on all matters related to the policy issues, resource-based exchange relationships, exclusions of some groups, shared values and consensus on preferred policy outcomes (Rhodes, 1997). Related concepts are advocacy coalitions (Sabatier and Jenkings-Smith, 1993), issue networks (Heclo, 1978) discourse coalitions (Hajer and Wagenaar, 2003), configurations (Termeer and Kessener, 2006), communities of meaning (Yanow, 2003) or epistemological communities (Fischer, 2003). These concepts all emphasise the shaping of communities by mutually reinforcing stable institutional and ideological structures. The policy community is a virtual 'space where relevant actors discuss policy issues and persuade and bargain in pursuit of their interests. During the course of their interaction with the other actors, they often give up or modify their objectives in return for concessions from others' (Howlett and Ramesh, 2003: 53). Policy communities may have different levels and vary in their degree of stability and integration (Kickert *et al.*, 1997; Scharpf, 1978). Rhodes distinguishes between several types of policy networks ranging along a continuum from tightly integrated policy communities to loosely integrated issue-networks (Rhodes, 1997).

Agricultural policy networks, in general, have served as notorious examples of tightly integrated policy communities (Montpetit, 2003). In the Netherlands, farmers and policy-makers used to share a common set of values, rules, and structures. For technical development the community had established a successful co-operation between research facilities, information services, and educational institutions. Finance, delivery, and product processing were organised largely along co-operative lines. Agricultural policy was made in a cosy iron triangle, consisting of the Ministry of Agriculture, the farmers' organisations, and the agricultural specialists from Parliament (De Vries, 1989; Van Dijk *et al.*, 1999). Self-regulation was the point of departure for the policy. All were proud of the increasing production volume and the improving living standards of Dutch farmers. The agricultural community was so focused on itself that it took a long time to notice the declining support from society for agriculture and to share with them the concerns about over-production, animal welfare and environmental problems. Strong pressure from outside was needed to open the policy community and remove the sector's blindness to other values, such as environment and animal welfare (Grin *et al.*, 2004).

Since the 1980's the agricultural policy community has been changing and new policy arrangements are being introduced, of which horizontal and vertical chain co-ordination are important ones. Within this context we consider the introduction of these new arrangements for horizontal and vertical chain co-ordination as interventions in existing policy communities. In the following two sections we analyze the possible negative effects of these new arrangements on the existing relationships and policy networks (Sabatier and Jenkins-Smith, 1993). We analyze how trust relations that are enacted in the policy communities may be damaged by the arrival of these new arrangements. We argue that new chain co-ordination arrangements could potentially result in less flexibility in dealing with consumer demands in the future. In our view, these potentially negative effects of chain co-ordination on relationships have thus far been neglected. Two questions will be central: (1) what are the potential effects of chain co-ordination on existing trust relations. We deal with this question in Section 3, and (2) what are the effects of the introduction of chain co-ordination on the existing co-operative structures (Section 4)?

3. Potential effects of chain co-ordination on trust relations

Supporters of chain co-ordination always point at the positive effects of chain co-ordination on trust relations. And indeed, the social capital literature teaches us that participation in policy communities could breed trust relations. Positive engagement in social networks, such as community welfare organisations, mutual aid societies, labour unions, political parties, and economic co-operatives, leads to trust (Fukuyama, 1995; Putnam, 1993, 2000, 2004). Once trust is established, it tends to increase through every interaction. It is a self-reinforcing mechanism, in which co-operation leads to social trust and social trust reinforces co-operation. In terms of Hirschman, trust is a moral resource 'whose supply may well increase rather than decrease through use' (Hirschman, 1984).

The history of the Dutch agricultural policy community illustrates clearly how this trust mechanism has worked for some decades. Especially in the period 1945-1975, most farmers, food processing industries, politicians, farmer representatives, and public officials established strong trust relations through co-operation. Their positive engagement in a variety of co-operative organisations constantly reinforced their trust (Breeman, 2006; Van der Ploeg, 1999; Van Dijk *et al.*, 1999). Hence, the co-operative structures gave farmers not only leeway to increase bulk production, as we mentioned in the last section, but also led to strong and viable trust relations.

We concur with the supporters of chain co-ordination that the introduction of chain co-ordination could lead to trust among consumers in food safety and among the actors that take part in the chain co-ordination. Obviously, this is a good thing. However, we argue below that chain co-ordination could also lead to the erosion of the initial enacted trust relations in the existing policy community and that, in due time, through chain co-ordination trust may even be lost altogether.

3.1 Losing enacted trust

Following Hirschman's statement that the supply of trust may well increase rather than decrease through use, we assume that as long as actors remain actively involved in a policy community interpersonal trust is ensured as well. Trust is enacted in running policies, co-operative structures, and various types of collaborative activities. However, the introduction of chain co-ordination can affect these co-operative structures, and the way people frame problems and solutions. For instance, one of the concepts of vertical chain co-ordination is, 'chain inversion', meaning a shift in focus away from the primary producers (as leading actors of the production chain) to those organisations that are closer to the consumers. This means that power relations are turned upside down and once influential actors could lose their key positions to other players. Moreover, some actors will be included and others excluded. Obviously, these changes affect the way policy communities are structured, how they operate, and how the self-reinforcing trust building mechanisms work. If the co-operative structures change, it will also change the way trust is fostered.

The exclusion of actors as well as the changing power relations may affect trust relations right from the start. Those actors who were previously involved may feel sidelined or even betrayed. Farmer spokespersons, who were, for instance, tightly incorporated in the iron policy triangle may become cynical and lose faith in the policy community altogether. Farmers, who were initially the primary focus of the food production chain, may lose their trust in the policy community. Exclusion could also lead to problems of trust among third parties who want to get involved later on. Closed networks imply that the involved actors will make no extra efforts to explore new relationships or to engage in trust building activities, thereby leaving third party actors frustrated outside the production chains. Hence, not only could chain co-ordination exclude creativity, it could also result in frustration and envy among the excluded actors. Destruction of trust through changing co-operative structures, such as the introduction of chain co-ordination or the exclusion of former partners, occurs in all kind of different policy domains. Consider, for instance, the case in which companies A and B have been working together successfully for years and now company A wants to obtain an ISO 22000 certification. Trust and co-operation could erode if company B will not (for instance, because of reasons of administrative efficiency) or cannot fulfil the requirements for the ISO 22000 standard. They might end their co-operation and lose some of their mutual trust.

3.2 Cannibalising trust

Another important possible consequence of the introduction of chain co-ordination is its potential cannibalising effect on trust relations. Cannibalising is a result of the formalisation of the relations between the included chain actors, and comes about only after a longer period of successful co-operation.

In his book *Trust and Rule* Tilly discusses the consequences when trust-networks become integrated in systems of rule (Tilly, 2005). He argues that through integration interpersonal trust relations will be replaced by some kind of governing system, in which legal and other policy instruments will be introduced to secure extant relationships. Tilly depicts (political) rulers as 'greedy predators' who want to incorporate as many social trust networks as possible in their systems of rule. The result of this integration is that it eats away the trust relations upon which the initial partners had established a (policy) community. In short, ruling takes the place of trust. This cannibalising trust mechanism could also be observed in the 1960's and 1970's when the government increasingly concerned itself with welfare organisations. Before then, most hospital and orphanages were for instance part of charity communities and usually financed by religious institutions. However, when the government started financing these institutions, they introduced a whole new type of governing system with detailed regulations and monitor systems, depleting the trust relations.

We posit that chain co-ordination will, in due time, show the same kind of effects. Initially, the relations between the included partners are based on mutual trust. This is time and again emphasised by supporters of chain co-ordination. However, the rules and regulations are building up from different sides. Inside the production chains, participants use various private legal instruments to secure trade relations and ascertain fair competition. They make, for instance, different agreements about product quality (labelling) and processing procedures. Outside the chains, public agencies produce public laws to safeguard food safety. An example in this case is the EU system of E-numbers through which the use of food supplements is regulated. The EU has also established an all-encompassing tracking and tracing system that should contribute to food safety. This hazard analysis and critical control points (HACCP) system is introduced to ensure all appropriate hygiene requirements during the food, production, processing, and distribution.

These systems of rule could lead to the erosion of interpersonal trust in the policy community. We postulate that regulations make trust relations redundant, because they will no longer be used, and, following Hirschman (1984) if trust is no longer enacted, the sub-system will be depleted from interpersonal trust all together.

4. Potential effects of chain co-ordination on co-operative structures

In addition to the potential negative effect on trust relations, chain co-ordination could also have negative effects on the co-operative structures of the existing policy community. The basic problem here is that chain co-ordination can result in the selective activation of only a part of the existing policy community. Co-ordination is 'extremely expensive in terms of its demands upon the capacity for information processing and conflict resolution within the system', and therefore selective activation of actors within the broader policy community is essential to run policy processes smoothly (Scharpf, 1978: 364). Crucial then is the correct identification of essential participants, as well as the lack of opposition from those who possess the required resources to frustrate co-ordination (Scharpf, 1978). In other words, one should select and mobilise the correct actors and de-activate the less important actors.

The selection of actors, through vertical and horizontal chain co-ordination may indeed result in stable policy networks, and in smooth-running policy processes, but in due time, as we will argue in this section, they may also lead to major disadvantages, such as the exclusion of creative and innovative actors, organisational blindness, and the suppression of variety.

4.1 Selection and exclusion

Selective activation of some actors implies the exclusion of others. This also results in the selection of certain values and norms and the dismissal of others. Only the cognitive frames of those actors that are 'activated' become part of the chain's institutions; other deviant norms and values can be ignored. Consider questions like, when does food qualify as good? Food that is considered good in the European Union may not be considered as good in the United States. Bio-friendly farmers will detest industrial eggs, while the industrial farmer will praise the quality of his eggs. In a broader context, environmentally damaging production techniques may qualify food as bad. Banana production, for example, that leads to the destruction of the rainforest will disqualify the food, although the quality of the bananas may be sufficient. In the same way, techniques, which damage the public health or the welfare of animals, may also disqualify the products (e.g. *foie gras*). Hence, to distinguish good food from bad depends on which values of which actors are taken into consideration in the agreements and protocols of the horizontal or vertical chains. Hence, chain co-ordination establishes certain cognitive frames and excludes other frames (Termeer and Kessener, 2007).

Moreover, the selection of leading policy goals and frames could have enormous consequences on the roles of certain key actors as well as on the power relations within the existing policy communities. As said before, chain inversion may turn power relations upside down. Chain inversion is assumed to improve market co-ordination, but it also makes the food industry and retail organisations more powerful, at least from the farmers' point of view. Obviously, this has raised interesting questions about the changing relations between farmers and the retail industry and about the changing cognitive frames in the chains.

The exclusion of actors may result in the barring of alternative policy goals. This is one of the central dilemmas of chain co-ordination: actor selection is necessary to efficiently establish certain primary aims, but it blocks other, equally desirable policy goals. If, chain co-ordination is aimed at increasing food quality rather than food security, the consequence may very well be that actors who were essential to establishing food security will be excluded from the policy community. Co-operation in the agricultural production networks in the Netherlands was initially aimed at securing national food supply. This meant that predominately actors who were able to produce bulk food were included in the policy community. These actors may now be excluded from modern production chains and one can wonder whether or not the shift from food security to food quality could jeopardise food security altogether. Do we lose actors that are strong players in efficient bulk production?

We believe, however, that the exclusion of actors is not the only and not even the most serious consequence of selective activation. Even more important is the sidelining and erosion of the former vital co-operative structures. Chain co-ordination may result in the erosion of these institutionalised co-operative relationships, which go back several decades. Through these co-operative relations, farmers were able to improve their production techniques and produce food quickly. Hence, farmer co-operatives assured food security and they could be counted as early examples of chain co-ordination. If now, in a worst case scenario, we encounter new problems with food security and the need for bulk food re-emerges, would we still be able to activate these production networks anew and organise bulk production?

4.2 Organisational blindness

Another possible side effect of chain co-ordination is organisational blindness. The policy community theory suggests that in the course of time, the interactions of actors within a policy community tend to become denser and tighter. This could lead to group thinking, positive feedback mechanisms, and organisational self referentiality, which means that an organisation starts focusing on itself only. These three effects refer to a similar kind of process, and all three lead to organisational blindness. In due time, the actors in a policy community start using the same cognitive frameworks, listen only to each other, and do not communicate with outsiders. Community members engage in the same kind of acts and use similar language. Various group processes reinforce these practices, often promoting internal cohesion as an identity marker with respect to other communities (Yanow, 2003: 237). These practices could not only generate some advantages but could also cause organisational blindness, making reform and innovation difficult.

As described above, the former Dutch agricultural policy community had been rather blind to issues such as environment and animal welfare. Since the 1980's it has been evolving into a more open community with less blindness. The introduction of vertical and horizontal chain co-ordination is an important aspect in this change process, but we believe that new forms of organisational blindness are bound to occur. If indeed production chains become denser and develop strong internal cohesion, we consider it likely that new forms of organisational blindness will arise. For all, chain co-ordination has the characteristics to develop into tightly coupled communities with strong internal cohesion because it tends to be selective to its members, build strong organisational boundaries, and suppress variety (Yanow, 2003; Weick and Sutcliff, 2001). Marsh and Smith (2000) convincingly show that policy communities and networks, such as the chain co-ordination arrangements are very good at resisting external pressure (see also Montpetit, 2003).

4.3 Blocking innovation

We finally argue that vertical and horizontal chain co-ordination may constrain innovation. This is worrying because fast technological developments, increasing global competition and stringent consumer demands severely pressurise companies to innovate. Innovation literature claims that the innovation process should be considered as the outcome of interaction between a variety of actors within firms, between firms, and between firms and other societal organisations such as universities, educational facilities, financing organisations, and government agencies (Nooteboom and Stam, 2008). The creation of inter-organisational network relations is considered as a key factor to sustained competitive advantages (Omta and van Rossum, 1999).

With the establishment of chain co-ordination, we expect tensions to arise between concepts like open innovation and shifting strategic alliances on the one hand and the more stabilised and closed relationships within production chains on the other. The consequences, however, are ambiguous; tensions can challenge and energise but they may also block or damp down inter-firm innovations and run the risk of becoming locked into insufficient variety (Nooteboom and Stam, 2008).

Furthermore, we also expect that chain co-ordination may hinder innovation if we consider innovation from a more abstract point of view. Van Gunsteren conceptualised innovation as a continuous learning process of generating variety, selection, generating new variety, and so on (Van Gunsteren, 2006). The idea here is that outsiders could be more creative than the included parties and thus encourage innovation. We hypothesise that chain co-ordination can delay or even block this innovation process in a twofold way. First, selective activation and the exclusion of outsiders through closed chain co-ordination limit the variety in a policy community directly. And second, since chain co-ordination is based on pre-selected criteria for judging and evaluating new developments, all innovations will be judged accordingly and thus disregard new ones. To put it simply, in the course of time, chain co-ordination may block out creative persons and blind chain participants to innovative ideas.

Rinnooy Kan, chairman of the Social and Economic Council, which is one of the most important advisory councils to the Dutch government, explained the problem of suppressing variety in a speech to the Dutch National Farmers' Association (29-02-2008). He said it was regrettable that the products of innovative farmers, who invest in animal welfare, environmental techniques, and the improvement of animal food, are still being processed as if they were bulk products. Hence, variety is blocked by the old institutional co-ordination networks of the processing industry and does not allow for diversification and, so he argued, is frustrating innovation (Rinnooy Kan, 2008). Contrasting with this image has been the Koppert-Cress company, who has been successful in bringing new products on to the market (micro-vegetables or 'cress') by aiming its marketing strategy entirely at chefs and restaurant owners. Koppert-Cress is an outsider, which bypassed the processing industries and retail sector in order to introduce its innovative products, and to force the retail sector only later to sell his products. However, this company has an extremely eccentric innovative product, which is not accessible for the general farmer, hence it is an exception to the rule.

5. Conclusions

Based on a theory driven analysis we highlighted in this chapter some of the possible negative effects of chain co-ordination. We believe that the positive effects have been amply discussed in the literature and that the negative effects of chain co-ordination have been neglected. However, we also believe that it is difficult to predict whether or not these negative effects will occur throughout the course of events and whether they will balance out the more positive consequences of chain co-ordination. In every progression of chain co-ordination we expect contradicting trends, which lead to ambiguous outcomes. Therefore, in this last section we provide a more balanced conclusion.

5.1 Ambiguous outcomes

In the early stages of the introduction of any chain co-ordination, both vertical and horizontal, we said that chain co-ordination may constrain innovation, exclude former partners, and turn power relations upside down, which could lead to distrust, cynicism, and frustration. Closed communities could however also result in strong collective intentions, shared values and common goals. The literature about social capital teaches us even that these kinds of collective intentions strengthen collaboration and intensify collective trust (Putnam, 1993, 2004). We expect that both effects are not mutually exclusive, but may occur at the same time and could contradict each other, resulting in ambiguous results.

We also foresee ambiguous effects of chain co-ordination in later stages. Social capital literature says that co-operation will increase the levels of trust. As said before, trust stimulates co-operation, whereas, in turn, successful collaboration strengthens trust. Co-operation in production chains may indeed result in strong cohesive resilient trust relations, in which people share common values and goals. However, at the same time we also envisage new problems in the later stages, because of the exclusion of third parties. While the included actors share strong resilient relations, creative outsiders may become frustrated at being left out. Once the contracts are signed, and the quality control systems are running, changes are difficult. The closed networks do not let newcomers in easily. Here too, we expect both effects to come about simultaneously.

Finally, we also see ambiguous effects occurring over time and production chains become denser and grow into strong institutions. This means that behaviour will be formalised and fixed in rules and regulations. From an organisational perspective, institutionalisation contributes to desirable effects, such as organisational stability, security, and continuity. It may make behaviour transparent and controllable. However, the formalisation of relationships may at the same time result in an erosion of trust relations. The regulatory frameworks could make trust relations redundant. With trust relations no longer in place, people will rely more and more on the rules, which indeed will make the creation of interpersonal trust even more difficult.

5.2 Lock-in effects

Additionally, when chain co-ordination grows into strong institutions, participants could be trapped inside. People's private ideas and values can change over time, but institutions are not easily transformed. Institutions consist of fixed social paradigms, values, goals, rules of conduct, and cognitive frames (March and Olsen, 1989). Once these are set, they provide stability and continuity, but they also constrain change. This goes both for institutions that focus on food security and those that focus on food quality.

Securing a nation's food production seems to be a different issue than establishing a food industry that produces a wide variety of high quality food. First, it refers to different values; food security is one of our primary necessities of life, whereas food quality pertains to values that are typical for a modern welfare society. It also needs different policies; food security is best established through a focus on efficient (bulk) production, whereas food quality needs product diversification, hi-tech production techniques, and quality control systems. The goals, values, policies and production techniques may be different, but in both cases networks of actors and institutions are required to make things happen.

In the past, the Dutch agricultural system was aimed at the high-speed production of bulk food. It has taken years to change the mind-set of the actors in the policy community to focus on food quality and diversification. If, however, through institutionalisation of the chain co-ordination the orientation becomes fixed on food quality, the Netherlands might lose its focus on food security, or even lose the capacity for bulk production altogether (see also Chapter 7 by Goverde, in this volume). Food security has become a non-issue in the Netherlands and even in Europe, but not so in the United States. The US regard agriculture still as a strategic economic sector that needs special attention. In the contribution of Grossman to this book (Chapter 10), one can read that food security is still an important issue; it pertains both to domestic security and to international food aid. To conclude, our observations in this chapter, lead us to plead for a thorough re-examination of chain co-ordination. Although chain co-ordination can have beneficial effects it can also have some negative ones and can cause important lock-in effects that determine the aims of food production for a long time.

5.3 Further research

The propositions we brought up in this chapter are theory driven and require empirical analyses. Since chain co-ordination is being introduced, we could already start analysing whether or not actors feel excluded and if the change in the power relations has resulted in frustrations and cynicism. Another question that remains to be answered is whether vertical or horizontal co-ordination is more harmful to trust relations and co-operative structures. More interesting may be an analysis of how chain co-ordination arrangements can be developed in such a way that its positive and negative effects on trust relations are balanced and viable and resilient trust relations are cultivated. Cultivating trust between the key actors in the food policy community is essential for achieving the aim of food quality *and* food security. Preventing lock-in situations is important for remaining resilient to unexpected developments in the markets, and for adapting to changing production and consumer demands.

References

Breeman, G., 2006. Cultivating Trust: How do Public Policies become Trusted. PhD thesis, OGC, Leiden, the Netherlands.

De Vries, J., 1989. Grondpolitiek en Kabinetscrises. Vuga, Den Haag, the Netherlands.

Fukuyama, F., 1995. Trust: The Political Virtues and the Creation of Prosperity. Penguin Books, London, UK.

Fischer, F., 2003. Reframing Public Policy: Discursive Politics and Deliberative Practices. Oxford University Press, Oxford, UK.

Grin, J., Felix, F. and Bos, B., 2004. Practices for Reflexive Design: Lessons from a Dutch Programme on Sustainable agriculture. International Journal of Foresight and Innovation 1/2: 126-148.

Hajer, M.A. and Wagenaar, H. (eds.), 2003. Deliberative Policy Analysis: Understanding Governance in the Network Society. Cambridge University Press, Cambridge, UK.

Hirschman, A.O., 1984. Against Parsimony: Three Easy Ways of Complicating Some Categories of Economic Discourse. The American Economic Review May 74: 89-96.

Heclo, H., 1978. Issue Networks and the Executive Establishment. In: Beer, S.H. and King, A. (eds.). The New American Political System. American Enterprise Institute for Public Policy Research, Washington, DC, USA.

Hendriks, C.M. and Grin, J., 2007. Contextualizing reflexive governance: the politics of Dutch transitions to sustainability. Journal of Environmental Policy and Planning 9: 333-350.

Hofstede, G.J., 2003. Transparency in Netchains. In: Harnos, S., Herdon, M. and Wiwczaroski, T.B. (eds.). Information Technology for a Better Agri-food Sector, Environment and Rural Living. Proceedings EFITA2003 Conference. Debrecen University, Debrecen, Hungary, pp. 17-29.

Howlett, M. and Ramesh, M., 2003. Studying Public Policy: Policy Cycles and Policy Subsystems. Oxford University Press, Oxford, UK.

Kalfagianni, A., 2006. Transparancy in the Food Chain (PhD-Thesis). University of Twente, Enschede, the Netherlands.

Kickert, W.J.M., Klijn, E.H. and Koppenjan, J.F.M. (eds.), 1997. Managing Complex Networks Strategies for the Public Sector. Sage Publishing, London, UK.

Kooij, P., Bieleman, J., Van der Woude, R., Karel, E. and Gerding, M., 2000. De Actualiteit van de Agrarische Geschiedenis. Nederlands Agronomisch Historisch Instituut, Groningen/Wageningen, the Netherlands.

Lazzarini, S.G., Chaddadm, F.R. and Cook, L.M., 2001. Integrating supply chain and network analyses: The study of netchains. Journal on chain and network science 1: 1.

March, J.G and Olsen, J.P., 1989. Rediscovering Institutions: the Organisational Basis of Politics. The Free Press, New York, NY, USA.

Marsh, D. and Smith, M., 2000. Understanding Policy Networks: Towards a Dialectical Approach. Political Studies 48: 4-21.

Montpetit, E., 2003. Life Sciences and Policy Networks in the European Union. Working paper no. 03/03, Université de Montréal, Montréal, Canada.

Nooteboom B. and Stam, E. (eds.), 2008. Micro-foundations for Innovation Policy 2008. Amsterdam University Press, Amsterdam, the Netherlands.

Omta, S.W.F. and Van Rossum, W., 1999. The Management of Social Capital in R&D Collaborations. In: Leenders, R.Th.A.J. and Gabbay, S.M. (eds.). Corporate Social Capital and Liability. Kluwer Academic Publishers, Boston, MA, USA, pp. 356-376.

Putnam, R.D., 1993. Making Democracy Work: Civic Traditions in Modern Italy. Princeton University Press, Princeton, NJ, USA.

Putman, R.D., 2000. Bowling Alone: The Collapse and Revival of American Community. Simon and Schuster, New York, NY, USA.

Putman, R.D., 2004. Better Together: Restoring the American Community. Simon and Schuster, New York, NY, USA.

Rhodes, R.A.W., 1997. Understanding Governance: Policy Networks, Governance, Reflexivity, and Accountability. Open University Press, Milton Keynes, UK.

Rinnooy Kan, A.H.G., 2008. Innovatie in de Agrarische Sector: Speech by Rinnooy Kan, chair of the SER, 29 februari 2008. SER, Rosmalen, the Netherlands.

Sabatier, P. and Jenkins-Smith, H., 1993. Policy Change and Learning: An Advocacy Coalition. Westview Press, Boulder, CO, USA.

Scharpf, F.W., 1978. Interorganizational Policy Studies: Issues, Concepts and Perspectives. In: Hanf, K. and Scharpf, F.W. (eds.). Interorganizational policymaking: Limits to coordination and central control. Sage Publishing, London, UK.

Scharpf, F.W., 1997. Games real actors play. Westview Press, Boulder, CO, USA.

Tilly, C., 2005. Trust and Rule. Cambridge University Press, Cambridge, UK.

Termeer, C.J.A.M. and Kessener, B., 2006. Revitalizing stagnated policy Processes: Using the configuration approach for research and intervention. Journal of Applied Behavioural Science 43: 256-272.

Tracy, M., 1989. Governement and Agriculture in Western Europe, 1880-1988. Harvester Wheatsheaf, Hemel Hempstead, UK.

Van der Ploeg, J.D., 1999. De virtuele boer. Van Gorcum, Assen, the Netherlands.

Van Dijk, G., Klep, L.F.M. and Merkx, A.J., 1999. De Corrosie van een IJzeren Driehoek: Over de Omslag rond de Landbouw. Van Gorcum, Assen, the Netherlands.

Van Gunsteren, H., 2006. Vertrouwen in Democratie. Van Gennep, Amsterdam, the Netherlands.

Weick, K.E. and K.M. Sutcliffe, 2001. Managing the Unexpected - Assuring High Performance in an Age of Complexity. Jossey-Bass, San Francisco, CA, USA.

Yanow, D., 2003. Accessing Local Knowledge. In: Hajer, M. and Wagenaar, H. (eds.). Deliberative Policy Analysis: Understanding Governance in the Network Society. Cambridge University Press, Cambridge, UK.

Chapter 9

Food safety governance from a European perspective: risk assessment and non-scientific factors in EU multi-level regulation[369]

Anna Szajkowska

1. Introduction

Food security encompasses several dimensions: the availability of food, the safety of food and the cultural acceptability of food.[370] Upon its initiation at the height of the cold war, the European Economic Community (now the European Union: EU) made it its first priority to attain self-sufficiency regarding food supply. At the macro level, this has been achieved magnificently. The EU no longer depends on imports of food, but has become one of the major net exporters in the world.

This does not mean, however, that food security is not an issue in the EU. Authorities both in the EU and in the Member States see it as their responsibility to ensure the safety of food and to protect human health. Regulation interfering with market mechanisms to protect interests such as health and the environment is defined as risk regulation (Hood *et al.*, 2001: 4).[371] Scientific and industrial developments result in the emergence of a 'risk society', where risks are no longer limited to certain localities or groups and increasingly escape the laypersons' perception (Beck, 1992). Confronted with pervasive risks, like toxins in foods, radiation or genetic engineering in everyday life, modern society depends on external knowledge to manage them both at institutional and personal levels.

Regulation in the field of food safety relies heavily on science. Risk analysis is a methodology incorporating technical expertise into the decision-making process, where risk managers take regulatory decisions based on the scientific risk assessment provided by experts. Theoretically, scientists refrain from any value judgements in this process. They deliver 'neutral' information, which serves as a basis for risk managers to decide on the acceptable level of risk imposed on society as a whole.

---

[369] A modified version of this chapter has been accepted for publication in the European Journal of Risk Regulation. It is published here by kind permission of the editors and publisher of the journal.

[370] On the concept of adequacy encompassing food in a quantity and quality sufficient to satisfy the dietary needs of individuals, free from adverse substances, and acceptable within a given culture see: General Comment 12 on the right to adequate food, available at: http://www2.ohchr.org/english/bodies/cescr/comments.htm.

[371] Some authors refer to 'social regulation' to describe this process; see Joerges (1997).

Excessive reliance on scientific opinions in risk governance may lead to a paradoxical situation where people are becoming more and more conscious of risks and, at the same time, decisions concerning these risks are taken further and further away from the political participation of citizens and their control (Lidskog, 2008). Clearly, science is not able to take responsibility for all decisions for society. Apart from scientific risk assessment, decision makers must take into account different interests and concerns of a social, economic or ethical character. Previous chapters address the duties of governments. But who exactly are the duty-bearers in the multi-level regulatory environment of the EU? Are policy-making processes able to reconcile both the scientific and democratic legitimacy of risk governance in the field of food safety?

In the risk analysis methodology established by Regulation 178/2002 laying down the general principles and requirements of food law, establishing the European Food Safety Authority and laying down procedures in matters of food safety (the so-called General Food Law, hereinafter also referred to as the GFL),[372] non-scientific concerns are recognised as 'other legitimate factors', considered at the risk management stage. Despite the growing controversy about whether and to what extent a legislator may decide against the findings of risk assessment by taking into account non-scientific factors, such as consumer risk perception, animal welfare, traditional methods of production etc., the role of these factors in EU food safety governance is not clearly formulated.

The extent to which risk managers can deviate from science in highly technical areas of risk regulation is in fact the question of how much discretion is conferred on the public authorities. The risk analysis methodology set out in the General Food Law does not operate in a regulatory vacuum. EU and national food safety regulation must be interpreted in light of the Treaty rules concerning the free movement of goods.[373] Hence, ultimately, the manner in which such discretion is exercised comports with the Treaty framework relating to the functioning of the internal market and the standard review applied by the European judiciary.

This chapter examines how the internal market mechanism and Treaty provisions relating to the free movement of goods determine the way 'other legitimate factors' are incorporated into the decision-making process in EU multi-level food safety regulation. It starts by describing shortly the division of competences between the EU and the Member States in this area and standard-setting mechanisms in the context of the internal market. Then, an overview of the main elements of the food safety scientific governance established by the General Food Law will be given. Next, we discuss the principle of risk analysis in light of the standard of judicial review applied to EU measures aimed at the protection of human health

[372] O.J. 2002, L 31/1.

[373] See Case C-47/90, *Delhaize*, [1992] ECR I-3669, at 26; Case C-315/92, *Clinique*, [1994] ECR-317, at 12.

and, subsequently, we analyse the discretion applied to national measures, with a special focus on the relationship between EU and national scientific opinions. Lastly, we look at the role of 'other legitimate factors' referring to two possibilities resulting from their inclusion in the decision-making process: standards ensuring a level of protection that is lower than that recommended by risk assessors, and a level of protection that is higher than that in scientific opinions. Although the focus of this chapter is the risk analysis model established in the General Food Law, we refer to examples from different areas of risk regulation, including legislation for hazardous substances and environmental standards. The chapter shows that, although the risk analysis methodology established by the General Food Law applies to all food safety legislation, the role of science and other legitimate factors determined by the context of the internal market and the scope of judicial review is different for EU and national measures.

2. Division of powers between the EU and the Member States

2.1 Treaty framework for EU food safety regulation

In EU multi-level food safety regulation, the division of competences between the Member States and the Union is shaped mainly by the Treaty provisions relating to the functioning of the internal market. In principle, the free movement of goods is ensured through the principle of mutual recognition, developed in the *Cassis de Dijon* judgement.[374] This principle breaks up national standards, precluding a Member State from prohibiting the sale of a product which does not conform to domestic regulation, but has been lawfully marketed in another Member State. As a consequence, it is no longer necessary to harmonise all food standards. From the adoption of the 1985 New Approach to Harmonisation further detailed regulation of compositional rules was abandoned (EC, 1985).

However, the Treaty on the Functioning of the European Union (TFEU) in Article 36 (ex Art. 30 EC) provides exceptions to the free movement of goods, *inter alia*, on grounds of the protection of health and life of humans, animal or plants.[375] Thus, in the area of food safety, where the main purpose of regulation is the protection of human life and health, Member States can invoke Article 36 TFEU as a basis for their legislation and be exempt from mutual recognition. As a consequence, harmonisation of national food safety laws remains the only way to assure the free movement of foodstuffs and national competences are pre-empted by Article 114 TFEU (ex Art. 95 EC). Nearly all legislation relating to food safety is now

[374] Case 120/78, *Rewe-Zentral AG* v. *Bundesmonopolverwaltung für Branntwein*, [1979] ECR 649.
[375] The other exceptions are: public morality, public policy or public security; the protection of national treasures possessing artistic, historic or archaeological value; and the protection of industrial and commercial property.

harmonised at the EU level, whereas technical food standards remain mostly within the competence of the Member States (Szajkowska, 2009).

2.2 Multi-level regulation through Article 114 TFEU

Because food safety regulation is to a large extent harmonised at the EU level, Article 114 TFEU in fact delimits competences between the EU and Member States in this field. The Article envisages three possibilities for national measures to derogate from harmonised European regulation. These possibilities are set out in paragraphs 4, 5 and 10.

Paragraph 4 allows Member States to *maintain* more stringent national provisions existing prior to harmonisation of laws at the EU level on grounds of major needs listed in Article 36 TFEU, or referring to the environment or the working environment.[376] This opt-out clause is similar to the pattern of minimum harmonisation, setting a threshold that standards have to meet, but giving Member States the possibility to retain or introduce more stringent measures.

Paragraph 5 provides the possibility to introduce *new* national measures after harmonisation.[377] This possibility, however, is limited to the environment or the working environment: paragraph 5 does not refer to Article 36 TFEU, and therefore the introduction of new national measures derogating from harmonised EU legislation in the area of food safety is not possible. The new national measures must be based on new scientific evidence relating to the protection of the environment or working environment on grounds of a problem specific to that Member State and arising after the adoption of the harmonisation measure.

National provisions derogating from EU harmonisation measures are subject to the Commission's approval (Art. 114(6) TFEU). They must be necessary and proportionate to the objective pursued and cannot constitute a means of arbitrary discrimination, a disguised restriction on trade or an obstacle to the functioning of the internal market.[378]

---

[376] The paragraph runs as follows: '4. If, after the adoption of a harmonisation measure by the European Parliament and the Council, by the Council or by the Commission, a Member State deems it necessary to maintain national provisions on grounds of major needs referred to in Article 36, or relating to the protection of the environment or the working environment, it shall notify the Commission of these provisions as well as the grounds for maintaining them.'

[377] The paragraph reads: '5. Moreover, without prejudice to paragraph 4, if, after the adoption of a harmonisation measure by the European Parliament and the Council, by the Council or by the Commission, a Member State deems it necessary to introduce national provisions based on new scientific evidence relating to the protection of the environment or the working environment on grounds of a problem specific to that Member State arising after the adoption of the harmonisation measure, it shall notify the Commission of the envisaged provisions as well as the grounds for introducing them.'

[378] See, e.g. Dec. 1999/5/EC, O.J. 1999, L 3/13, at 15; Dec. 1999/830/EC, O.J. 1999, L 329/1, at 18.

Paragraph 10 introduces the possibility to include safeguard clauses in harmonisation measures. The safeguard clauses permit Member States to restrict or suspend trade in foodstuffs for one or more non-economic reasons referred to in Article 36 TFEU. The national safeguard measure must be temporary and subject to a Union control procedure. However – unlike the other two possibilities discussed above – this provision is addressed to the EU legislature and does not confer on the Member States any powers *per se*. In addition, apart from safeguard clauses, which are actually contained in almost all food safety legislation, harmonisation may allow other derogations or leave some aspects of the subject matter to the Member States. This possibility will be discussed in Section 6.3.

By introducing the opt-out clauses the Single European Act took a more flexible approach to the creation of the internal market. In the beginning, the new harmonisation provisions, especially paragraph 4, raised fears of a threat to the free movement of goods (Pescatore, 1987). Consider the conditions set out in paragraphs 4 and 5. The requirement that national measures be based on new scientific evidence on grounds relating to a problem specific to that Member State does not exist in paragraph 4. In his opinion in the case *Denmark* v. *Commission*, Advocate General Tizzano stressed that paragraph 4 needed to be interpreted strictly, requesting justifications consisting in a situation specific to the Member State concerned, otherwise it would open the possibility to maintain or introduce a derogation from the principle of unity of the market, on the basis of a 'unilateral assessment of the need for raising the standard'.[379] He went on to conclude that:

> 'to allow such a claim would mean transforming Article 95(4) EC into a veritable permanent opt-out clause from any harmonisation directive, in stark contrast with the principles and purposes of the system and with the logic which, in the protection of the general interest, inspires the division of powers between the Community and the Member States.'[380]

The Court did not confirm these arguments, ruling that 'neither the wording of Article 95(4) EC nor the broad logic of that article as a whole entails a requirement that the applicant Member State prove that maintaining the national provisions which it notifies to the Commission is justified by a problem specific to that Member State'.[381]

In practice, however, very few Member States have used Article 114(4) TFEU to maintain their national provisions.[382] The limited number of notifications may be explained by the fact that the very threat of invoking the opt-out clause compels

[379] Opinion of A.G. Tizzano of 30 May 2002 in Case 3/00.
[380] Opinion of A.G. Tizzano of 30 May 2002 in Case 3/00, at 76.
[381] Case 3/00, *Denmark* v. *Commission*, [2003] ECR I-2643, at 59.
[382] For an overview of national derogations see Onida (2006).

the Member States to seek a harmonisation measure that ensures a level of health protection satisfactory for all States (Hanf, 2001: 12). In fact, Article 114(3) TFEU explicitly stipulates that in the field of health, safety and consumer protection the Union institutions are obliged to take as a basis a high level of protection, taking into account any new development based on scientific facts.

2.3 Race to the top[383]

The mechanism applied to EU standards follows the legal approach defined by Article 114 TFEU. In Grey's words, in the EU harmonisation system, '[e]ach national act calls for a Community act to resolve the situation' (Grey, 1990: 111). In many cases, the initially differentiated standards lead to a uniform regulation: the internal market mechanism eventually brings the Union legislation in line with the national standards. As a consequence, national derogations help develop stricter Union environmental and food safety standards and lead to increased involvement in the field of the environment, food safety, and consumer protection. The European Economic and Social Committee in its opinion on the relation between the articles referring to the internal market and environmental protection stated:

> 'The practical implementation of these articles involves both approaches at the same time, i.e. harmonization and 'non-harmonization'; by not harmonizing and opting for a higher level of protection, Member States force a constant upwards adjustment of environmental standards. This leads to a halfway house situation where there are technical barriers to trade, but these barriers may help to reach the ultimate aim - environmental protection' (1998, 1.4).

This phenomenon was described by David Vogel in 1995 in relation to the history of United States automobile emission standards. In 1970, the Clean Act Amendments expressly allowed California to introduce stricter standards than those required for the rest of the US. In 1990, the US emission standards became stricter following California's standards. Congress once again permitted California to set stricter standards, and gave the other states the possibility to chose either national or California standards. A few years later 12 states asked the federal government permission to introduce California's standards. The new standards became the basis for setting new minimum federal requirements. Thus, instead of a race to the bottom to enhance competitiveness and attract investment, standards are eventually becoming a uniform regime in line with the strictest requirement (Vogel, 1995: 259).

The 'California effect' can be observed in the EU regulation on creosote, a wood preserving agent. Creosote is principally used for railway sleepers, poles for electricity transport, hydraulic constructions or fences. It is highly toxic and its

[383] The title refers to Otsuki *et al.* (2001).

release to the environment can cause harm to animals and wildlife. Exposure to creosote is also dangerous to human health and can be carcinogenic. Directive 94/60/EC amending Directive 76/769/EEC on the marketing and use of certain dangerous substances harmonised the use of creosote and laid down restrictions on its marketing.[384] Creosote was not permitted for wood treatment if it contained benzo[a]pyrene above a concentration of 0.005% by mass and water extractable phenols at a concentration greater than 3% by mass. By way of derogation, in industrial installations creosote containing benzo[a]pyrene of a concentration of less than 0.05% was allowed.

Despite the EU restrictions, four Member States – the Netherlands, Germany, Sweden, and Denmark – decided to maintain their national provisions (on the basis of Article 114(4) TFEU). In this case, following the national requests for derogations, the Commission ordered additional studies and asked the Scientific Committee on Toxicity, Ecotoxicity, and the Environment (SCTEE) to evaluate the findings. The Committee concluded that there was a cancer risk to humans from creosote containing benzo[a]pyrene at a concentration of less than 0.005% by mass and wood treated with such creosote, although the magnitude of the risk could not be estimated with certainty. In consequence, the Commission approved all four national derogations in light of the precautionary principle.[385]

The review of the studies on health effects of creosote and approval of the national measures resulted in a re-consideration of the EU harmonisation measure in line with the national derogations. In 2001, the Commission amended Directive 76/769/EEC.[386] The marketing and use of creosote was prohibited, except for industrial applications, where the levels previously applied to all wood treatment (i.e. creosote containing benzo[a]pyrene at a concentration of less than 0.005% by mass and water extractable phenols at a concentration of less than 3% by mass) were still allowed.

In 2002, the Netherlands notified the Commission once again about its stricter national provisions on the use of creosote, derogating from the 2001 Directive in that the ban on creosote included industrial applications. The SCTEE confirmed that the environmental risks highlighted in the scientific studies submitted by the Netherlands, relating to the extensive use of creosote-treated wood for riverbank protection, were justified. The Commission approved the measure and once again announced its intention of examining the harmonisation measure in light of the scientific evidence provided by the Netherlands, in accordance with Article 114(7) TFEU.[387]

384 O.J. 1994, L 365/1.

385 Dec. 1999/832/EC, O.J. 1999, L 329/25; Dec. 1999/833/EC, O.J. 1999, L 329/43; Dec. 1999/834/EC, O.J. 1999, L 329/63; Dec. 1999/835/EC, O.J. 1999, L 329/82.

386 Dir. 2001/90/EC, O.J. 2001, L 283/41.

387 Dec. 2002/884/EC, O.J. 2002, L 308/30, at 80.

To evaluate whether national derogations are necessary and proportionate, and whether they do not constitute a means of arbitrary discrimination or a disguised restriction on trade, the Commission refers to scientific assessment. In the creosote case, an external expert was asked to carry out an additional environmental impact study. The Commission, however, emphasised that, according to the Treaty provisions, the responsibility of producing scientific evidence lies with the requesting Member State and seeking additional justification for the maintenance of the national measures by the Commission cannot constitute a precedent for the future.[388] We will see further in this chapter that – in most cases – the Commission confines itself to asking EU scientific institutions to give an opinion on the scientific evidence submitted by the national authorities.

This leads us to the core issue in the application of national derogations and the functioning of the internal market in general, *viz.* possible conflicts between national and Union scientific opinions. The Commission, in principle, relies on the advice of its own scientific committees or EU agencies, whereas Member States invoke studies produced by national scientific institutions. These tensions are especially visible in the field of food safety, where the serious mishandling of the BSE crisis in the nineties shattered public trust in the EU scientific advice and food safety system.[389] A series of actions aimed at improving scientific governance of food safety and regaining consumers' trust resulted in the creation of a system of principles and requirements of food law, new procedures in matters of food safety and a new scientific institution: the European Food Safety Authority (EFSA). The next part of this chapter will briefly discuss the main elements of this scientific food safety governance established by Regulation 178/2002.

3. Scientific food safety governance

3.1 European Food Safety Authority

The European Food Safety Authority (EFSA) was established by the General Food Law as an important element of the new EU policy to improve food safety and restore confidence in the food supply in the EU (EC, 2000). The Authority provides communication on food risks and independent scientific advice to support the Commission, European Parliament and Member States in taking risk management decisions. Accordingly, Scientific Committees attached to the Commission and

[388] See, e.g. Dec. 1999/833/EC, O.J. 1999, L 329/43 at 32-34.

[389] A special Committee of Inquiry set up by the European Parliament severely criticised the Commission for protecting the market and putting business interests over public health concerns (European Parliament, 1997). On the BSE crisis see Vos (2000) and Krapohl (2005).

delivering scientific advice in the field of food safety were reorganised and their role was taken over by EFSA.[390]

EFSA carries out risk assessment underpinning food safety legislation in accordance with the principles of excellence, independence and transparency. In the EU legislation introduced after the General Food Law risk assessment is centralised and entrusted to EFSA.[391] Although the Authority does not have any regulatory powers, its scientific opinions are crucial for drawing up food safety measures within the framework of the risk analysis methodology. Where the draft measure is not in accordance with EFSA's opinion, the Commission must give reasons for its decision.[392] Thus, in exercising its administrative discretion, the Commission has to give careful consideration to the scientific opinion of the Authority. Some authors suggest that this provision may make it difficult for the Commission to diverge from EFSA's conclusions (Alemanno, 2007: 194). As the analysis of several examples later in this chapter will show, this especially concerns national derogations.

One of the basic principles of EFSA's mission is to network with national authorities carrying out similar tasks and to facilitate the sharing of knowledge.[393] EFSA is not empowered with a decisive voice in the case of a conflict between national and EU scientific opinions. Instead, it is obliged to be vigilant in identifying at an early stage any diverging scientific opinions between the Authority and Member States, and to co-operate with a view to either resolving the divergence or preparing a joint document explaining the differences.[394] Thus, in the network of scientific excellence, EFSA was created to become an authoritative point of reference, whose reputation in scientific matters would 'put an end to competition in such matters among national authorities in the Member States' (Byrne, 2002).

Another procedure designed to deal with diverging scientific opinions is mediation introduced in Article 60 GFL. According to this provision, where a Member State is of the opinion that a measure taken by another Member State is either incompatible with the General Food Law or likely to affect the functioning of the internal market, the Commission may request EFSA to give an opinion on the contentious scientific issue. This procedure is aimed at reaching a 'mutual scientific opinion'

[390] These Committees were: Scientific Committee on Food, Scientific Committee on Animal Nutrition, Scientific Veterinary Committee, Scientific Committee on Pesticides, Scientific Committee on Plants, and Scientific Steering Committee.

[391] See, e.g. Reg. 1829/2003 on genetically modified food and feed, O.J. 2003, L 268/1; Reg. 1924/2006 on nutrition and health claims made on foods, O.J. 2007, L 12/3; Reg. 1331/2008 establishing a common authorization procedure for food additives, food enzymes and food flavourings, O.J. 2008, L 354/1.

[392] See, e.g. Art. 7(3) Reg. 1331/2008, O.J. 2008, L 354/1.

[393] Art. 36(3) GFL and Reg. 2230/2004 laying down detailed rules with regard to the network of organizations operating in the fields within EFSA's mission, O.J. 2004, L 379/64.

[394] Art. 30 GFL.

and resolving scientific conflicts before the matter is brought before the Court of Justice.

3.2 Risk analysis methodology: science and other legitimate factors

The General Food Law establishes general principles and requirements of food law that apply to all stages of production, processing and distribution of food and feed.[395] 'Food law' is defined as laws, regulations and administrative provisions, governing food in general, and food safety in particular, whether at Community or national level.[396] Therefore, the principles established in the GFL apply to all legislation in this field at all levels of European governance.

One of the general principles governing EU food safety regulation is the principle of risk analysis, set out in Article 6 GFL. Risk analysis is a widely recognised methodology used for developing food safety standards. The risk analysis approach is used, *inter alia*, in the Codex Alimentarius Commission (CAC) – an inter-governmental body established by the Food and Agriculture Organisation (FAO) – and the World Health Organisation (WHO) to develop international food standards to ensure fair practices in trade and consumer protection in relation to the global trade in food (CAC, 2008).

The risk analysis principle requires that measures adopted by the EU and the Member States should generally be science-based.[397] Risk analysis for regulatory actions can be described as a two-step process.[398] During the risk assessment stage, scientists evaluate risks related to a product or technology.[399] The next step is risk management: a political process where decision makers utilise risk assessment findings to develop policy alternatives and select the most appropriate measures to reduce or eliminate risks.[400]

Risk assessment provides an essential input to risk management. Science alone, however, cannot constitute a basis for food safety decisions. Establishing policy in

[395] Art. 4(1) GFL.
[396] Art. 3(1) GFL.
[397] The risk analysis methodology is limited to measures aimed at the protection of human health (food safety law). Technical food standards not having as their objective the protection of human health are excluded from the scope of application of risk analysis. See also Szajkowska (2010). On risk analysis see National Research Council (1983), American Chemical Society (1998) and FAO (1997).
[398] The definition of risk analysis consists of three components: risk assessment, risk management and risk communication. Risk communication is the interactive exchange of information concerning risks throughout the risk analysis process among risk assessors, risk managers, but also consumers, food businesses and other stakeholders (Art. 3(13) GFL). This element of risk analysis, however, is outside the scope of this study.
[399] Risk assessment consists of four steps: hazard identification, hazard characterisation, exposure assessment and risk characterization (Art. 3(11) GFL). On the application of risk assessment to food safety see McKone (1996).
[400] Art. 3(12) GFL.

a democratic manner requires taking into account a broad array of societal values and preferences. The creation of public policy in the field of food safety has to reconcile the European patchwork of diverse cultures and traditions concerning food with the single market,[401] which has always been a particularly contentious issue (Ansell and Vogel, 2006).

Typically such conflicts in the democratic technological society have been described as the science-democracy dichotomy. The politics of food safety creates a paradox of reliance on scientific expertise considered as a device legitimating decision making, which at the same time leads to distancing society from political participation in risk governance by failing to respond to other social concerns. Thus, efforts at 'democratising expertise', where scientific opinions are supplemented by social, ethical or economic concerns expressed by the civil society, also have to be interpreted in a broader context of achieving legitimate risk governance by combining science and public participation in the decision-making process (EC, 2001; Christoforou, 2003; De Marchi, 2003; Grundmann and Stehr, 2003; Liberatore and Funtowicz, 2003; Millstone, 2007).

This conflict between scientific considerations and other values translates into the question of how much discretion public administration enjoys in taking regulatory decisions in highly technical areas of risk management. In this regard, Fisher argues that the disputes over standard-setting in risk regulations should be regarded as a discussion about the nature of public administration and the choice between two paradigms: rational-instrumental and deliberative-constitutive (Fisher, 2007).

The rational-instrumental model is the *technocratic* approach, which limits administrative discretion to a minimum and sees science as a primary validation of standard setting. The deliberative-constitutive theory reflects the *democratic* approach, granting public administration 'substantial and problem solving' discretion in regulating risks. In this model, public administration is not merely a 'transmission belt' constrained by a strict rational methodology and exercising discretion in an easy-to-assess and clearly defined way, but a body actively engaged in standard-setting and, depending on circumstances, including different types of information in the decision-making process (Fisher, 2007: 28-31).

The General Food Law recognises that other legitimate factors, such as societal, economic, traditional, ethical, environmental concerns, and the feasibility of controls should be taken into account when drawing up food safety legislation.[402] The concept of 'other legitimate factors' is also included in the definition of risk

[401] As an example, Vogel describes the reaction to the ruling of the Court of Justice concerning the German 'Reinheintsgebot' (Case 178/84, *Commission* v. *Germany*, [1987] ECR 1227). A petition to maintain the German purity decree was signed by 2.5 million German citizens (Vogel, 1995: 40).
[402] Rec. 19 GFL.

management[403] and in Article 6 GFL, which stipulates that, in order to achieve the general objectives of food law established in Article 5 GFL, risk management should take into account the results of risk assessment, other factors legitimate to the matter under consideration and the precautionary principle.

Risk assessment in the General Food Law is confined to the evaluation of a hazard defined as 'a biological, chemical or physical agent in, or condition of, food or feed, with the potential to cause an adverse health effect'.[404] It does not take into consideration factors other than natural science, such as cultural, economic, social, religious, psychological or ethical issues.[405] The Regulation merely states that other legitimate factors come into play during the risk management stage. The General Food Law does not provide, however, any indication of how they should be incorporated in the decision-making process and to what extent measures based on 'other legitimate factors' can deviate from the conclusions of scientific risk assessment.

4. Standard of judicial review of EU health protection measures

4.1 Discretion

The level of discretion that public administration enjoys is coupled with the standard review applied by courts (Jasanoff, 1994: 53). The level of intensity of the scrutiny applied by the judiciary in reviewing legislative or administrative acts based on complex technical assessments is a particularly sensitive area. On the one hand, courts do not have a basis for 'second-guessing administrators on such highly technical issues'. On the other hand, although the Court of Justice does not express the desire to carry out scientific assessments on its own, the exercise of administrative discretion in technical areas cannot be totally excluded from judicial review.

In general, the Court of Justice grants the EU institutions broad discretion in drawing up legislation aiming at removing barriers to trade and ensuring a high level of protection of the environment and of human health. This discretion especially applies to determining the level of protection deemed acceptable for society.[406] In the *Cheminova* case concerning the Commission's withdrawal of authorisations for malathion – an insecticide sprayed on certain crops – the Court of First Instance (now renamed the General Court) stated:

[403] Art. 3(12) GFL.
[404] Art. 3(14) GFL.
[405] On the inclusion of these factors see Scientific Steering Committee (SSC), 2003.
[406] See, e.g. Case 157/96, *National Farmers' Union*, [1998] ECR I-2211, at 61.

> '[I]f the Commission is to be able to pursue effectively the objective assigned to it, account being taken of the complex technical assessments which it must undertake, it must be recognized as enjoying a broad discretion.'[407]

In the same judgement, the Court set forth a standard of such a review. The legality of a measure can be affected if the authorities have not complied with the procedural requirements, if the facts established by the authorities are not correct, if there has been a manifest error of appraisal of those facts or a misuse of powers.[408] Where the EU institutions enjoy wide discretion, the review of administrative procedure is an important element of the judicial review. In this regard, the judicature must establish whether the evidence is accurate, reliable and consistent, and whether it contains all the information necessary to draw conclusions in such complex situations.[409]

In the *Pfize*r case the Court of First Instance confirmed that, in cases involving complex assessments, the discretion also applies to a certain extent to the establishment of the facts:

> 'The Community judicature is not entitled to substitute its assessment of the facts for that of the Community institutions, on which the Treaty confers sole responsibility for that duty.'[410]

The discretion is particularly wide for measures adopted under conditions of scientific uncertainty, where – in accordance with the precautionary principle – the institutions may take protective measures without having to wait until the reality and seriousness of risk to health become fully apparent.[411] In the *Commission* v. *France* judgement, concerning national measures requiring a prior authorisation scheme for processing aids and foodstuffs whose preparation involved the use of processing aids, the Court has reiterated that, to the extent that uncertainties exist in the current state of scientific research, it is for the Member States to determine the level of protection they deem appropriate.[412] Yet, a correct application of the precautionary principle requires 'identification of the potentially negative consequences for health' and a 'comprehensive assessment of the risk to health based on the most reliable scientific data and the most recent results of international research'.[413] A national measure constituting an obstacle to trade can be adopted

---

[407] Case T-326/07, *Cheminova v. Commission*, [2009] ECR 0000. at 106.

[408] Case T-326/07, *Cheminova v. Commission*, [2009] ECR 0000 at 107. See also Joined Cases T-74/00, T-76/00, T-83/00, T-84/00, T-85/00, T-132/00, T-137/00 and T-141/00, *Artegodan and Others* v. *Commission*, [2002] ECR II-4945, at 200.

[409] Case C-405/07 P, *Netherlands* v. *Commission*, [2008] ECR I-8301, at 55.

[410] Case T-13/99, *Pfizer Animal Health SA* v. *Council*, [2002] ECR II-3305, at 169.

[411] Case 157/96, *National Farmers´ Union*, [1998] ECR I-2211, at 63.

[412] Case C-333/08, *Commission* v. *France*, [2010] ECR 0000, at 85.

[413] Case C-333/08, *Commission* v. *France*, [2010] ECR 0000, at 92.

only if 'the real risk alleged for public health appears sufficiently established on the basis of the latest scientific data available at the date of the adoption of such decision.'[414] The judgement indicates that also the discretion applied to measures adopted under the precautionary principle is different depending on whether the measures are adopted at EU or national level.

The *Commission* v. *France* judgement explicitly referred to the risk analysis framework established by Regulation 178/2002 in the context of evidence that national authorities have to provide to justify their measures. Processing aids, except for extraction solvents,[415] are not harmonised at the EU level, hence the French legislation was considered under Articles 34 and 36 TFEU. In this regard, the Court held that:

> 'A Member State cannot justify a systematic and untargeted prior authorisation scheme...by pleading the impossibility of carrying out more exhaustive prior examinations by reason of the considerable quantity of processing aids which may be used or by reason of the fact that manufacturing processes are constantly changing. As is apparent from Articles 6 and 7 of Regulation No 178/2002, concerning the analysis of risks and the application of the precautionary principle, such an approach does not correspond to the requirements laid down by the Community legislature as regards both Community and national food legislation and designed to achieve the general objective of a high level of health protection.'[416]

Thus, the risk analysis methodology is interpreted by the Court in connection with a high level of health protection, requiring in-depth assessments of the risks and high-scientific thresholds for national derogations.

4.2 Cost-benefit analysis: proportionality *stricto sensu*?

Emiliou distinguishes three elements of the principle of proportionality applied in the EU context: a rationality test, necessity test and proportionality *stricto sensu* (Emiliou, 1996). The rationality test requires that the measure is an adequate means to achieve a purported end. At this stage Courts determine whether, e.g. a measure that claims to protect public health is not a disguised restriction on trade. The necessity test goes beyond the simple rationality test and consists in checking if there is a less trade-restrictive means to achieve the same objective.[417] Finally, proportionality *stricto sensu* juxtaposes the costs of introducing a measure and the benefits, and analyses whether the means are proportionate to the objective

[414] Case C-333/08, *Commission* v. *France*, [2010] ECR 0000, at 89.
[415] Dir. 88/344/EEC, O.J. 1988, L 157/28.
[416] Case C-333/08, *Commission v. France*, [2010] ECR 0000, at 103.
[417] See Case 104/75, *De Peijper*, [1976] ECR 613, at 16-17.

of the measure.[418] In practice, however, the division is not consistent, and the degree and intensity of review by Courts may differ.

In accordance with the distinction made by Emiliou, Trachtman argues that proportionality *stricto sensu* is in fact a cost-benefit analysis with a certain margin of appreciation (Trachtman, 1998). The margin of appreciation gives more room for all considerations that have to be taken into account during the decision-making process, as opposed to merely asking whether the costs outweigh the benefits. In the *Pfizer* judgement the Court has recognised that 'a cost-benefit analysis is a particular expression of the principle of proportionality in cases involving risk management' and applied the three elements of the proportionality test:

> '[T]he principle of proportionality...requires that measures adopted by Community institutions should not exceed the limits of what is appropriate and necessary in order to attain the legitimate objectives pursued by the legislation in question, and where there is a choice between several appropriate measures, recourse must be had to the least onerous, and the disadvantages caused must be not be disproportionate to the aims pursued.'[419]

Using the risk analysis terminology set out in the General Food Law, the cost-benefit analysis as interpreted by the European judiciary can be interpreted as the weighing of different policy alternatives, taking into account science and other legitimate factors that influence the content of risk management measures. In *Pfizer*, the Court took into account disadvantages caused by banning the use of an antibiotic – virginiamycin – as a growth promoter in animals to assess whether they are proportionate to the objective pursued, i.e. the protection of human health linked to the probability of transfer of antibiotic resistance developed in animals to humans. The Court referred to various aspects including animal welfare (whether the use of the antibiotic prevents certain diseases and reduces the mortality rate in animals),[420] economic factors related to the use of growth hormones in animal farming (possible increased costs to society of such changes in farming methods), and the effects of this agricultural practice on the environment.[421]

Although the Court devoted considerable space to these factors, the review remained rather marginal as Court justified the wide discretion concerning measures protecting human health with the political responsibility. Moreover, according to settled case

[418] See Case 302/86, *Commission* v. *Denmark*, [1988] ECR 4607 (*Danish bottles*). In this case, the ECJ found that 'the system for returning non-approved containers is capable of protecting the environment and, as far as imports are concerned, affects only limited quantities of beverages compared with the quantity of beverages consumed in Denmark... In those circumstances, a restriction of the quantity of products which may be marketed by importers is disproportionate to the objective pursued' (21).
[419] Case T-13/99, *Pfizer Animal Health v. Council*, [2002] ECR II-3305 at 411.
[420] Case T-13/99, *Pfizer Animal Health v. Council*, [2002] ECR II-3305 at 420-429.
[421] Case T-13/99, *Pfizer Animal Health v. Council*, [2002] ECR II-3305 at 464-474.

law, public health must take precedence over economic considerations.[422] Hence, it is unlikely that, in assessing a risk management measure involving the weighing of different political and economic factors, and protecting a broad public interest, the courts will go beyond the weak interpretation of the proportionality principle.

Although when assessing the proportionality of EU measures protecting human health Courts do not seem to make a distinction between basic acts (e.g. the directive on the marketing and use of certain dangerous substances laying down restrictions on the use of creosote) and implementing measures (the decisions withdrawing the authorisation of virginiamycin and malathion), the picture becomes more complicated for national measures derogating from EU legislation, which may require more intensive scrutiny. The next part of this chapter will discuss the standard of review applied to national derogations in comparison with the discretion enjoyed by the EU institutions in the light of the risk analysis methodology.

5. Science in national derogations

5.1 National versus EU scientific opinions

It is apparent that the necessity and proportionality of the national derogation cannot be judged in isolation from the existing harmonised legislation, which is supposed to set, according to Article 114(3) TFEU, a high standard of protection, taking into account recent scientific developments. In this context, criteria for assessing the proportionality of national measures seem to be even more stringent than those underlying Article 36 TFEU.[423] A Member State must not only prove that a derogating measure is necessary, proportionate and does not constitute a means of arbitrary discrimination or a disguised restriction on trade, but also that the level of protection ensured by the harmonisation measure is not sufficient in light of scientific facts and/or that the measure contains an error of fact or of assessment.[424]

In practice, this requirement means that Member States invoking Article 114(4) TFEU must still either prove that their population is in a specific situation of risk compared to the EU as a whole[425] or challenge the scientific risk assessment on

[422] Case T-13/99, *Pfizer Animal Health v. Council*, [2002] ECR II-3305 at 471, and case law referred to therein.
[423] See Opinion of A.G. Tizzano of 30 May 2002 in Case 3/00 at 99.
[424] Case 3/00, *Denmark v. Commission*, [2003] ECR I-2643 at 93.
[425] See Opinion of A.G. Tizzano of 30 May 2002 in Case 3/00. In this regard he states that '[i]f it were a problem common to all or the majority of Member States, it would presumably already have been resolved by the directive, but if that were not so, it would be necessary to verify whether the conditions for challenging the directive were met, given that the directive must ensure not just general protection but a high level of protection; in any case, the problem would be of a general nature and it is not therefore possible to understand why it should be resolved only for the fortunate citizens of a single meticulous Member State, to the detriment of the uniform application of the harmonised rules and hence of the functioning of the internal market' (77).

which the EU measure is based to show that the harmonised legislation does not ensure sufficient protection. As discussed above, the specific situation in the Netherlands related to the use of creosote was the basis for approval of the national derogation in 2002. Similar criteria were applied by the Commission in assessing the legality of Swedish measures concerning the use of certain colours and sweeteners in foodstuffs.[426] The national studies provided by the Swedish authorities were confronted with the evaluations of the Scientific Committee of Food (SCF) on which the harmonisation measure was based. The Commission concluded that the Swedish studies did not show that the colours in question posed any particular health problem for the Swedish population compared with other Member States and that all the risks highlighted in the national reports were sufficiently dealt with in the SCF reports. It may indeed seem difficult to demonstrate, to quote Tizzano, 'that a directive negotiated over a period of years can have overlooked information so crucial as to justify the derogation', and that 'this information (or even new scientific evidence) can then emerge in the extremely short period of time between approval of the directive and the request to maintain national derogations'.[427]

Confronted with divergent scientific opinions of different institutions, the Court has expressed on several occasions confidence in international scientific research, and in particular the findings of the Scientific Committee of Food and later of EFSA.[428] The limited scrutiny is generally confined to analyzing whether the scientific basis underpinning the contested legislation adopted by the EU institutions is obviously wrong or lacks a scientific basis. It is very difficult, however, to challenge the scientific basis itself. In the *Cheminova* case, e.g. the Court assessed whether the results of a study, suggested by EFSA and undertaken by the company, which established no genotoxicity potential for a specification of isomalathion of 0.2%, could exclude the uncertainties pointed out in the EFSA report. The Court concluded that the new study could not affect the legality of the decision, because, according to EFSA report, 'further genotoxicity studies [had] to be provided'. In the Court's opinion, the mere use of the plural clearly indicated that the results of the study provided by Cheminova would not be sufficient.[429]

5.2 National versus EU interpretations of a scientific opinion

Interestingly in this context, the Danish derogations concerning the use of sulphites, nitrites and nitrates did not refer to national studies, but to the opinions of the Scientific Committee on Food - the same opinions on which the contested EU measure was based. Directive 95/2/EC established a list of additives other than

[426] Dec. 1999/5/EC, O.J. 1999, L 3/13.
[427] Opinion of A.G. Tizzano of 30 May 2002 in Case 3/00, at 82.
[428] See, e.g. Case 178/84, *Commission v. Germany*, [1987] ECR 1227, at 44.
[429] Case T-326/07, *Cheminova v. Commission*, [2009] ECR 0000, at 146.

colours and sweeteners authorised in food.[430] The list of additives was based on the scientific evaluations of the SCF. Before the entry into force of the directive, the Danish legislation contained a list of permitted food additives, including the conditions of use of nitrates, nitrites and sulphites. The Danish authorities intended to maintain the provisions regulating these three additives after transposing the directive.

Sulphites are used as preservatives and antioxidants in, *inter alia*, wine, jam, biscuits and dried fruit. In larger amounts they can, however, be harmful to health and provoke allergic reactions, especially for asthma sufferers. The SCF carried out the toxicological assessment of sulphites in 1981 and gave an opinion on sulphites used as food additives in 1994, establishing an acceptable daily intake (ADI) for sulphur dioxide at a maximum of 0.7 mg/kg of body weight and recommending that the use of sulphites be limited as much as possible. According to the Danish government, their use had to be further restricted because any additional authorisation of sulphite use would lead to the acceptable daily intake being exceeded. The national provisions authorised the use of sulphites in only 16 categories of the 61 permitted by the EU harmonisation measure.

The Danish provisions also prohibited the use of nitrates (except for one product) and authorised the use of nitrites in meat within limits lower than half those provided for by the EU measure. Nitrites and nitrates are also used as preservatives, mostly to prevent the growth of *botulinum* bacteria. Nitrites in meat, however, can be transformed into nitrosamines, which are considered carcinogenic. The SCF issued two opinions on the risks and technological need for the addition of nitrites and nitrates: in 1990 and 1995. Both opinions recommended to reduce the amounts of these preservatives as far as possible. In addition, the SCF confirmed the carcinogenicity of nitrosamines and stated that it is not possible to determine a level below which they could be safe to human health.

In the view of the Danish authorities, the EU harmonisation measure, permitting an excessive use of nitrites and nitrates, did not take into consideration the recommendations of the SCF, which indicated a correlation between the addition of nitrates and the formation of nitrosamines. In the view of the Danish authorities, the second opinion of the SCF criticised the quantities of additives authorised by the directive as being unnecessary and disproportionate and stated that if the levels of nitrate residues were as high as those authorised by the harmonisation measures, the ADI would be exceeded.

The Commission did not approve the Danish legislation as not justified by the need to protect public health (sulphites) and excessive in relation to this aim

[430] O.J. 1995, L 61/1.

(nitrites and nitrates).[431] Subsequently, the Danish Government repealed the contested national provisions and decided to challenge the Commission decision in the Court of Justice, giving the Court the opportunity to rule for the first time on the application against the Commission's refusal to authorise the maintenance of national derogations.[432]

The evidence submitted by Denmark did not contain either new scientific evidence or proof that the Danish population was in a specific situation of risk concerning the consumption of nitrites compared to the European population. The Commission claimed that Directive 95/2 sufficiently took into account the findings of the SCF because the 1995 opinion repeated the same conclusion which had been contained in the opinion issued before the adoption of the directive. Hence, the national derogation was based solely on a different interpretation of the scientific opinion given by the SCF, which – according to the Danish authorities – was not sufficiently taken into account in setting the limits for nitrites and nitrates.

The Court, however, did not explicitly refer to the inadequacy of protection of the EU measure, but went on to state that, pursuant to Article 114(4) TFEU, a Member State may maintain in force national derogations on the basis of an assessment of risks made unilaterally by that Member State, if the measures are necessary to attain a level of protection that is higher than the harmonisation measure:

> '[T]he applicant Member State may, in order to justify maintaining such derogating national provisions, put forward the fact that its assessment of the risk to public health is different from that made by the Community legislature in the harmonization measure. In the light of the uncertainty inherent in assessing the public health risks...divergent assessments of those risks can legitimately be made, without necessarily being based on new or different scientific evidence.'[433]

The Court thus confirmed that Member States can decide on their acceptable level of risk, taking into account the same scientific evidence on which the EU risk management decision is based. The necessity of the national measure, however, must be justified in the light of the existing EU measure – therefore, the national authorities must show that the EU measure is insufficient to attain the objective set out by the national legislator. Being reluctant to adjudicate on diverging scientific and technical issues and consistently assigning a central role to EU scientific bodies, the Court should indeed look more favorably at a national derogation based on a divergent interpretation of the EU scientific evidence than at a national derogation based on a national risk assessment. Another issue which is relevant for

---

431 Dec. 1999/830/EC, O.J. 1999, L 329/1.

432 Case 3/00, *Denmark v. Commission*, [2003] ECR I-2643.

433 Case 3/00, *Denmark v. Commission*, [2003] ECR I-2643 at 63.

the interpretation of scientific findings in this context is the possibility to invoke the precautionary principle. The creosote example discussed above shows that, in the face of scientific uncertainty, Member States have the right to maintain more restrictive measures to ensure a higher level of protection. Divergent EU and national risk assessment may create uncertainties *per se*, but the Danish case concerning the use of nitrites and nitrates illustrates that a national measure may be justified by a different interpretation of the same scientific evidence on which the EU measure is based, or even parts of that risk assessment, which, in the view of the national authorities, provide sufficient basis to conclude that uncertainty as to a risk to human health exists.[434]

6. Other legitimate factors in EU and national legislation

6.1 Other legitimate factors lowering the level of health protection

The Danish authorities succeeded in proving that the national measures in so far as they concerned the use of nitrites and nitrates as food additives were legitimate because the harmonisation measure did not ensure sufficient protection of human health. In consequence, the Court annulled the Commission Decision which rejected the Danish provisions concerning the use of those additives.

According to Article 114(7) TFEU, when a Member State is authorised to mantain or introduce national provisions derogating from a harmonisation measure, the Commission is obliged to examine immediately whether to propose an adaptation to that measure. This is in line with the requirement that the Commission take a high level of protection as the basis for its measures concerning human health, environmental or consumer protection, in particular taking account of any new development based on scientific facts.

The ruling delivered in March was discussed at the next meeting of the Standing Committee on the Food Chain and Animal Health in April 2003. The Commission informed the Committee that it was planning to ask EFSA to give an urgent opinion on the minimum levels of nitrites and nitrates in meat products necessary to achieve the preservative effect and to guarantee microbiological safety and asked Member States to provide national data on the technological need and real amounts of those additives used in foods.

EFSA in its opinion confirmed that lowering the maximum levels of nitrites and nitrates was necessary and that, instead of residual amounts, the ingoing amounts of nitrites and nitrates should be monitored (EFSA, 2003). Accordingly,

[434] See also on this subject, Case T-13/99, *Pfizer Animal Health v. Council*, [2002] ECR II-3305, where the precautionary measures were introduced despite the conclusions of the risk assessment stating no risk to human health.

the Commission proposed an amendment to the Directive on food additives other than colours and sweeteners, introducing the new method for establishing the permitted level of nitrites and nitrates. The proposal, however, allowed exemptions from this rule for several meat products from the U.K., including Wiltshire bacon and ham, justified by their traditional method of production. For these products, the maximum residual level would be maintained.[435] Other Member States also asked for similar derogations for their traditional products, while some Member States were opposed to introducing specific exemptions based on factors other than scientific risk assessment (European Parliament, 2004). In the amended directive, these exemptions were eventually included for several traditionally manufactured meat products specifically named, as well as for products produced in a similar manner, which could if necessary be categorised in accordance with the comitology procedure.[436]

Despite the amendments to Directive 95/2 changing the authorisations for nitrites and nitrates as a follow up to the judgement of the Court of Justice and EFSA's opinion, Denmark once again decided to maintain, on the basis of Article 114(4) TFEU, its stricter national law. In November 2007 the Danish government notified the Commission of its intention not to transpose the amended provisions insofar as they concerned the addition of nitrite to meat products. In comparison with the amended directive, the Danish provisions laid down lower maximum added amounts for several types of meat products and did not allow exceptions referring to the traditional methods of production, forbidding the placing on the market of products for which no ingoing amounts could be established. Denmark highlighted that these provisions had been in place for many years and had successfully maintained a very low rate of food poisoning cases caused by sausages compared to other Member States.

Like in the previous request to maintain the national measures, Denmark referred to the opinions of the Community's scientific bodies. The Danish government considered the national provisions in full compliance with the recommendations of the SCF made in 1990 and 1995, as well as EFSA's risk assessment from 2003, since none of these opinions suggested any exceptions to the principle of controlling maximum 'added amounts' of nitrites rather than residual levels.

In the light of the facts provided by Denmark, EFSA, asked by the Commission to issue a scientific opinion on the validity of the previous risk assessments, confirmed the recommended control of ingoing amounts, as well as the maximum added amounts for different meat products. As a result, the Commission concluded that the Danish provisions were based on the scientific evidence of the SCF and EFSA

[435] According to the Commission, for these products, due to the nature of the manufacturing process, it was not possible to control the ingoing amount of curing salts absorbed by the meat.
[436] Dir. 2006/52/EC, O.J. 2006, L 204/10.

and thus justified on grounds of protection of public health.[437] The decision was all the more easy when taking into account that the national derogations did not seem to constitute a serious obstacle to the functioning of the internal market. Denmark showed that imports of meat products from other Member States had even been increasing during the time when the stricter provisions were in force.

In this case, the Danish derogation was approved despite the fact that the Commission expressly stated that the harmonisation measure was adequate to ensure the microbiological safety of meat products.[438] Although harmonised legislation has to ensure a high level of health protection, this level does not necessarily have to be the highest that is technically possible.[439] In its risk management decision, the Commission decided not to pursue the strictest maximum levels for nitrites and took into account other factors; relating to specific traditional methods of production and the diversity of the meat products across the EU. Similarly, Regulation 178/2002 defines the aim and scope of food law as 'the assurance of a high level of protection of human health and consumers' interest in relation to food, taking into account in particular the diversity in the supply of food including traditional products, whilst ensuring the effective functioning of the internal market.'[440]

Although the General Food Law applies at both EU and national level, as we have shown in the previous cases, discretion permitted for the EU institutions and Member States varies. Mortelmans, in his analysis of the relationship between the Treaty provisions and secondary legislation, argues that the EU institutions are bound by the Treaty rules relating to free movement in the same way as the Member States, but enjoy a broader margin of discretion (Mortelmans, 2002). Referring to the risk analysis framework, in the case of conflict between the free movement rules, discretion is especially relevant to the extent to which other legitimate factors can be a basis in risk management decisions.

In the amended directive on the use of nitrites and nitrates as food additives, the EU legislature, in addition to the scientific opinion, took into account other legitimate factors in the risk management choice. This measure was easily challenged by the Member State requesting a derogation from the harmonisation measure and invoking the scientific opinion on which the harmonisation measure was based. The other legitimate factors, taking into account traditional methods of production and the diversity of products on the internal market, lowered the level of protection of the harmonisation measure. This situation is clearly not possible in the case of national derogations based on Article 114(4) TFEU, because they have to be more stringent than the EU harmonisation measure.

[437] Dec. 2008/448/EC, O.J. 2008, L 157/98. The approval was temporary, until 23 May 2010 (2 years).
[438] Dec. 2008/448/EC, O.J. 2008, L 157/98, at 41.
[439] Case T-13/99, *Pfizer Animal Health v. Council*, [2002] ECR II-3305 at 152.
[440] Art. 1(1) GFL.

In line with Article 114 TFEU, one would expect the mechanism of the internal market to bring the differentiated standards to a uniform regulation in accordance with the 'race to the top'. The food safety paradigm, however, is influenced by socio-cultural implications of trade in foods. If the EU regulator decided to follow the scientific opinions, traditional cured meat products would disappear from the market. Thus, permitting exceptions justified by specific methods of production in this case does not seem to hinder the functioning of the internal market, even if several Member States decide to opt out, along the lines marked by the scientific opinions of EFSA.

6.2 Other legitimate factors raising the level of health protection

Another possibility resulting from taking into consideration other values and interests in a regulatory procedure is a measure ensuring a higher level of protection compared to the recommendation of the scientific risk assessment on which it is based. This possibility in theory concerns both EU measures and national derogations. The extent to which other legitimate factors can be taken into account by decision makers, again, depends on the discretion framed by the Treaty provisions relating to the functioning of the internal market.

The EU ban on the use of growth hormones except for therapeutic or other purposes and the ban on the import or intra-EU trade in meat from cattle treated with these hormones is an example of an EU measure introduced to a large extent as a response to consumer concerns about the safety of meat treated with hormones, prompted by abuse of good veterinary practice, rather than conclusive scientific studies, although these non-scientific assumptions were concealed behind scientific uncertainty.[441] The relationship between this measure and the internal market was dealt with in the *Fedesa* case, a preliminary ruling concerning proceedings brought before a national court in the UK, where the applicants challenged the validity of the national regulations implementing Directive 88/146/EEC.[442] The applicants argued that the directive banning the use of hormones in meat was not underpinned by scientific evidence and hence frustrated the legitimate expectations of traders.[443] The Court did not uphold this claim, referring to the discretionary power conferred on the EU institutions, and did not discuss the existence of scientific evidence showing the safety of the five hormones, merely stating that:

[441] Dir. 81/602/EEC banned the use of substances having hormonal action, but the use of oestradiol-17β, progesterone, testosterone, trenbolone acetate and zeranol was left to be regulated in accordance with the individual regulatory schemes in Member States and other countries, pending further research. Dir. 85/649/EEC extended the ban to include these five hormones, but it was annulled by the Court of Justice on procedural grounds. The same provisions were re-introduced by Dir. 88/146/EEC. The ban was a reaction to 'hormone scandals' in Italy in the late 1970s. Premature development of schoolchildren was suspected to be linked to illegal hormones in meat from school canteens (Roberts, 1998: 386).

[442] Case C-331/88, *Fedesa*, [1990] ECR I-4023.

[443] Case C-331/88, *Fedesa*, [1990] ECR I-4023, at 7.

> '[F]aced with divergent appraisals by the national authorities of the Member States, reflected in the differences between existing national legislation, the Council remained within the limits of its discretionary power in deciding to adopt the solution of prohibiting the hormones in question, and respond in that way to the concerns expressed by the European Parliament, the Economic and Social Committee and by several consumer organizations'.[444]

In the *Fedesa* case, it was the EU measure that hindered free movement of goods, and not the opposite – more typical – scenario, i.e. a Member State against, and the Commission for, the internal market.[445] The discretion in responding to consumer perceptions and public concerns seems to be much more limited in the case of national measures. Consider the case *Land Oberösterreich* v. *Commission*, concerning the lawfulness of national measures banning genetic engineering. The Land Oberösterreich intended to prohibit the cultivation of genetically modified organisms (GMOs) and breeding of transgenic animals and notified its draft measures in accordance with Article 114(5) TFEU. The Commission rejected the request for derogation.[446] After the Court of First Instance (now renamed the General Court) dismissed the action brought by the Land Oberösterreich and Austria in which they sought annulment of the decision,[447] the applicants appealed to the European Court of Justice to set aside the judgement.[448]

Austria claimed that the ban was justified by the specificity of the province of Upper Austria, founded on the existence of small size farms and organic production in the area. The Land Oberösterreich provided a study suggesting that it was not possible for organic and conventional production to coexist alongside a vast GMO cultivation.[449] Thus, the Austrian request for derogation related more to socio-economic aspects than to scientific evidence concerning the protection of the environment. The Land Oberösterreich stated itself that the notified measure was 'prompted by the fear of having to face the presence of GMOs'.[450]

The Commission based the rejection on the opinion of EFSA, in which it concluded that the information provided by Austria did not contain any new scientific evidence which could justify the measure.[451] Moreover, in the view of the Commission, the small size of farms was a common characteristic for all Member States, and hence

---

[444] Case C-331/88, *Fedesa*, [1990] ECR I-4023, at 9.

[445] Mortelmans refers in this regard to the *BSE* case, where the UK argued against the Commission measure prohibiting exports of UK beef (Case C-180/96, *UK* v. *Commission*, [1998] ECR I-2288) (Mortelmans, 2002: 1338).

[446] Dec. 2003/653/EC, O.J. 2003, L 230/24.

[447] Cases T-366/03 and T-235/04, *Land Oberösterreich* and *Austria* v. *Commission*, [2005] ECR II-4005.

[448] Cases C-439/05 P and C-454/05 P, *Land Oberösterreich* and *Austria* v. *Commission*, [2007] ECR I-7141.

[449] Dec. 2003/653/EC, O.J. 2003, L 230/24, at 34.

[450] Cases T-366/03 and T-235/04, Land Oberösterreich and Austria v. Commission, [2005] ECR II-4005, at 67.

[451] Dec. 2003/653/EC, O.J. 2003, L 230/24, at 63-68.

did not constitute a problem specific to Upper Austria. Finally, the Commission rejected Austria's argument justifying the national derogation by recourse to the precautionary principle as too general and lacking substance. In this case, too, the Commission referred to EFSA's opinion, which did not identify any risk that 'would justify taking action on the basis of the precautionary principle at Community or national level.'[452]

The Court dismissed the appeals of Austria and the Land Oberösterreich. Obviously, the case concerned the lawfulness of national measures notified under Article 114(5) TFEU, which explicitly requires that a national derogation be based on new scientific evidence on grounds of a problem specific to that Member State. However, the ruling would not be different if the national authorities wished to maintain existing national measures on the basis of Article 114(4) TFEU, which does not refer to these conditions. In fact, the autonomy of Member States is limited under both opt-out clauses - both possibilities of derogation are closely linked to the assessment of scientific evidence put forward by the Member State.

According to Article 114(4) TFEU, national derogations must be justified on grounds listed in Article 36 TFEU or relating to the environment or the working environment. The narrow interpretation of the exception relating to the protection of health and life of humans, on which food safety measures are based, requires that the national authorities must carry out a comprehensive risk assessment. This in-depth, case-by-case assessment should appraise the degree of probability of harmful effects and the actual risk a foodstuff might pose to health.[453] This restrictive approach practically precludes Member States from invoking factors other than science, be they social, traditional or economic reasons, to justify their national derogations.

6.3 Flexible harmonisation

Because food safety regulations directly affect the functioning of the internal market, they are usually totally harmonised, depriving Member States of any power to retain their national legislation, apart from the possibility of derogation provided in Article 114(4) TFEU. EU harmonisation measures, however, may allow Member States to adopt more specific provisions, taking into account particular situations or problems existing in their territories.

This approach is an important element of the so-called 'hygiene package': the re-organised EU regulatory framework for food hygiene and safety, which became applicable in 2006. Flexibility is necessary to guarantee the viability of traditional products and small food producers. According to Article 10(3-8) of Regulation

---

[452] Dec. 2003/653/EC, O.J. 2003, L 230/24, at 73.

[453] Case C-333/08, *Commission v. France*, [2010] ECR 0000, at 101.

853/2004 laying down specific hygiene rules for food of animal origin,[454] Member States may introduce national measures adapting the general hygiene requirements to traditional methods of production[455] or to food businesses in regions that are subject to geographical constraints. The national measures, however, cannot compromise the objectives of food safety. Products to which specific national hygiene measures apply are in free circulation in the internal market. The procedure is transparent; Member States have to notify their measures to the Commission and other Member States and any disagreements are discussed by the Standing Committee on the Food Chain and Animal Health.

In determining the content of harmonisation measures the objective of the internal market is the most important. The flexible method of harmonisation, however, touches on the diversity of the internal market, by giving the Member States the possibility to apply adaptations, taking into account local factors. This method of harmonisation creates an opportunity for incorporating a wider set of societal concerns into the decision-making process at the national level.

7. Conclusion

The risk analysis methodology established in Regulation 178/2002 applies to both EU and national food safety measures. The methodology, however, has different implications for the EU and national risk managers in the context of the Treaty provisions relating to the internal market and judicial review of measures protecting human health. This difference is reflected in the role science and other legitimate factors play in the risk management process.

The EU institutions enjoy a broad discretion in setting food safety standards, which can be largely motivated by other legitimate factors. The inclusion of these factors can lead to laxer standards, justified, e.g. by the diversity in the food supply or traditional methods of production, but also to stricter measures, based, e.g. on consumer risk perception. The role of non-scientific assumptions in food safety measures, however, is not clearly defined and socio-economic considerations often tend to be camouflaged by scientific risk assessment or scientific uncertainties.

Contrary to EU measures, national exceptions to the rule of the free movement of goods are dealt with restrictively. The national authorities must demonstrate that the their legislation is necessary to protect human health and, to this end,

---

[454] O.J. 2004, L 226/22 (Corrigendum).

[455] 'Foods with traditional characteristics' are defined in Art. 7 Reg. 2074/2005 laying down implementing measures for certain products under Reg. 853/2004 as 'foods that, in the Member State in which they are traditionally manufactured, are: (a) recognised historically as traditional products, or (b) manufactured according to codified or registered technical references to the traditional process, or according to traditional production methods, or (c) protected as traditional food products by a Community, national, regional or local law.' (O.J. 2005, L 338/27).

they must carry out a risk assessment. The internal market mechanism and the necessity to resist national protectionism leave practically no possibility to justify national food safety measures on the basis of factors other than science. Moreover, in the case of diverging national and EU scientific opinions, the EU Courts tend to follow the scientific expertise provided by EU scientific institutions, with EFSA playing a central role.

The inclusion of socio-economic considerations in the decision-making process creates an opportunity to promote diversity of the internal market by taking into account local traditional, cultural and social implications, as opposed to technocratic food safety regulation based on 'sound' science. Striking a balance between the safety paradigm and other legitimate factors at the EU rather than national level is understandable from an internal market perspective, but raises concerns about the legitimacy of supranational risk regulation. In addition to the questions of transparency and procedural guarantees for public participation in the decision-making process, the challenge facing food safety governance at the EU level involves establishing a credible risk governance across a variety of different national cultures. A more flexible approach to harmonisation, giving more possibilities to include non-scientific factors in national measures, as well as a clear formulation of how science and a wide set of other legitimate factors interact in food safety policy-making, could improve the credibility of risk governance in this field.

References

Alemanno, A., 2007. Trade in Food: Regulatory and Judicial Approaches in the EC and the WTO. Cameron May, Cambridge, UK.

American Chemical Society, 1998. Understanding Risk Analysis: A Short Guide for Health, Safety, and Environmental Policy Making. Washington, DC, USA.

Ansell, C. and Vogel, D. (eds.), 2006. What's the Beef? The Contested Governance of European Food Safety. The MIT Press, Cambridge, MA, USA.

Beck, U., 1992. Risk Society: Towards a New Modernity. Sage Publishers, London, UK.

Byrne, D., 2002. EFSA: Excellence, integrity and openness. Inaugural meeting of the Management Board of the European Food Safety Authority, 18 September 2002, Brussels, Belgium.

Christoforou, T., 2003. The precautionary principle and democratizing expertise: a European legal perspective. Science and Public Policy 30: 205-211.

CAC (Codex Alimentarius Commission), 2008. Procedural Manual 18th edition. FAO/WHO, Rome, Italy.

De Marchi, B., 2003. Public participation and risk governance. Science and Public Policy 30: 171-176.

EFSA (European Food Safety Authority), 2003. The effects of Nitrites/Nitrates on the Microbiological Safety of Meat Products. The EFSA Journal 14: 1-31.

Emiliou, N., 1996. The Principle of Proportionality in European Law: A Comparative Study. Kluwer Law International, London, UK.

EC (European Commission), 1985. Completing the Internal Market. White Paper from the Commission to the European Council (Milan, 28-29 June 1985). European Commission COM (85) 310 final, Brussels, Belgium.

EC (European Commission), 2000. White Paper on Food Safety. European Commission COM (1999) 719 final, Brussels, Belgium.

EC (European Commission), 2001. Report of the working group: Democratising expertise and establishing scientific reference systems (Group 1b) R. Gerold, pilot; A. Liberatore, rapporteur. White Paper on Governance, Work Area 1 - Broadening and enriching the public debate on European matters, Brussels, Belgium.

EC (European Commission), 2002. Communication from the Commission concerning Article 95 (paragraphs 4, 5 and 6) of the Treaty Establishing the European Community. European Commission COM (2002) 760, Brussels, Belgium.

European Economic and Social Committee, 1998. Opinion on 'The single market and the protection of the environment: coherence or conflict (SMO)'. O.J. C 019, 21/1/1998.

European Parliament, 1997. Report on the alleged contraventions or maladministration in the implementation of Community law in relation to BSE, without prejudice to the jurisdiction of the Community and the national courts, A4-0020/97/A, PE 220.544/fin/A.

European Parliament, 2004. Draft Report on the proposal for a Directive amending Directive 95/2/EC on food additives other than colours and sweeteners and Directive 94/35/EC on sweeteners for use in foodstuffs (COM(2004)0650 - C6-0139/2004 - 2004/0237(COD)).

FAO (Food and Agriculture Organisation), 1997. Risk Management and Food Safety. Report of a Joint FAO/WHO Consultation - Rome, 27-31 January 1997. FAO Food and Nutrition Paper 65.

Fisher, E., 2007. Risk Regulation and Administrative Constitutionalism. Hart Publishing, Oxford, UK.

Grey, P., 1990. Food law and the internal market: Taking stock. Food Policy 15: 111-121.

Grundmann, R. and Stehr, N., 2003. Social control and knowledge in democratic societies. Science and Public Policy 30: 183-188.

Hanf, D., 2001. Flexibility clauses in the Founding Treaties, from Rome to Nice. In: De Witte, B., Hanf, D. and Vos, E. (eds.). The Many Faces of Differentiation in EU Law, Intersentia, Antwerp, Belgium, pp. 3-26.

Hood, C., Rothstein, H. and Baldwin, R., 2001. The Government of Risk: Understanding Risk Regulation Regimes. Oxford University Press, Oxford, UK.

Jasanoff, S., 1994. The Fifth Branch: Science Advisers and Policymakers. Harvard University Press, Cambridge, MA, USA.

Joerges, C., 1997. Scientific Expertise in Social Regulation and the European Court of Justice. In: Joerges, C., Ladeur, K.H. and Vos, E. (eds.). Integrating Scientific Expertise into Regulatory Decision-Making: National Traditions and European Innovations. Nomos, Baden-Baden, Germany, pp. 295-323.

Krapohl, S., 2005. Thalomide, BSE and the Single Market: A Historical-Institutionalist Approach to Regulatory Regimes in the European Union. EUI Working Paper LAW No. 2005/03.

Liberatore, A. and Funtowicz, S., 2003. 'Democratising' expertise, 'expertising' democracy: what does this mean, and why bother? Science and Public Policy 30: 146-150.

Lidskog, R., 2008. Scientised citizens and democratised science. Re-assessing the expert-lay divide. Journal of Risk Research 11: 69-86.

McKone, T.E., 1996. Overview of the risk analysis approach and terminology: the merging of science, judgment and values. Food Control 7: 69-76.

Millstone, E., 2007. Can food safety policy-making be both scientifically and democratically legitimated? If so, how? Journal of Agricultural and Environmental Ethics 20: 483-508.

Mortelmans, K., 2002. The Relationship between the Treaty Rules and Community Measures for the Establishment and Functioning of the Internal Market - Towards a Concordance Rule. Common Market Law Review 39: 1303-1346.

National Research Council, 1983. Risk Assessment in Federal Government: Managing the Process: National Academy Press, Washington, DC, USA.

Onida, M., 2006. The Practical Application of Article 95(4) and 95(5) EC Treaty: What Lessons Can We Learn About the Division of Competences Between the EC and Member States in Product-Related Matters? In: Pallemaerts, M. (ed.). EU and WTO Law: How tight is the Legal Straitjacket for Environmental Product Regulation? VUB Brussels University Press, Brussels, Belgium.

Otsuki, T., Wilson, J.S. and Sewadeh, M., 2001. A Race to the Top? A Case Study of Food Safety Standards and African Exports. The World Bank Policy Research Working Paper WPS2563.

Pescatore, P., 1987. Some Critical Remarks on the 'Single European Act'. Common Market Law Review 24: 9-18.

Roberts, D., 1998. Preliminary Assessment of the Effects of the WTO Agreement on Sanitary and Phytosanitary Trade Regulations. Journal of International Economic Law 1: 377-405.

Scientific Steering Committee, 2003. Final Report on Setting the Scientific Frame for the Inclusion of New Quality of Life Concerns in the Risk Assessment Process - 10 -11 April 2003. Scientific Steering Committee, European Commission, Brussels, Belgium.

Szajkowska, A., 2009. From mutual recognition to mutual scientific opinion? Constitutional framework for risk analysis in EU food safety law. Food Policy 34: 529-538.

Szajkowska, A., 2010. The Impact of the Definition of the Precautionary Principle in EU Food Law. Common Market Law Review 47: 173-196.

Trachtman, J.P., 1998. Trade and...Problems, Cost-Benefit Analysis and Subsidiarity. European Journal of International Law 9: 32-85.

Vogel, D., 1995. Trading Up: Consumer and Environmental Regulation in a Global Economy. Harvard University Press, Cambridge, MA, USA.

Vos, E., 2000. EU Food Safety Regulation in the Aftermath of the BSE Crisis. Journal of Consumer Policy 23: 227-255.

Part 3

Transnational law, resource complexes and food security

Chapter 10[456]

Farmland and food security: protecting agricultural land in the United States

Margaret Rosso Grossman

1. Introduction

In February 2010, the Food and Agricultural Organization of the United Nations observed that the world had entered 'a period of grave concern for the fate of the world's hundreds of millions of poor and hungry people' (FAO, 2010: 103). Both the global food crisis in 2006-2008 and the economic recession that followed have triggered increased hunger and malnutrition, especially in developing countries. The current crisis has focused the world's attention on 'agriculture, agricultural development, and global food security' and may well lead to 'more decisive action to eliminate hunger and food insecurity in the world' (FAO, 2010: 119, 122). Indeed, a special section on food security in a February 2010 issue of *Science* insists that '[f]eeding the 9 billion people expected to inhabit our planet by 2050 will be an unprecedented challenge' (Ash *et al.*, 2010: 797).

Agricultural production is critical for food security, and farmers in the United States (US) produce a large part of the world supply of food and feed. US agricultural production is more than adequate to feed the American population, and many products are exported to other nations, both commercially and as food aid.[457] While the US appears to be in no immediate danger of becoming food insecure, the global outlook is less certain. The US Department of Agriculture (USDA) estimated that 833 million people in the world were hungry in 2009, nearly 2% more than in 2008. Most of the hungry live in developing countries.[458] Food security problems in these countries are a result of several factors, including poor domestic yields, high food prices, and financial constraints, exacerbated by the global financial crisis (Shapouri *et al.*, 2009: v).

---

[456] This chapter is based on work supported by the National Institute of Food and Agriculture, US Department of Agriculture, under Hatch Project No. ILLU-470-309. The author thanks David Holesinger and Robert Romashko for their research assistance.

[457] An important issue, not addressed here, is the appropriate role of the US in global food security. Should the US provide exports and food aid, or should the US help developing countries to build 'sustainable, local food systems?' (See Taylor (2001: 189)).

[458] USDA estimates of hunger include transitory hunger, using annual data, and assume a 2,100 calorie-a-day diet (Shapouri *et al.*, 2009: 4). The US, too, has hungry people. USDA research published in November 2009 indicated that at some time during 2008, 49 million people (14.6% of households) experienced food insecurity, an increase from 36 million (11.1% of households) in 2007 (Nord *et al.*, 2008: 6).

The US Action Plan on Food Security (IWG, 1999) outlined ways in which the US planned to address the food security goals of the World Food Summit. The Action Plan focuses on broad issues, including economic security and trade. Most relevant for a consideration of agricultural land is the emphasis on sustainable food systems and environment. That is, one of the key priority strategies outlined in the Action Plan is the 'integration of environmental concerns into food security efforts' to ensure sustainable production. An important element of this strategy is 'developing and implementing flexible, environmentally sensitive agriculture... and land-use policies' (IWG, 1999: 2).

Focusing on domestic food production, the US Action Plan on Food Security insists that one of the critical issues for sustainable food production is 'farmland preservation because once land is developed for urban or industrial use, it rarely returns to agriculture'.[459] To preserve farmland, the US should implement farmland protection policies, monitor conversion of farmland, ensure that government policies do not encourage conversion, and support state and local government incentives for farmland preservation (IWG, 1999: 35).

Under the Action Plan, moreover, use of US agricultural land to produce food should remain sustainable.[460] Environmentally sensitive policies should be implemented 'to conserve soils, protect fragile lands, and protect watersheds' (IWG, 1999: 33). Land retirement, targeted at the most sensitive land, and environmentally sound management practices are elements of the sustainability strategy.

The Action Plan recognises that both the quantity and quality of US land affect food security, in the US and globally. Scientific assessment agrees. In light of the fact that '[f]uture world populations will require ever-increasing food supplies,' and that more than 99% of food calories are produced on land, it becomes clear that '[m]aintaining and augmenting the world food supply basically depends on the productivity and quality of soils' (Pimentel, 2006: 119).

The OECD recognised the growing demand for food and the global importance of sustainable agricultural production:

[459] Another is reduction of nonpoint source pollution of waters (IWG, 1999: 35).

[460] US statutes define sustainable agriculture (7 USC § 3103(19)):

'The term "sustainable agriculture" means an integrated system of plant and animal production practices having a site-specific application that will, over the long-term –

(A) satisfy human food and fiber needs;
(B) enhance environmental quality and the natural resource base upon which the agriculture economy depends;
(C) make the most efficient use of nonrenewable resources and on-farm resources and integrate, where appropriate, natural biological cycles and controls;
(D) sustain the economic viability of farm operations; and
(E) enhance the quality of life for farmers and society as a whole.'

> 'In the next half-century agriculture, worldwide, will be required to double its output if it is to meet the expected increased global demand for food and reduce hunger. The challenge is whether agriculture can efficiently produce the food to meet this growing world demand over time without degrading natural resources – productive soils, unpolluted air, clean and sufficient supplies of water, conserved habitats, biodiversity and landscapes – and do so in ways that are socially acceptable' (OECD, 2004: 11).

A recent US Department of State document on global food security identified barriers to global food production: 'increasing threats from climate change, scarce water supplies, and competition for energy resources' (US Department of State, 2009: 1).

In light of the importance of US food production for domestic and global food supplies, this chapter focuses on agricultural land in the US. After a brief overview of US production and expected global food demand, the chapter turns to programmes that protect both the quantity and the quality of US farmland. It discusses both federal and state programmes designed to protect agricultural land from conversion to non-agricultural uses. The chapter then reviews some of the more important US conservation programmes, both for land retirement and for working lands. Many of the federal laws discussed in this chapter are embedded in federal agricultural legislation, often referred to as Farm Bills. The most recent Farm Bill, the Food, Conservation, and Energy Act of 2008, emphasised conservation of farmland.

2. Agricultural production and global food demand

2.1 US agricultural production and exports

The US is the world's largest producer of many agricultural commodities. In 2005,[461] the US ranked first in world production of corn and soybeans and third in wheat production. The US was the largest producer of cattle, chicken, pig, and turkey meat, as well as other commodities. It ranked in the top three world producers in 44 commodities, including many fruits and vegetables, as well as grains and animal products (FAO, 2006b). The size of the agricultural harvest is evident from the value of agriculture in the US economy. In 2007, the value of US agricultural production was $329.2 billion; the value of crop production was $150.8 billion; livestock, $138.1 billion. Net farm income was $86.8 billion (US Census Bureau, 2009: Table 813).

The US produces a large share of the world's grain commodities. In 2008, for example, the US produced 307 million metric tonnes of corn for grain, 39.1% of

[461] Data from 2005 is the latest available on the FAO website that ranks commodity production by country.

world production. The 81 million metric tonnes of soybeans were 36.8% of world production. Wheat production of 68 million metric tonnes was only 10% of world production (US Census Bureau, 2009, Table 824). In 2007, the US was the world's leading producer (350 million tonnes) and exporter (44 million tonnes) of coarse grains (FAO, 2009a: 68). In 2010, however, the USDA expects US farmers to plant 5.7 million fewer acres (2.3 million ha) of eight major crops (corn, soybeans, wheat, cotton, sorghum, barley, oats, and rice) than in 2008 (Glauber, 2010: 1).

Indeed, many agricultural products leave the US for export to other parts of the world. The value of agricultural exports in 2008 was $115.4 billion, a number that reflects rising commodity prices. Grain and feed ($36.9 billion) and oilseeds and products ($23.7 billion) together accounted for over 52% of total agricultural exports (US Census Bureau 2009, Table 827). Exports of corn and soybeans in 2008[462] (44 and 32.9 million metric tonnes) were 57.8% and 44.5% of world exports. Wheat exports of 26.5 million metric tonnes, though smaller, still accounted for 20.3% of world exports (Table 824). The European Union accounted for 8.7% of 2008 US agricultural exports ($10 billion). Asia received 38.4% of total agricultural exports ($44.4 billion); Central and South America (7.3%) and Mexico (13.9%), together 21.2% ($24.5 billion). Africa received only 5% of the total ($5.8 billion) (Table 828).

Food imports account for a growing proportion of consumption in developing countries. In 1999, commercial imports accounted for 8.2% of all available food in 70 low-income countries; in 2007, commercial imports were 9.5%, with a projected reduction to 8.8% by 2018 (Shapouri *et al.*, 2009: 3). Some regions are especially dependent on food imports; in North Africa, grain imports vary annually, but 'average nearly half of total grain supplies'; in Sub-Saharan Africa, imports are more than 20% of grain supplies. Latin America and the Caribbean also import about half their grain supplies (Shapouri *et al.*, 2009: 15, 18, 20).

Though much of US production is exported through commercial markets, some products are exported in the form of food aid. Food aid does not account for a large proportion of consumption in developing countries, but it contributes to global food security. While in the aggregate food aid does not solve global hunger, it saves lives during food emergencies. One study indicated that in 70 of the poorest countries the amount of consumption supplied by food aid is only 4%, but in various countries and in times of strife, food aid is critical. For example, since 2000, about half of Eritrea's consumption has come from food aid. The US supplied over 60% of worldwide tonnes of food aid in the period from 2000-2002 (Shapouri and Rosen, 2004: 40, 43).[463] In fiscal 2008, the US AID provided

[462] Similarly, in 2007, the US was the world's largest exporter of maize and soybeans and sixth in wheat exports (FAO, 2009b, Exports: Major commodity exporters).
[463] Though US food and foreign aid are large 'in sheer dollar terms,' the US 'lags behind most other developed countries on a per-capita basis for foreign aid' (Bertini *et al.*, 2006: 71).

$2.06 billion in food aid to 26 developing countries. In several countries, the US purchased food from local sources (US AID, 2009).

US food aid has been criticised for its inefficiency and for the practice of monetisation.[464] In addition, critics indicate that some food aid is 'dumped' – that is, sold below the cost of production (Lilliston, 2007b: 7).[465] An influential group of independent agricultural leaders recommended changes in US policies for food aid, including increased local (instead of US) purchase of food for recipient countries and decreased monetisation of food aid (Bertini and Glickman, 2009: 104-106). In a 2009 consultation document, the US Department of State indicated its intent to continue giving in-kind food aid, but also to increase use of local and regional food procurement and food vouchers, 'when they make fiscal and development sense' (US Department of State, 2009: 10).

2.2 Global population growth and food demand

Projected world population growth will affect global demand for food. The world population reached six billion in 1999 (UN, 1999: 3), 6.5 billion by 2005, and 6.8 billion in 2009, with average annual increases of 79 million since 2005 (UN, 2009: 7). Continued population increases are projected, though at a somewhat slower rate. World population is expected to reach 7 billion during 2011; by 2050, world population will grow to 9.1 billion, with annual increases of about 31 million (UN, 2009: 4, 7).

Most of the projected population increases will occur in developing or least developed countries. According to the United Nations:

> 'Most of those additional 2.3 billion people will enlarge the population of developing countries, which is projected to rise from 5.6 billion in 2009 to 7.9 billion in 2050....
> In contrast, the population of the more developed regions is expected to change minimally, passing from 1.23 billion to 1.28 billion, and would have declined to 1.15 billion were it not for the projected net migration from developing to developed countries....'(UN, 2009: 5)

[464] A recent study from the Government Accountability Office found various challenges that reduce the 'amount, timeliness, and quality' of food aid. High costs and other difficulties in delivery cause inefficiencies, and other challenges and resource constraints often interfere with effective alleviation of hunger for vulnerable populations. Monetisation (selling food aid abroad to raise cash for development projects), the report indicated, 'is an inherently inefficient use of resources' (GAO, 2007: 2-6). A 2006 FAO report identified desired reforms of food aid, in part consistent with the GAO recommendations (FAO, 2006c).

[465] Dumping occurs in contexts other than food aid, when exports are sold at prices lower than production costs, thanks in part to export subsidies and export credits authorised by agricultural legislation. Dumping often harms poor farmers, who may be driven off the farm or who suffer from depressed global prices (Lilliston, 2007b: 4; see Murphy *et al.*, 2005).

Many in developing and least developed countries are poor. About 1.25 billion people, most in rural areas, live on less than $1 per day, and 3 billion live on less than $2 per day (Bertini *et al.*, 2006: 70).

The World Bank projects that, even after the recent financial crisis, developing economies will experience improved economic performance in the years leading to 2015. (The year 2015 is significant in the context of the Millennium Development Goal to decrease by half the proportion of people who are undernourished.) Poverty projections show an expected decline in persons living on less than $1.25 or $2 per day, though even by 2015, 2 billion people (one-third of the population in developing countries) may still live on less than $2 per day (World Bank, 2010: 41-42).

Improved economic growth in developing countries triggers increased demand for food. As the FAO noted in 2006, '[g]rowth in per capita incomes will contribute to hunger alleviation by reducing poverty and increasing per capita food demand' (FAO, 2006a: 13).[466] Higher incomes will increase demand for food, because 'overall calories consumed increase with income' (Offutt *et al.*, 2002: 43). Demand focuses first on staple food items (grain), but at higher income levels, diets include higher-value plant products (fruits, vegetables) and livestock products. Economists estimate that developing countries will make up 85% of increased demand for cereals and meat by 2020 (Offutt *et al.*, 2002: 43-44).

An annual increase of 1.2 to 1.3% in global cereal demand was projected for the next several decades, with most of the increased demand from developing countries. Growth in crop production has so far kept up with demand for agricultural products, with some variations globally.[467] Average per capita availability of food has increased, and some analyses suggest that productivity increases can be sustained, making food available to feed the increased global population expected in the next decades (Wiebe, 2003: 9).[468]

Though food production in developing countries has been increasing, cereal production in those countries is not expected to meet demand fully. Indeed, USDA researchers indicated that '[m]any developed and some developing countries are

[466] The relationship between poverty and hunger is complex. As FAO indicates, 'hunger is not only a consequence but also a cause of poverty, and...it compromises the productive potential of individuals, families and entire nations' (FAO, 2006a: 13). Growth in income does not always assuage hunger.
[467] The USDA noted: 'Between 1961 and 1999, FAO's aggregate crop production index grew at an average annual rate of 2.3 percent. Relatively rapid and steady annual increases in crop production were reported in Asia (averaging 3.1 percent) and Latin America (2.7 percent). Crop production generally grew more slowly and with greater variation in Sub-Saharan Africa and the developed regions. Total global cereals production grew about 2.3 percent per year....' (Wiebe, 2003: 8).
[468] Moreover, whether the economic and environmental costs of increased production will be acceptable remains an open question. In light of those costs, some analyses suggest that production will satisfy demand, albeit with 'considerably less excess capacity' (Wiebe, 2003: 9).

close to their maximum scientific and technical potential for growing crops,' so growth of agricultural production may slow (ERS, 2005: 22).[469] Moreover, in 70 developing countries, the widespread food insecurity (related to the economic recession) in 2008 and 2009 is expected to continue over the next decade, affecting more people (Shapouri *et al.*, 2009: 103). Thus, food imports are expected to increase significantly. Even before the recession, authoritative estimates suggested that by 2020, 60% of cereal imports for developing countries would come from the US (Pinstrup-Andersen *et al.*, 1999: 5; Wiebe, 2003: 7).

2.3 Meeting global food demand: agricultural land

The ability of agriculture to meet global food demands depends in large part on two important elements: the availability of agricultural land and the sustainability of that land. Water, too, is a critical element for agriculture, but this chapter does not address important current concerns about the availability of sufficient water and the quality of surface and ground waters.

The FAO published sobering statistics about the availability of land to feed growing world populations: 'Arable land per person is shrinking, from 0.38 hectares in 1970 to 0.23 in 2000, with a projected decline to 0.15 hectares per person by 2050' (FAO, 2001: 1). Some potential farmland could be added to the global farmland base. The FAO observed that unused potential farmland is suitable 'in varying degrees' for rain-fed crop production, but 'only a fraction of this extra land is realistically available for agricultural expansion in the foreseeable future, as much is needed to preserve forest cover and to support infrastructural development. Accessibility and other constraints also stand in the way of any substantial expansion' (FAO, 2002: 4).

Much of the potential cropland is located in Latin America and sub-Saharan Africa; in other regions where food will be needed, little additional land is available for production. Even where land is available, environmental costs, including erosion and loss of biodiversity, may accompany expansion of farmed cropland (Wiebe, 2003: 10). Crop production in developing countries will increase to meet growing food demand, but intensified production will stress an already fragile environment,[470] and projected expansion of arable land (estimated at 12%, or 120 million hectares) is likely be achieved through deforestation (FAO, 2001).

---

[469] The potential role of biotechnology in meeting increased food demand is subject to debate. For some considerations, see Pew (2004).

[470] Intensified agricultural production carries risks in the form of environmental damage from polluted water, salinised lands, soil erosion, and lower fertility. Scientists noted that 'modern agricultural land-use practices may be trading short-term increases in food production for long-term losses in ecosystem services, including many that are important to agriculture' (Foley *et al.*, 2005: 570-71). The short-term increase in material goods (including food) may 'undermine' the ecosystem, 'even on regional and global scales' (Foley *et al.*, 2005: 572).

Yields, of course, are related to the quality of cropland (Wiebe, 2003: 2, 4, 19-27), and human activity that causes degradation of farmland reduces yields. Since 1945, almost 40% of world cropland has suffered degradation; erosion is the main cause, especially in Africa and Latin America, where the majority of cropland has been degraded to some extent (Wiebe, 2003: 15). Irrigated land (20% of irrigated land in developing nations) suffers damage from salinity or excess water (FAO, 2001). Estimates of the impact on productivity vary, but it seems clear that over the years land degradation has resulted in significant losses in productivity in some geographic areas. Losses have sometimes been masked by yield growth due to improved technology and higher use of inputs like fertilisers. In the future, however, projected slowing of yield increases may make yield losses from degradation more alarming (Wiebe, 2003: 15-18).

The impact of 'land degradation...on productivity may affect food security in some areas both through losses in aggregate production...and through losses in income' (Wiebe, 2003: iii). Lack of data make it difficult to quantify the effect of land quality and land degradation on productivity and food security. But both loss of cropland and reduced yields due to degradation can affect food security, particularly in developing countries (Wiebe, 2003: 2, 43-47). 'Given that crop yields are projected to increase more slowly in percentage terms than food demand over the next several decades, even small degradation-induced losses of productivity raise concerns' (Wiebe, 2003: 18).

The discussion above indicates that a relatively small amount of potential farmland is available worldwide and that degradation of agricultural land is widespread and affects global crop production. Both facts suggest that protection of the quantity and the quality of US farmland is desirable, both for Americans and for citizens of other nations. Protection of agricultural land becomes even more prudent in the light of developments that increase demand for crops. For example, ethanol and biodiesel production is expected to affect the availability and the price of grains for food and feed, and the reality of climate change will affect agricultural production and require adaptive measures.

2.4 The claim of renewable fuels on crop production

The US, like the EU, encourages the replacement of fossil fuels with renewable fuels (biofuels).[471] Since 1992, alternative fuels have played a role in US energy policy, and US law governs blending of renewable fuel into motor vehicle fuel to reduce reliance on imported petroleum and to limit emissions of greenhouse gases. The Energy Policy Act of 2005 required the establishment of a renewable

[471] The EU mandates that by 2020, 10% of transportation fuels should be biofuels (Office of the Chief Economist, 2010: 21). A report published in the UK criticises this 10% target, citing the effects of biofuels on global food prices and increased poverty (ActionAid, 2010).

fuels programme, and the Energy Independence and Security Act of 2007 (EISA) increased renewable fuel requirements. 'Renewable fuel' is fuel produced from renewable biomass and used to replace fossil fuel in transportation fuel. EISA defines three subcategories of renewable fuel: advanced biofuel (renewable fuel other than ethanol from cornstarch, with a 'lifecycle GHG emission displacement of 50%'), cellulosic biofuel (60% GHG emission displacement), and biomass-based diesel (50% GHG emission displacement) (Clean Air Act, 42 USC § 7545(o)).

Transportation fuels must contain specific volumes of renewable fuel, in amounts that increase each year. The required volume of renewable fuel increased from 4 billion gallons (15.14 billion litres) in 2006 to 12.96 billion gallons (49.06 billion litres) in 2010 and will reach 36 billion gallons (136.3 billion litres) in 2022, with an increasing component (21 billion gallons (79.5 billion litres) by 2022) of advanced biofuels. The renewable fuel obligation applies to refineries and importers. In February 2010, the Environmental Protection Agency finalised its renewable fuel regulations (40 CFR part 80, subpt M). In a commentary about those regulations, the EPA estimated that by 2022, net farm income would increase significantly; corn exports would decrease by 8% and soy exports by 14%; and the cost of food would increase by $10 per person (US EPA, 2010: 7).

At least two dozen programmes, administered by five federal agencies, govern biofuels, and many offer incentives (e.g. tax credits) or assistance for producers and others (Yacobucci, 2009). For example, a provision of the 2008 Farm Bill established the Biomass Crop Assistance Program, which authorises financial assistance to producers who will supply biomass to a biomass conversion facility.

At present, ethanol production in the US depends on corn as its primary feedstock,[472] and biodiesel relies on soybeans. By 2006, ethanol was the third largest user of US corn, following livestock feed and exports (RFA, 2006: 14). In 2009-2010, ethanol production used almost one-third of US corn, and biodiesel production (about 400 million gallons (1.5 billion litres)) used 13-15% of US soybean oil (Office of the Chief Economist, 2010: 58, 60). But USDA projected that corn-based ethanol production will expand less rapidly in the near future, allowing increased corn exports (albeit a declining world market share) as global meat consumption increases (Office of the Chief Economist, 2010: 58). USDA expects limited growth of soy production and a declining market share as soybean and soy meal exports decline due to South American competition (Office of the Chief Economist, 2010: 60).[473]

[472] The current conversion rate is 2.7 gallons (10.22 litres) of ethanol per bushel, but many plants do better (Baker and Zahniser, 2007: 68). Sorghum is also an ethanol feed stock.

[473] USDA predicts a decline in soybean export market share from 45% in 2009-2010 to 38% in 2019-2020 (Office of the Chief Economist, 2010: 60). The share of Brazilian soy exports is expected to increase from 33% to 45% of world soy trade between 2006 and 2016 (D-G Agri, 2006: 5-6).

Opinions differ about the effect of production of biofuels from crops like corn and soybeans. Though the USDA asserted that biofuel production has affected world food prices only slightly, international organisations disagree (Bertini and Glickman, 2009: 111). The 2007 EISA mandate that the US use 36 billion gallons (136.3 billion litres) of renewable fuel by 2022 may be 'insensitive to market forces [and] can easily be criticised as a threat to global fuel supply' (Bertini and Glickman, 2009: 111).

In fact, some fear that US grain production will be inadequate to serve both existing markets and the increased demand for ethanol production.[474] Experts projected that 'the magnitude of increase in grain-ethanol production is expected to have a large impact on commodity prices, agricultural profitability, and global food security' (CAST, 2006: 1). Others assert that additional corn needed for ethanol production is likely to raise grain prices and reduce US corn exports, making less corn available for global food and feed (Baker and Zahniser, 2007: 69).[475] Moreover, the 'demand for corn acres to produce ethanol,' may decrease land available for soybean production for biodiesel and other purposes (CAST, 2008: 5). Moreover, shifts in agricultural production and the use of marginal cropland to grow feed stocks for biofuels may have environmental consequences related to increased use of nitrogen fertilisers, run-off of nutrients, and increased erosion (Malcolm *et al.*, 2009).

UN-Energy considered the implications of bioenergy for food security and concluded that '[t]o the extent that increased demand for biofuel feedstock diverts supplies of food crops (for example, maize) and diverts land from food crop production, global food prices will increase' (UN-Energy, 2007: 31). Globally, however, bioenergy is influenced by policies that govern energy, environment, agriculture, and trade, and its effect on food security is complex (UN-Energy, 2007: 35).

Representatives of the biofuels industry minimise these concerns. The Renewable Fuels Association insisted that ethanol production from corn has little effect on farmland. In 2007, land used to grow crops for ethanol produced in the US was 0.6% of global cropland, and in 2008 was still less than 1% (RFA, 2009: 20-21). The RFA also noted that ethanol production has little effect on US farmers' ability to feed the world or on the price of food. RFA cited studies that connect only 2-3% of the increase in global food prices to ethanol and noted that food inputs are a small proportion of retail food prices (RFA, 2009: 26-27). The National Corn

[474] Questions about land use change (e.g. cultivation of forest and other virgin land) surround ethanol, and its production also has implications for water supplies and land prices.

[475] Some data suggests, however, that '55 percent of U.S. corn exports are directed to other wealthy countries,' while less than one percent goes to the eleven 'most undernourished countries' (Lilliston, 2007a: 6). Still, a well-known US environmentalist insisted that 'the emerging competition between the 800 million automobile owners who want to maintain their mobility and the world's 2 billion poorest people who want simply to survive will be on center stage' (Brown, 2006).

Growers Association (NCGA) noted that ethanol production also creates co-products, including distillers grain or corn gluten feed and meal.[476] These co-products, particularly distillers dried grains with solubles, are suitable as feed for some (but not all) livestock. Most US corn, used domestically or exported, feeds livestock rather than humans, and 2007 exports of distillers grains were 2.4 million metric tonnes (of 15 million produced) (NCGA, 2008: 3-5).

The real impact of biofuels on food production and international food security remains unclear. The US Government Accountability Office, which considered the effects of increased use of biofuels, concluded that 'ethanol production has raised demand for corn and contributed to higher corn prices. This has had several effects on US agriculture, including an increase in acres planted to corn, a reduction in acres planted to other crops, an increase in crop production on lands that were formerly used for grazing or idled, and an increase in feed costs for livestock' (GAO, 2009: 35). The effects of increased production of biofuels and advanced biofuels, required under federal law, remain uncertain. Though production of new energy crops (cellulosic biomass) may increase demand for land, some can grow on marginal land, and USDA programmes provide incentives for production of alternative crops for biofuels. Moreover, global food demands may help to keep productive farmland in food and feed crops (GAO, 2009: 45-46, 48).

3. Agricultural land in the United States

The United States has a total land area of about 2.3 billion acres (931 million ha). More than 60% of US land – almost 1.4 billion acres (567 million ha) – is privately owned, and 99% of US cropland is privately owned. The federal government owns about 28% of US land (635 million acres (257 million ha), mostly in Alaska and the West). State and local governments own about 9%, and the Bureau of Indian Affairs holds about 2%. (Lubowski *et al.*, 2006b: 35-36).

The USDA's Economic Research Service (ERS) has compiled estimates of major land uses since 1922, with consistent data compiled since 1945. In 2002, the latest year for which data have been compiled, the ERS calculated that about 1.2 billion acres (486 million ha) (almost 52% of total land) was used for all agricultural purposes. This includes cropland, grassland pasture and range, grazed forest land, and land in farm homes, roads, and lanes. In the 48 contiguous states, forest use constituted 29.5% of total land; grassland pasture and range, 30.8%; and cropland, 23.3%. Urban use accounted for only 3% of total US land (Lubowski *et al.*, 2006a,b).

[476] A 56 pound (25.4 kg) bushel of corn yields 18 pounds (8.2 kg) of distillers grain (dry grind ethanol process) or 13.5 pounds (6.1 kg) of corn gluten feed and 2.6 pounds (1.8 kg) of high-protein corn gluten meal, plus corn oil (wet mill process) (NCGA, 2008: 5).

3.1 Farmland loss

Since 1840, the US has taken a regular census of agriculture, at first decennially, and now every five years. Census Bureau data focuses on 'land in farms,' a different measure than that used by ERS.[477] Though ERS and Census Bureau numbers differ, the trend is clear: land used for agriculture has decreased since the mid-twentieth century. According to census data, land in farms in the US has decreased significantly, from about 1,142 million acres (462 million ha) in 1945, to 1,017 million acres (411.6 million ha) in 1974, to 938.3 million acres (380 million ha) in 2002, and to 922.1 million acres (373 million ha) in 2007 (ESA, 1947: 5; NASS, 2004: 6; NASS, 2009: 7, Table 1). As farmland acres decreased, urban land increased from 15 million acres (6.07 million ha) in 1945 to about 60 million acres (24.3 million ha) in 2002 (Lubowski *et al.*, 2006b: 5).[478]

As a component of farmland, cropland has also decreased significantly in recent decades. Cropland includes cropland used for crops (harvested, failed, or fallowed), idle cropland, and cropland pasture.[479] Total cropland in 1949 was 478 million acres (193 million ha); by 2002 the total was only 442 million acres (178.9 million ha), the lowest level since 1945, when consistent major land use data was first compiled.[480] Cropland actually used for crops in 1949 was 383 million acres (155 million ha); in 2002, only 340 million acres (137.6 million ha).[481] Forty million acres (16.2 million ha) were idle in 2002 (Lubowski *et al.*, 2006b: 5, 12).

As these numbers indicate, farmland has been converted to other uses at an alarming rate. Between 1945 and 1992, the loss of farmland was about 4.2 million

[477] 'Land in farms,' as defined by the Census Bureau does not include grazing lands. (Lubowski *et al.*, 2006b: 1-2) In 2004, the US had 205,000 commercial farms (sales above $250,000); these farms, about 10% of total farms, produced about 75% of total output. Commercial farms owned only 29% of total farmland. Rural residence farms (1.4 million, or two-thirds of all farms) owned almost one third of farmland, but produced only 8% of total output. Intermediate farms (528,000) owned 31% of farmland and produced 16% of total farm output (USDA, 2006b).

[478] Census of Agriculture data (after 1945, for most land use) includes all 50 states. USDA analysis often focuses on the 48 contiguous states, because Alaska has little cropland, and Hawaii grows unique crops. The land area in the contiguous 48 states is about 1.9 billion acres (769 million ha).

[479] The relationship between cropland used for crops and idle cropland is influenced by federal acreage-reduction programmes and other factors. Cropland pasture (62 million acres (25.1 million ha) in 1962) varies in quality. Some is rotated between pasture and cropland; other marginal land is not planted (Lubowski *et al.*, 2006b: 13, 16).

[480] Cropland data shows a considerable variation from census to census, rather than a gradual decline, perhaps because clearing, drainage, and irrigation helped to put additional land in production, temporarily offsetting loss of cropland to non-agricultural uses (Tweeten, 1998: 240). Cropland used for crops, for example, increased in 1978 and 1982, but thereafter declined significantly (Lubowski *et al.*, 2006b: 12). Moreover the dynamic nature of land use change means that land may leave the cropland inventory, and other land may then be used for growing crops (Vesterby, 2001: 18).

[481] NASS has somewhat different figures: 440 million acres (178 million ha) in 1974 declining to 434.2 million acres (175.7 million ha) in 2002 (NASS, 2004: 6) and to 406.4 million acres (164.5 million ha) in 2007 (of which 309.6 million acres (125.3 million ha) were harvested) (NASS, 2009: 7).

acres (1.7 million ha) per year (Tweeten, 1998: 240). Based on the Natural Resources Inventory, a regular assessment of US resources, the USDA estimated that between 1992 and 2001, the US suffered an average annual loss of around 2.2 million acres (890,000 ha) of farmland – double the average rate of loss in the previous decade (Nickerson and Barnard, 2006: 213).[482] Between 1992 and 1997, 11.2 million acres (4.5 million ha) were lost; between 1997 and 2002, total cropland was reduced by 14 million acres (5.7 million ha) (Bertini *et al.*, 2006: 49), with further losses by 2007 (NASS, 2009: 7). Thus, farmland losses remain significant.

Conversion of cropland to urban uses is often irreversible and is therefore a cause for concern.[483] In recent years, about 20% of new urban land was converted from cropland (Lubowski *et al.*, 2006a: 6-7). Indeed, loss of cropland is often concentrated in areas with large or growing urban populations (e.g. the north-east and the south-east) (Tweeten, 1998: 240). Farms in the counties most influenced by urbanisation (and thus subject to conversion) produce a majority of the nation's dairy products, fruits, and vegetables (AFT, 2003: 1).[484]

The best cropland, prime farmland that is flat and well drained,[485] is also easiest to develop; moreover, developers can afford the land more easily than farmers (AFT, 2003: 1). Therefore, loss of prime farmland is significant. For example, between 1992 and 1997, when 11.2 million acres (4.5 million ha) of farmland land were converted to non-agricultural uses, 3.2 million acres (1.3 million ha) of that land was prime farmland. That is, each year, about 645,000 acres (261,000 ha) of prime farmland were lost to urban uses (Libby, 2002: 3).

3.2 Effect of farmland loss

As a recent policy report indicated, 'One of the greatest threats to agriculture is the loss of farmland to nonfarm use' (Bertini *et al.*, 2006: 53). Continued loss of productive farmland will affect US agricultural production and the food available for domestic use and for export.

[482] USDA authors note that 'this annual rate represents barely 0.2 percent of the Nation's 1.03 billion acres [417 million ha] of cropland, grassland, pasture, and rangeland, and suggests little threat to the Nation's capacity to produce food and fiber' (Nickerson and Barnard, 2006: 213).

[483] Some economists emphasise the 'option value' of cropland; that is, 'avoiding conversion of cropland to low-value irreversible non-crop uses,' to ensure that it will be available to meet future food and open space needs (Tweeten, 1998: 239).

[484] Urbanisation has significant implications for vegetable production, because about 61% of land used for vegetable production is in metropolitan areas. Vegetables occupy less than 1% of US cropland (Heimlich and Anderson, 2001: 41). Of course, urban encroachment is only one factor in farmland loss, and some research has suggested that lack of economic viability in the farming sector, rather than urban encroachment, could be the main reason for loss of cropland (Tweeten, 1998: 243).

[485] Prime farmland is '[l]and that has the best combination of physical and chemical characteristics for producing food, feed, forage, fiber, and oilseed crops and is also available for these uses' (NRCS, 2007).

The American Farmland Trust, an NGO that focuses on farmland protection, stated that 'American farmland is at risk. It is imperiled by poorly planned development, especially in urban-influenced areas, and by the complex forces driving conversion' (AFT, 2003: 1). The extent of that risk, however, is still being debated, and reactions to the loss of farmland vary.

Because the US produces significantly more food than it consumes, loss of farmland is unlikely to result in national food shortages. Development of farmland, researchers indicate, is 'not considered a threat to the Nation's food production overall' (NRCS, 2001). In the 2001 edition of a USDA publication called Agricultural Resources and Environmental Indicators, a USDA economist noted that '[l]osing farmland to urban uses...does not threaten total cropland or the level of agricultural production, which should be sufficient to meet food and fiber demands' (Vesterby, 2001: 18). In the 2006 edition, however, USDA economists insisted that '[e]xcessive loss of cropland to urban uses could lessen the production of food and fiber and the supply of rural amenities' (Lubowski *et al.*, 2006a: 6). Though short-term food shortages are unlikely, anticipated population growth, especially in developing countries, will make US farmland critical for meeting global demand.

Agricultural land offers far more to society than its ability to produce food and fiber. Beyond food security, justifications exist to preserve productive cropland. As one thoughtful economist noted, 'food security is only one dimension of farmland protection' (Tweeten, 1998: 237). He observed that

> '[p]olicies to preserve farmland ideally recognize the diverse interests in farmland as follows: local populations wish to preserve amenity value of open space, states wish to preserve an economic base, and the national government in cooperation with other countries seeks food security. Often, these interests overlap. A legal framework for public policy that coordinates these interests while allowing markets to work where possible will best serve the public' (Tweeten, 1998: 248).

In addition, farmland itself is multifunctional; it creates rural amenities, including benefits to society from open space, scenic vistas, recreational opportunities, and wildlife habitat. Recent discussions about farmland and the multifunctionality of agriculture in the context of agricultural trade indicate that society values the many amenities offered by agricultural land use, including open space, natural beauty, rural heritage, and various natural resource services (Cardwell *et al.*, 2003).

Urbanisation affects these important values and others, including 'rural lifestyles, watershed protection, local rural economies, unique farmlands, sprawl, infrastructure costs, and others,' which vary by location (Vesterby, 2001: 18). Moreover, conversion of land to non-agricultural uses is 'a critical issue because it can lead to fragmentation of agricultural and forest land; loss of prime farmland, wildlife habitat, and other resources; additional infrastructure costs for communities and regional authorities; and competition for water' (NRCS, 2001). The focus on food production here does not discount these other important reasons for protecting farmland.

4. Programmes to protect agricultural land from conversion

Land use planning and regulation in the US are governed primarily by state and local governments.[486] Local (that is municipal and county) provisions predominate, with recent interest in an enhanced state role, especially in states with significant growth or especially sensitive land.

Though federal statutes – for example, many environmental laws – affect the use of land,[487] those measures do not form a comprehensive system of national land use law (Kayden, 2000: 445-47). In fact, in 1970, Congress considered, but did not pass, a National Land Use Policy Act, which would have encouraged statewide land use plans, provided federal planning grants, and engaged planning officials at all levels in 'a process of consultation and information exchange on matters of land-use' (Kayden, 2000: 448). The law would have increased federal involvement in land use, but did not authorise federal land use planning.

Federal interest in farmland, its ownership, and its uses has a long history. A few federal programmes focus on protection of farmland from conversion to non-agricultural uses, but most federal involvement protects the quality, rather than the quantity, of farmland.

4.1 Federal programmes

In the 1970s, serious concerns about the extent of farmland conversion precipitated the National Agricultural Lands Study, which produced an inventory of US agricultural land, explored the extent of conversion, and studied programmes to preserve farmland (USDA and CEQ, 1981). Many of the farmland protection statutes and programmes discussed later in this chapter trace their development to the late 1970s and early 1980s, during the era of the National Agricultural Lands Study.

[486] A number of factors explain the absence of federal land use planning. The Constitution does not mention land use planning expressly, though it is likely that Congress would have authority to regulate land use. But the size of the US and the overwhelming variety in local conditions would challenge national planners. The importance of private property ownership and a preference for local control help to explain the limited role of the federal government (Kayden, 2000: 450-53).
[487] Federal laws govern the use of federally-owned land, which makes up about 28% of the total land area in the US. Federal land includes, for example, national parks, wilderness areas, forest, grazing land.

Also in the 1970s, questions about the effect of foreign ownership of farmland led to the 1978 passage of the Agricultural Foreign Investment Disclosure Act. This law does not restrict foreign investment. Instead, it requires foreign investors to report their holdings of US agricultural land. Foreign ownership is around 1.6% of privately-owned agricultural land (0.92% of all US land), but much of the foreign-held acreage has been owned by US corporations with foreign investors. Foreign holdings are concentrated in forest and pastureland, with less cropland in foreign ownership (Blevins *et al.*, 2009).

Several federal laws work more directly to protect farmland. For example, the Farmland Protection Policy Act, enacted in 1981, recognises the importance of US farmland and its role in providing food for the US and its export markets. The FPPA is intended to ensure that federal programmes do not contribute to the unnecessary and irreversible conversion of farmland and that they comply with state and local farmland protection policies. Using criteria developed by USDA, federal government agencies must identify the adverse effects (including conversion) of their programmes on farmland and consider alternative actions to avoid or minimise those adverse effects. The FPPA imposes requirements on federal agencies whose projects affect farmland, but does not give citizens the legal right to challenge federal activities that affect farmland adversely.

The FPPA has not been particularly effective, and farmland continues to be consumed by federally supported programmes (e.g. transportation and utility infrastructure). Policy experts suggest that clear USDA enforcement authority is necessary to protect important agricultural lands against encroachments supported by other federal agencies (Bertini *et al.*, 2006: 55).

US agricultural legislation has included provisions intended to protect farm and ranch lands, and the Food, Conservation, and Energy Act of 2008 (the 2008 Farm Bill) continued several programmes. The Farmland Protection Program[488] is a co-operative programme that provides matching funds for purchase of conservation easements on privately-owned productive farmland. Eligible land has prime, unique or other productive soil or harbours historical or archaeological resources. State or local agencies or private farmland protection organisations work with landowners to acquire conservation easements, and the federal government contributes up to 50% of the value of the easement. After sale of the conservation easement, a landowner may not convert the land to non-agricultural use and must implement a conservation plan for any highly erodible cropland.

Between 1996 and 2005, an earlier version of the programme has helped to protect 449,000 acres (182,000 ha), most with permanent easements, on 2,290

[488] A 1996 Farmland Protection Program became the 2002 Farm and Ranch Lands Protection Program, but is now again called Farmland Protection Program.

farms. Payments to producers in 2005 were $112 million (USDA, 2006a: 7). Some commentators believe that the programme 'will help protect the nation's farmland and will slow the pace of farmland loss' (Eitel, 2003: 604), but others believe that limitations of the programme, especially insufficient funding, hamper its effectiveness.

The Grassland Reserve Program (GRP) provides incentives to keep grazing land from being converted to cropland by helping private landowners to restore and conserve vulnerable grassland. Intended for working grazing operations, the GRP will enhance biodiversity and protect grass and shrub land from conversion. To participate in the GRP, landowners enter 10-20 year rental contracts or grant permanent easements to the USDA, state agencies or private conservation organisations. Landowners receive annual rental payments or easement payments plus a share of restoration costs; they may continue to use the land for grazing, but not for crop production.

In the interests of global food security, some would prefer that the federal government take the lead in protecting farmland, because states may lack a compelling incentive to protect land for food security. In the 1990s, the US accounted for about 10% of world food production; an average state provided 2% of US production and only 0.2% of world production. Thus, the average state has a small impact on global food supplies and perhaps little incentive to protect farmland to protect those supplies. This small impact is 'unlikely to motivate the state's citizens to pay the high cost of preserving farmland from development to promote third-world food security.... Hence, preserving farmland to promote world food security requires a national or, better yet, an international effort' (Tweeten, 1998: 248).

More recently, policy experts have recommended renewed federal assistance in protecting farmland. That is, the federal government should 'support and encourage effective land use planning,' in part 'through the provision of funding incentives in support of local land use planning.' Effective land use planning is critical 'to sustain farmland and ranch land, the most fundamental agricultural assets' (Bertini *et al.*, 2006: 53-54).

4.2 State programmes

States have taken an active role in the protection of farmland. As the discussion above suggested, however, the main goal of state efforts to preserve farmland may not be food security. As an economist noted, 'Farmland preservation by any one state has no perceptible impact on international food supplies. Hence, the food security dimension of farmland preservation will only be served by combining that dimension with other objectives of farmland preservation' (Tweeten, 1998: 249). State farmland protection programmes do help to continue production of food and fiber, but they also further other interests, including orderly development, open space, and enhancement of rural economies.

State governments, acting under the police power, have the authority to regulate the use of land within their borders. The police power allows states to regulate to protect public health, safety, and general welfare. Regulation of land, including protection and preservation of agricultural land, is an exercise of the police power. Constitutional protections limit the police power, however, and land use regulation violates the US Constitution when it interferes so much with property rights that it 'takes' property without just compensation.

4.2.1 Zoning

Zoning is a common exercise of the police power. Some states (e.g. Oregon,[489] Vermont, Hawaii, Florida, and others) have enacted statewide land use and growth management plans, usually with local participation. Although these programmes differ in design, long-range policies adopted at the state level normally protect farmland, which may be developed only with approval from the appropriate public body.

More commonly, states have enacted enabling laws that authorise local governments to adopt land use plans and related zoning regulations. Acting under these enabling laws, local governments adopt comprehensive land use plans, implemented through zoning ordinances that apply to land within their boundaries (and sometimes to closely adjacent land) (See Kayden, 2000). Zoning ordinances typically include a text that creates districts with permitted land uses and a map that assign uses to geographic districts.

In the US, zoning is the most common technique for preventing conversion of agricultural lands to non-agricultural uses (Paster, 2004: 292). Though every state has a zoning enabling act and most states authorise zoning of agricultural land, zoning does not protect all farmland. In 1981, 270 counties in the US had zoning to protect agricultural land; by 1995, that number had increased to almost 700 jurisdictions in 24 states (AFT, 1998). Thus, much agricultural land remains vulnerable to development.

Zoning of agricultural land (sometimes called agricultural protection zoning) takes several forms. Exclusive agricultural zoning generally prohibits most non-farm development on land zoned for agriculture. This approach is unpopular with farmers and therefore little used. Non-exclusive agricultural zoning, which limits

[489] Voters in Oregon passed Measure 37, a land-value compensation law (Oregon, 2004). This measure, upheld by the Oregon Supreme Court, triggered challenges against farmland zoning, because it required payments to owners when certain land use restrictions reduced the value of their property. Measure 49 modified Measure 37. It provides relief only for land use regulations that restrict residential uses or agricultural practices and also reduce the value of property (Oregon, 2007).

or discourages non-farm development, is more common. Several approaches have been used.

Area-based zoning ordinances restrict development of farmland by allowing only a limited number of building lots, with the number calculated by reference to the size of the agricultural land parcel. Fixed-area zoning allows a specified number of building lots per number of acres (e.g. one lot per 20 or 50 acres (8 or 20 ha)). Sliding-scale zoning usually allows a proportionately smaller number of lots as the land area increases (e.g. one lot per 5 acres (2 ha), but only four lots for 60 acres (24 ha)); it may permit more dense development on less productive farmland. Development may be clustered so that the remaining land can be farmed more efficiently. This approach can foster development on marginally-productive land, while preserving prime farmland for production (Grossman, 1992: 177; Paster, 2004: 292-293).

Large-lot (or minimum lot size) zoning permits non-farm development – usually a residence – as of right on large lots (e.g. 5 to 640 acres (2 to 259 ha)). Large-lot zoning can keep economically viable parcels in farming, provided that lot sizes are large enough. Small minimum lot sizes (e.g. 5 acres (2 ha)), however, may result in expensive subdivisions and increased urban sprawl. Cluster (or open-space) zoning concentrates development in a single location, leaving the remainder of the land available for agricultural production. Often used for residential subdivisions, cluster zoning may be more suitable for protecting environmentally sensitive land than for large-scale agriculture (Paster, 2004: 293-297).

Some zoning ordinances include provisions for transferable development rights (TDRs), which can help to compensate farmers for declines in market value when land is zoned for agricultural use. A TDR is a legal interest in property – the right to develop. Under conditions set by a zoning ordinance, TDR programmes allow landowners to separate the right to develop land from underlying land ownership and to transfer that right. Development rights originate on a 'sending' parcel and, through a private market, are sold for use on a 'receiving' parcel, often located in a designated zone. The buyer of development rights can build at a density higher than ordinarily permitted by zoning, and a conservation easement restricts the sending land's right to develop. In theory, farmland can be protected permanently. Though TDR programmes exist in at least 100 counties and municipalities, relatively few acres of farmland have been protected; a survey reported only about 130,000 acres (52,600 ha) by January 2008 (AFT, 2008c).

When preceded by careful land use planning, zoning is among the most effective and least expensive ways to protect agricultural land. Because of the perception that farmers bear the cost of agricultural zoning, primarily through lower market value, zoning may be unpopular politically and difficult to enact.[490] Moreover, once enacted, zoning does not offer permanent protection. Land use plans and zoning ordinances are vulnerable to amendment, and variances that weaken protection are sometimes available to avoid individual hardship. Thus, zoning alone is unlikely to preserve farmland for future generations (Grossman, 1992: 178).

4.2.2 Agricultural districts

Agricultural districts or agricultural areas respond to the problem of undisciplined urban development. In exchange for an agreement to keep land in agricultural use for a prescribed period (e.g. 10 years), agricultural districts offer protections to agricultural operations. Laws in at least 16 states, most enacted since 1965, authorise agricultural district programmes.

Though the statutes vary somewhat, they tend to follow a rather similar pattern (See generally Grossman, 1985: 708-15; Illinois, 2008). The following description is based, in the main, on Illinois law. Participation is voluntary, and landowners can propose to create an agricultural district of at least a minimum size (e.g. in Illinois, at least 350 acres (142 ha) or, in urban counties, at least 100 acres (40 ha)). The prescribed public body, often the county board, gives public notice of the proposal, holds a public hearing, and decides whether the viability of active farming and other factors justify formation of the district. If the proposed district lies within the zoning authority of a municipality, the municipality may object. If the proposal is approved, the district is formed for 10 years, with subsequent renewal periods (after review) of 8 years. Landowners may petition to add land to the district, to withdraw land from a district, or to dissolve the entire district. Once an agricultural district has been formed, land in the district can be used only for agricultural production, broadly defined, but land can be bought and sold freely.

Farmers in an agricultural district typically receive several types of benefits in exchange for land use restrictions. For example, local governments cannot enact ordinances that unreasonably restrict or regulate farming practices or farm structures in an agricultural district, unless their purpose is public health and safety. Public agencies cannot burden agricultural district landowners with a disproportionate share of benefit assessments or special *ad valorem* taxes to finance public services. Moreover, state agencies are encouraged to maintain viable farming in agricultural districts. Some statutes restrict the use of agricultural district land for public development projects, but most do not prevent government acquisition of protected farmland through eminent domain. In a few states, agricultural land

[490] Indeed, some have argued that 'downzoning' (a zoning change to a less intensive land use) to protect open space or farmland provides benefits to the public; therefore, the public at large should compensate landowners for the loss of value when regulatory restrictions reduce the value of rural property (e.g. Richardson, 2003).

must be part of an agricultural district to qualify for property tax abatement and right to farm protection from nuisance suits (see below).

Because districts are both voluntary and temporary, they do not prevent the conversion of farmland. Instead, they keep agriculture in place until farmland is needed for urban growth. Thus, they foster orderly and efficient development, rather than permanent protection of farmland.

4.2.3 Development restrictions

Development restrictions (e.g. conservation easements or purchase of development rights) can protect agricultural land by separating the right to develop from the underlying land ownership. They prevent development, while the farmer maintains a property interest and the right to use the land for agricultural purposes. Conservation easements can be sold or donated; development rights are often purchased by a government agency. Private land trusts and other qualified conservation organisations hold conservation easements on agricultural and other valuable land.

Conservation easements are voluntary legal agreements between landowners and a government agency or conservation organisation. Flexible devices, they limit the use of land to protect significant resources on the property. Once agreed, conservation easements are conveyed as a deed restriction, either permanently or for a term of years. Conservation easements on agricultural land keep farmland available for future generations by limiting conversion to non-agricultural uses. They do not require that the land be farmed, but some require that farmers apply conservation measures. Moreover, they do not usually protect farmland from government taking by eminent domain (Grossman, 1985; AFT, 2008a).

Every state authorises voluntary donations of conservation easements on agricultural land. Donation of an easement can trigger tax benefits; its value, of course, must be determined by appraisal. The donation of a permanent easement for a conservation purpose (including preservation of farmland, when pursuant to a government conservation policy) may qualify as a deductible charitable gift under US tax law (26 USC § 170(h)). Estate taxes may be reduced because the remaining land value has been reduced (§ 2031(c)). Local property tax assessments, calculated by reference to land value, may also be reduced.

Since the mid-1970s, a number of state and local laws have authorised the purchase of agricultural conservation easements (PACE). PACE programmes in 23 states (of 27 with authorised programmes) have acquired easements to protect agricultural lands. Sources of funding, including appropriations, bonds, and private contributions, vary among states; a number of states use matching funds authorised by the federal Farmland Protection Program (discussed above). As of May 2009, these state programmes had acquired 11,117 conservation easements or restrictions to protect 1.89 million acres (764,855 ha) of agricultural land (AFT, 2009b). In addition, 19 states have a total of at least 77 local PACE programmes, the majority at the county level. Under these programmes, as of January 2009, 2,746 easements protected about 342,000 acres (138,500 ha) (AFT, 2009a).

PACE programmes seem to be growing in popularity, perhaps because they offer a voluntary means of preserving farmland permanently, as well as the ability to use limited funds to protect farmland that is particularly valuable or under development pressure. Conservation easements on the land most vulnerable to development are expensive, so PACE programmes cannot protect all farmland subject to development pressure (Heimlich and Anderson, 2001: 60-64).

4.2.4 Right-to-farm laws

Right-to-farm laws, enacted in every state since the late 1970s, responded to rapid loss of farmland to non-agricultural uses. They are intended to conserve and protect agricultural land for production of food and other agricultural products. Right-to-farm laws limit the loss of agricultural resources by discouraging nuisance suits against agricultural operations.

Most right-to-farm laws protect farmers from public or private nuisance litigation that results from changes in surrounding land use (e.g. a residential development built near an existing farm). Conditions for protection vary among the states. A typical right-to-farm statute provides that a farm that has operated for at least a year cannot become a nuisance because of changed circumstances, if the farm was not a nuisance when it began operation and if its operation is not negligent. In effect, these statutes legislate a 'coming to the nuisance' defence to nuisance suits. Some laws protect specific agricultural activities, and others ('feedlot statutes') protect livestock operations. Other requirements often apply, and some statutes offer additional protection (e.g. from local regulations that might restrict farming operations). Some statutes discourage nuisance suits by providing that a prevailing farm defendant may recover costs and expenses, including attorney fees, for defending the nuisance case (See generally Grossman and Fischer, 1983; Centner, 2006).

In a few instances, courts have found that right-to-farm laws are unconstitutional takings of the private property rights of neighbours. Under one statute, for example, the court held that a right-to-farm law's restrictions on the right to sue in nuisance created legislative easements on neighbouring property, resulting in an unconstitutional *per se* taking of property interests (Bormann, 1998). In another decision, the court held that a statute protecting animal production from nuisance suits violated provisions of the state constitution (Gacke, 2004). Although these decisions raise questions about right-to-farm laws in other states, especially when those statutes offer expansive immunity to agricultural operations, most state right-to-farm laws continue to apply (See Centner, 2006).

Right-to-farm laws discourage nuisance suits and allow continued production, even when nearby residents may object to activities perceived as nuisances. But they do not actually preserve farmland (Grossman, 1992: 180). Indeed, '[r]ight-to-farm laws are only effective in preventing involuntary conversion against a landowner's wishes; they do little to protect farmland when an owner desires to convert' (Cordes, 2002: 421).

4.2.5 Preferential tax assessment
Preferential (or differential) tax assessment of farmland is a state-law incentive for keeping land in farms. Local real estate taxes, assessed annually on the value of property at its highest and best use, finance government functions, including public schools. Preferential assessment reduces the burden of these significant taxes by assessing land in agricultural use on the basis of its lower value as farmland; they ensure that property taxes more accurately reflect the level of government services provided to agricultural landowners. Though state law governs preferential assessment, local governments implement the programmes and bear the burden of reduced tax revenue (AFT, 2008b: 6).

State preferential assessment laws vary. Some include only farmland; others include forest or open space. Statutes determine the methods of determining eligibility (e.g. a minimum size or level of production) and formulas for assessing value. Some states apply a deferred tax or a rollback penalty if land is converted to a non-qualifying use; others impose no penalty. A few states have laws that allow 'circuit breaker' tax credits to eligible farmers, sometimes connected with zoning or contractual restrictions on land use (AFT, 2008b: 6).

Preferential tax assessment, authorised in 49 states,[491] is popular with farmers; those with land under development pressure (and therefore with the highest market value) benefit most from these programmes. A government study estimated that US farm operators receive a total tax subsidy of $1.1 billion per year through preferential assessment (Heimlich and Anderson, 2001: 60). Preferential assessment helps to maintain farm property taxes at manageable levels and may discourage sale of farmland for development. But despite the high cost, these incentives alone do not keep farmland in agricultural use.

5. Agricultural land quality

Loss of farmland to non-agricultural uses is not the only threat to agricultural productivity. Intensive farming practices may result in degradation of land, which reduces productivity. The OECD has recognised the environmental harm from intensified production.

[491] Michigan authorises circuit breaker tax credits, rather than preferential assessment; participants must enter restrictive agreements to prevent conversion of farmland (AFT, 2008b: 6).

> 'Increasing food demand – together with policies encouraging production – and technological and economic changes have often led to a marked intensification of agriculture (more output per unit of land or labour) and farming on environmentally sensitive land, which in some cases has led to environmental harm. These detrimental effects include mainly water and air pollution, but also the loss of wildlife, habitats and landscape features. Soil degradation and water depletion are also serious concerns in some areas....' (OECD, 2004: 12)

Though land degradation in the US is generally less severe than in some other geographical areas, the protection of productive soils in the US remains critical. As USDA research noted, '[a]s rising populations and incomes increase pressure on land and other resources around the world, agricultural productivity plays an increasingly important role in improving food supplies and food security,' both in the US and abroad (Wiebe, 2003: iii).

The US has long emphasised the conservation of productive soils. Since the 1930s, the federal government has provided financial assistance (cost-sharing) for conservation practices,[492] and the USDA has provided technical conservation assistance to landowners, most recently through the Natural Resources Conservation Service, supplemented by technical service provided by certified non-federal entities. As the USDA noted,

> '[h]istorically, the bulk of conservation program funding focused largely on maintaining the productivity of cropland. Assistance primarily focused on vegetative, engineering, and crop management measures to control soil erosion. Strip cropping, terracing, drainage, crop rotation, contouring, pasture management, tree planting, and other measures became part of farm conservation plans' (USDA, 2006a: 2).

Since 1985, when environmental interest groups played a significant role in drafting federal agricultural legislation, conservation has played an even more important role in US agricultural policy. Current US law includes more than 25 conservation programmes, designed to protect the fertility of farmland (Bertini *et al.*, 2006: 50).[493] The 2008 Farm Bill includes fifteen different conservation programmes, including new programmes and continuation or simplification of existing programmes. Though most are voluntary, significant incentives help to induce participation. Agricultural legislation imposes conservation requirements (cross compliance) for producers who receive farm programme payments and

[492] Some of the material in this section is based on Grossman (2005).

[493] States also have programmes to encourage, and sometimes to require, sustainable farming.

authorises financial incentives (environmental subsidies) for conservation and sustainable land use practices.[494]

5.1 Conservation compliance on highly erodible cropland

Conservation compliance, introduced in 1985 farm legislation, protects the long-term capacity of US farmland to produce food and fiber. Producers who receive federal farm support must implement measures to conserve highly erodible land, and they must also protect wetlands.[495] These measures are voluntary, at least in theory, because farmers need not comply if they choose not to accept federal farm payments. Conservation compliance on cropland is most relevant for food security and is the focus here.

Federal conservation provisions for highly erodible land (HEL)[496] are designed to keep HEL in production, to protect soil productivity, and to reduce off-site damage from erosion.[497] They apply to a producer 'who in any crop year produces an agricultural commodity on a field on which highly erodible land is predominate' (16 USC § 3811(a)). The producer is not eligible for specific government support,

[494] Compliance mechanisms, like regulation, 'prescribe limits on producer actions and provide for penalties (loss of farm program benefits in the case of noncompliance). Like taxes and fees, however, violation does not imply illegal activity and maximum penalties are limited and known in advance.' In contrast, environmental subsidies (available under other federal programmes) confer a benefit for 'taking an action that enhances...environmental performance' (Claassen *et al.*, 2004: 3-4).

[495] Since 1780, nearly half of US wetlands have been drained, primarily for agricultural use; the largest losses occurred since 1885 (Hansen, 2006). The rate of conversion of wetlands has declined since the 1950s, and a number of legal rules protect wetlands. Since 1985, federal agricultural legislation has discouraged conversion of wetlands. Under Wetland Conservation (or so-called 'swampbuster') provisions, producers are ineligible for certain federal farm programme benefits if they have produced commodities on converted wetlands after 1985 or have converted wetlands for crop production after 1990.

About 92 million acres (37 million ha) of wetlands could be subject to swampbuster, but USDA researchers estimate that fewer than 13 million acres (5.3 million ha) are adjacent to existing cropland, and some of these may be too expensive or difficult to convert profitably. Thus, swampbuster protects between 1.5 million acres (607,000 ha) and 3.3 million acres (1.3 million ha) vulnerable to conversion. This relatively small number of acres is worth protecting, especially because USDA efforts (through the Conservation Reserve and Wetlands Reserve) restored fewer than 3 million acres (1.2 million ha) between 1990 and 2003 (Claassen *et al.*, 2004: 22-24).

[496] Highly erodible land has an erodibility index of 8 or more. The erodibility index is defined by the ratio of inherent erodibility (tons/acre/year on clean-tilled land) to the soil loss tolerance (tons/acre/year without long-term loss of productivity) (Claassen *et al.*, 2004: 2). If a field has predominately highly erodible soils, conservation measures must be applied to the whole field (Claassen *et al.*, 2004: 14). A field is predominately HEL if one third or more of the total area or 50 or more acres (20.2 ha) are highly erodible (7 CFR 12.22).

[497] Soil erosion is a 'slow insidious process' often induced by human activities that expose soil to rain or wind. It diminishes the quality of soil and reduces the productivity of agricultural land (Pimental, 2006: 119, 131).

loans, and conservation programmes[498] unless he or she applies a conservation plan using approved conservation systems. Violations of the HEL requirements normally result in loss of federal farm programme payments for all commodities and on all land farmed by the producer. In case of violations in 'good faith and without an intent to violate' (16 USC § 3812(f)), programme benefits may be reduced, instead of denied.

Conservation systems required for HEL must be technically and economically feasible, based on local conditions, cost-effective, and must not cause undue economic hardship to producers. Conservation requirements are flexible, and many approved systems are inexpensive. Most common practices are conservation tillage (which leaves at least 30% of residue on the soil surface at planting time), seasonal crop residue management (crop residue remains on the surface; land is tilled before planting), and conservation cropping (less profitable crops or cover crops for part of the season) (Claassen *et al.*, 2004: 19-21). Special rules apply to HEL farmed before 1985 and native vegetation converted to cropland after 1996 (7 CFR 12.23).

About 98% of farms with highly erodible cropland applied approved conservation systems on their land (Claassen *et al.*, 2004: 11, 13). USDA researchers estimated that on farms receiving government payments, 'an erosion reduction of 294.6 million tonnes per year' could be attributed to conservation compliance, that is, about 25% of all reductions in erosion between 1982 and 1997 (Claassen *et al.*, 2004: 19). Non-market benefits to society from conservation compliance, including better soil productivity and improved quality of surface water, are estimated to be $1.4 billion per year (Claassen *et al.*, 2004: 17).

5.2 Conservation incentives

US agricultural legislation has long offered financial and technical incentives for voluntary conservation measures, and a number of important new conservation programmes have been enacted since 1985. Under current law, these programmes include both land retirement measures and programmes for 'working land.'[499] USDA statistics, using 2007 data, indicate that farms with income less than $50,000, rather than large commercial operations, received about 55% of payments ($3 billion

[498] Conservation compliance applies to payments under a number of federal farm programmes: direct payments under commodity programmes, disaster programmes (including market loss assistance), important conservation programmes, and most federal government loan programmes (16 USC § 3811(a); Claassen *et al.*, 2004: 8-10),

[499] Under WTO measures, domestic support provided by US agri-environmental policies for land retirement and working land has at most a minimal impact on production and trade. These programmes are generally considered exempt from limitation as 'green box' policies under the Uruguay Round Agreement on Agriculture. For more information on trade aspects of domestic support, including multifunctionality and non-trade concerns, see chapters by Grossman (chapters 2, 5) and Cardwell (chapter 6) in Cardwell *et al.* (2003).

total) under major conservation programmes (Conservation Reserve, Wetlands Reserve, and Environmental Quality Incentives Program), and almost 19% of total government payments (ERS, 2009).

5.2.1 Land retirement

Retirement of environmentally sensitive cropland is an important component of US conservation policy. Much of that idle cropland has been entered into the Conservation Reserve Program (CRP), enacted in 1985 and reauthorised through 2012.[500] From the outset, both environmentalists and advocates of supply control supported the CRP, which provides environmental benefits and reduces commodity production.

The CRP is voluntary, and maximum authorised enrolment is 32 million acres (12.9 million ha) through 2012 (reduced from 39.2 million acres (15.9 million ha) through 2009). Eligible land includes highly erodible cropland, marginal pasture land, and specific other land that will provide environmental benefits. Recent enrolments have focused on high-priority field practices (e.g. filter strips), rather than whole fields (UDSA, 2006a: 17).

When USDA announces a sign-up period, interested persons may submit bids to enrol their land.[501] Enrolled land must balance equitably the conservation purposes of soil erosion control, water quality, and wildlife habitat. Using a formula called the Environmental Benefits Index, the Farm Service Agency (part of USDA) ranks offers for contracts and selects land providing maximum environmental benefits for the CRP.

---

[500] Of the 40 million acres (16.2 million ha) of farmland idled in 2002, 85% was part of the CRP, and one million acres (404,686 ha) were enrolled in the related Wetlands Reserve Program (Lubowski *et al.*, 2006b: 14).
The Wetlands Reserve Program was enacted in 1990 to protect farmed or converted wetlands through permanent easements, 30-year easements, restoration cost-share agreements, or a combination of these options. Priority for enrolment goes to wetlands that maximise wildlife values and have the least likelihood of re-conversion at the end of the easement period. Owners of WRP land must grant an easement to the United States, implement a conservation plan, and record a deed restriction under state law. In exchange, they receive compensation for the easement plus cost-share payments for conservation measures and technical assistance. Because the programme is small, with an enrolment of 2 million acres (0.8 million ha) in 2008 and a maximum enrolment of about 3 million acres (1.2 million ha), it has little impact on crop production.

[501] Continuous sign-up applies to small areas of land to be set aside for purposes like filter strips, riparian buffers, windbreaks, etc. About 4 million acres (1.62 million ha) (of which 3.29 million acres (1.6 million ha) were under contract in January 2010) are reserved for this programme, which yields higher rental payments. In January 2007, about 20% of funding supported 'buffer practices associated with working land' (e.g. filter strips, riparian buffers) on 10% of CRP land (Claassen, 2007: 31).
The Conservation Reserve Enhancement Program, with co-operation from state and local entities, identifies and retires environmentally sensitive land with high priority for the state or nation. A farmable wetland programme, expanded in 2008, may enrol up to 1 million acres (404,686 ha); about 209,000 acres (85,000 ha) were enrolled by 31 January 2010 (Sullivan *et al.*, 2004: 5, 8; FSA, 2010).

Under the CRP, agricultural owners, operators, or tenants enter into 10 to 15 year contracts with the federal government. In exchange for annual rental payments, they agree to retire eligible land from agricultural use, to implement an approved conservation plan, and to plant an approved vegetative cover on the CRP land. The federal government pays 50% of the cost of establishing conservation practices required by contract.

As of January 2010, total CRP enrolment was 31.19 million acres (12.62 million ha), with 733,248 contracts on 411,388 farms and annual average rental payments of $53 per acre ($45 per acre for most land, enrolled in the general programme) (FSA, 2010). The USDA expected to pay about $1.7 billion in CRP payments during fiscal 2010, plus another $169 million for cost-share and other payments (FSA 2010; USDA, 2009).[502] The USDA held a general sign-up period in August 2010.

The CRP has helped to reduce erosion by taking marginal cropland out of production[503] and establishing permanent vegetative cover. Non-market benefits from reduced erosion through the CRP were estimated at $694 million per year (Claassen *et al.*, 2004: 16-17). The 2008 Farm Bill reduced the size of the Conservation Reserve; moreover, between 2010 and 2013, CRP contracts on about 18.7 million acres (7.6 million ha) will expire (FSA, 2010). The USDA may re-enrol some of the most environmentally sensitive land, but when CRP contracts are not renewed, land can be returned to production. Much of the land currently enrolled in the CRP could be farmed, although some is marginal, and important environmental benefits and habitat would be lost.[504] Demand for corn for ethanol feed stock may tempt farmers whose contracts expire to take their land out of the CRP.

Some CRP land offers an important resource for global food security, but not all CRP land is ideal for crop production. In 2004, a USDA administrator indicated that world crop supply and demand considerations are factors in government decisions to accept cropland into the CRP (Committee on Agriculture, House of Representatives, 2004: 11-14, 66). The 2008 Farm Bill's reduction in CRP acreage may reflect concern for food production.

[502] In 2005, farm operators and landowners received CRP payments of $1.8 billion, almost two thirds of total conservation payments (USDA, 2006a: 6, 8).
[503] Some have raised concerns that taking millions of acres out of production would affect nearby communities adversely. A USDA study, however, concluded that 'the adverse economic impacts of the CRP are generally small and fade over time.' Though some farm-related businesses suffered, other businesses expanded, and expenditures on recreation increased in many rural areas (Sullivan *et al.*, 2004: iv).
[504] Highly erodible cropland coming out of the CRP must meet the standard applied to other HEL in the area, and producers have 2 years to install structural measures to comply (7 CFR 12.23).

5.2.2 Working land

Land retirement has accounted for a large majority (85%) of conservation payments since enactment of the CRP in 1985. Beginning in 2002, however, US conservation policy has focused on 'working land' – land in agricultural production. The 2002 Farm Bill allocated 40% of conservation spending for working land, with reduced emphasis on land retirement (Bertini *et al.*, 2006: 52). Many programmes under the 2008 Farm Bill also focus on working land. The 2002 and 2008 Farm Bills authorised higher spending for conservation. Congress has not always appropriated the full amount authorised for conservation programmes, and budget reductions to agriculture have affected conservation programmes disproportionately (Cattaneo *et al.*, 2005).

a. Environmental Quality Incentives Program

The largest USDA conservation programme for working lands is the Environmental Quality Incentives Program (EQIP), created in 1996. The 2008 Farm Bill extended EQIP through 2012 and increased authorised funding each year, reaching $1.75 billion in fiscal 2012. In fiscal 2008, USDA entered 48,116 EQIP contracts that obligated $943 million (NRCS, 2010a).[505]

EQIP is intended 'to promote agricultural production, forest management and environmental quality as compatible goals, and to optimise environmental benefits' (16 USC § 3839aa). Among other purposes, the programme helps producers to comply with regulatory requirements concerning soil, water and air quality, wildlife habitat, and surface and groundwater conservation.[506] The programme is also intended to avoid the need for regulation by helping producers to meet environmental quality criteria, to provide assistance to install and maintain conservation practices, and to help streamline conservation planning and regulatory compliance.

To carry out these intentions, EQIP authorises contracts, lasting from one to 10 years, with producers who agree to implement eligible environmental and conservation practices. 'Practice' is defined to include, among other things, site-specific structural practices, land management practices, and comprehensive nutrient management planning practices. Environmental practices relating to livestock production receive 60% of EQIP funding. Producers may receive payments for income foregone and cost-share payments, as well as technical assistance. Cost-share payments are 75% (90% for beginning or limited resource farmers) of the cost of conservation practices. Participants must submit and implement

[505] In 2005, 63,800 producer contracts on 94.5 million acres (38.2 million ha) resulted in total payments of $444 million (USDA 2006a: 6). Researchers questioned the allocation of EQIP funds to states and suggested that EQIP funds may not be going to states that suffer the most serious environmental effects from agricultural production (GAO, 2006).

[506] By providing support to help producers to meet environmental regulatory requirements, EQIP may not implement the 'polluter pays' principle (See generally Grossman (2006)).

a conservation plan that describes practices and environmental purposes and comply with other requirements. Payments for most EQIP participants are limited to $300,000 in any 6-year period for all contracts entered during fiscal years 2009 through 2012.

b. Conservation Stewardship Program
The 2002 Farm Bill introduced the Conservation Security Program, which paid producers on working land to adopt or maintain[507] conservation practices that help to protect or improve the quality of soil, water, air, energy, plant and animal life, and for other conservation purposes (Even, 2005: 197-204). Three tiers of conservation contracts imposed increasingly stringent requirements, and practices eligible for compensation had to exceed minimum conservation compliance (highly erodible land and wetlands protection) required for receipt of federal farm programme payments.[508] Though the programme was intended to be an entitlement programme, budgetary considerations meant that it was implemented gradually, with eligible watersheds identified each year.

The 2008 Farm Bill enacted a substitute programme, the Conservation Stewardship Program, designed to encourage farmers to improve and maintain current conservation activities and undertake additional conservation practices. Implemented by NRCS, the CSP is a voluntary 'payment for performance' programme open to all producers who meet eligibility requirements. Programme acres are allocated by state, and over 12 million acres (4.86 million ha) may be enrolled annually through 2017.[509]

Annual payments reward new and existing conservation activities. NRCS has identified more than 100 conservation activities that qualify for payment. Applicants use a Conservation Measurement Tool (CMT) to determine whether they meet minimum requirements for eligibility and to calculate the conservation effects of their existing and proposed conservation activities. The CMT, which calculates

[507] Research suggests that payments to those who already have a good stewardship record leave fewer resources to encourage new conservation practices (Cattaneo *et al.*, 2005). Land enrolled in the CRP, WRP, and Grasslands Reserve Program and some newly cultivated land was not eligible for CSP (16 USC § 3838a).

[508] This limitation is analogous to the approach to agri-environment payments under Council Regulation No. 1257/1999, as amended. 'Agri-environmental...commitments shall involve more than the application of usual good farming practice.... They shall provide for services which are not provided for by other support measures, such as market support or compensatory allowances' (art. 23(2)). Effective 1 Jan. 2007, this provision was replaced by art. 39 of Council Regulation No. 1698/2005: 'Agri-environment payments cover only those commitments going beyond the relevant mandatory standards [in Regulation No. 1782/2003] as well as minimum requirements for fertiliser and plant protection product use and other relevant mandatory requirements established by national legislation....' (art. 39(3)).

[509] Expenditures under CSP should average $18 per acre nationally, including all payments to farmers, technical assistance, and expenses of enrolment and participation. CSP excludes payments for animal waste storage or treatment and cost-free conservation activities. Resource-conserving crop rotations may result in additional payments (16 USC § 3838g).

conservation performance points, helps to rank applicants for participation (NRCS, 2010b). Eligible producers who are selected for participation enter a renewable 5-year stewardship contract, which requires implementation of a conservation stewardship plan for the whole operation. Participants' payments, based on their conservation performance points, are unique to their operation and reflect environmental benefits from the practices they carry out.[510]

c. Other conservation programmes

Other federal conservation programmes that offer financial assistance have more narrow goals or apply only in limited regions. Some focus on rural natural resources – like forests and wildlife habitat – instead of on cropland. These programmes are also voluntary and offer incentives for participation. They may provide conservation opportunities for landowners who do not qualify for other USDA programmes.

For example, the Wildlife Habitat Incentives Program (WHIP) protects habitats and species, especially those of national or regional significance, on private agricultural and forest land. States identify priorities that guide enrolment of land, and selected landowners enter WHIP agreements for 5 to 15 years. Under these contracts, landowners establish and improve wildlife habitat with the help of USDA technical assistance and cost-share payments. In fiscal 2009, NRCS enrolled more than 812,000 acres (328,600 ha) under 3,700 contracts.

Agricultural Management Assistance provides cost-share payments to improve conservation management and to mitigate risk through diversification or resource conservation. It applies in only 16 states with historically low participation in Federal Crop Insurance Program. Another programme, Soil and Water Conservation Assistance, provides cost share and incentive payments to eligible producers who prepare a conservation plan and enter a cost-share agreement. The programme is available nationally, but not all land qualifies.

5.2.3 Observations

Despite the many conservation programmes that focus on land retirement and conservation on working lands, improvement in US conservation efforts is possible. A recent policy report indicated that 'while current conservation programmes... have enhanced land productivity and protected some of the most environmentally sensitive land, the programmes are fragmented and lack a broader vision. They are ill equipped to address the wide range of agriculture-related environmental challenges we face' (Bertini *et al.*, 2006: 50). The 2008 Farm Bill helped to improve

[510] Under the payment for performance aspect of the CSP, annual land use payments are calculated by a formula: Annual Payment = (Land Use Acres) x (Existing + New Activity Performance Points) x (Land Use Payment Rate) (NRCS, 2010b). Payments may not exceed $40,00 per year or $200,000 over 5 years.

US conservation programmes to encourage and reward stewardship of valuable agricultural land.

6. Conclusion

More than half of the total land area of the US is devoted to agricultural use, and cropland constitutes 23.3% of land in the 48 contiguous states (and 19.5% in the 50 states). Production from that land is abundant, and US crops contribute significantly to the global food supply. The continued availability of US farmland and the conservation of its productive capacity are critical, both for the US itself and for other nations.

Though 'impending food scarcity' does not demand protection of every acre of farmland, 'many...support state and local farmland protection efforts with the intuitive sense that arable cropland has limits and the prudent course is to...encourage farmers to continue farming' (Libby, 2002: 8). 'Evidence of food abundance may be overwhelming,' noted a US expert on farmland, 'but people still have the nagging feeling that profligate abuse of farmland today will return to haunt us in the future. That perception is genuine and will affect farmland policy in the U.S. and elsewhere' (Libby, 2000: 2).

Moreover, protecting the most productive farmland has assumed increasing importance. US land retirement programmes, like the CRP, have generally taken the more fragile cropland out of production, leading to improved environmental quality (Claassen, 2007: 30). But if loss of higher quality farmland and global food demands eventually require production on those fragile lands, environmental quality will suffer. Much retired land is by nature less suitable for crops, and putting the land back into production may lead to land degradation and environmental impairment.

The ability to meet the increasing global demand for food will depend, too, on the 'quality of productive resources and on the market incentives, policy measures, and research investments that influence how those resources are used' (Wiebe, 2003: 9). As an environmental scientist observed, in an era when the world population is increasing, but per capita production of cereal grains has decreased and annual increases in yields of those grains are declining, the conservation of soil is becoming ever more critical (Pimental, 2006: 132).

To some extent, protecting farmland 'is really about risk preference' (Libby, 2002: 8). That is, a thoughtful policy now can ensure that future generations, both in the US and globally, will have food security. Moreover, the 'consequence of underestimating the amount of farmland needed to feed future generations would likely be greater than the consequence of saving "too much" farmland' (Libby, 2002: 8). The American Farmland Trust agrees: 'With a rapidly increasing world population and expanding global markets, saving American farmland is a prudent investment in world food supply and economic opportunity' (AFT, 2003: 1). The president of the American Farmland Trust recently insisted that 'food security is national security' (Scholl, 2009). Indeed, protecting US farmland aids global food security and might be considered a sensible application of the precautionary principle (See Libby, 2002: 8).

Thoughtful agricultural leaders recommended this wise approach for US land policy:

> 'Several principles should be enshrined in land use policy, beginning with compelling national need. It should be recognized that sustainability and competitiveness are dependent on an adequate supply of high-quality productive land. Policy should encourage landowners to keep the best land in agricultural production rather than having it developed or idled. It should also encourage the most efficient use of resources in environmental programs' (Bertini *et al.*, 2006: 54).

US agricultural policy should continue to protect farmland, especially prime farmland, and to encourage conservation practices that will ensure the sustained productivity of that farmland. In so doing, the United States can continue to share its abundant harvest of food with other nations.

References

ActionAid UK, 2010. Meals per gallon: The impact of industrial biofuels on people and global hunger. London.

AFT (American Farmland Trust), FIC (Farmland Information Center), 2009a. Fact Sheet: Status of Local Pace Programs. Available at: http://www.farmlandinfo.org.

AFT, FIC, 2009b. Fact Sheet: Status of State Pace Programs.

AFT, FIC, 2008a. Fact Sheet: Agricultural Conservation Easements.

AFT, FIC, 2008b. Fact Sheet: The Farmland Protection Toolbox.

AFT, FIC, 2008c. Fact Sheet: Transfer of Development Rights.

AFT, FIC, 2003. Fact Sheet: Why Save Farmland?

AFT, FIC, 1998. Fact Sheet: Agricultural Protection Zoning.

Ash, C., Jasny, B.R., Malakoff, D.A. and Sugden, A.M., 2010. Feeding the Future. Science 327: 797.

Baker, A. and Zahniser, S., 2007. Ethanol Reshapes the Corn Market. Amber Waves 5 (Special Issue): 66-71, updated from Apr. 2006.

Bertini, C. and Glickman, D. (co-chairs), 2009. Renewing American Leadership in the Fight Against Global Hunger and Poverty. Report Issued by an Independent Leaders Group on Global Agricultural Development, Chicago Council on Global Affairs. Chicago, IL, USA.

Bertini, C., Schumacher, G. and Thompson, R.L. (co-chairs), 2006. Modernizing America's Food and Farm Policy: Vision for a New Direction. Report of the Agriculture Task Force, Chicago Council on Global Affairs. Chicago, IL, USA.

Blevins, P.A., Johnson, L. and Smith, D., 2009. Foreign Holdings of U.S. Agricultural Land Through February 29, 2008. Farm Service Agency, USDA, Washington, DC, USA.

Brown, L.R., 2006. Exploding U.S. Grain Demand for Automotive Fuel Threatens World Food Security and Political Instability. Earth Policy Institute, 3 Nov. 2006. Available at: http://www.earth-policy.org/Updates/2006/Update60.htm.

Cardwell, M.N., Grossman, M.R. and Rodgers, C.P. (eds.), 2003. Agriculture and International Trade: Law, Policy and the WTO. CABI Publishing, Oxon, UK.

Cattaneo, A., Claassen, R., Johansson, R. and Weinberg, M., 2005. Flexible Conservation Measures on Working Land: What Challenges Lie Ahead? ERS, USDA, ERR No. 5. Washington, DC, USA.

Centner, T.J., 2006. Governments and Unconstitutional Takings: When Do Right-to-Farm Laws Go Too Far? Boston College Environmental Affairs Law Review 33: 87-148.

Claassen, R., 2007. Emphasis Shifts in U.S. Conservation Policy. Amber Waves 5 (Special Issue):29-34, updated from July 2006.

Claassen, R., Breneman, V., Bucholtz, S., Cattaneo, A., Johansson, R. and Morehart, M., 2004. Environmental Compliance in U.S. Agricultural Policy: Past Performance and Future Potential. ERS, USDA, AER No. 832. Washington, DC, USA.

Committee on Agriculture, House of Representatives, 2004. Hearing to Review the Implementation of the Conservation Title of the Farm Security and Rural Investment Act of 2002, 108th Congress, 2d Session (15 June), Serial No. 108-32. Washington, DC, USA.

Cordes, M.W., 2002. Agricultural Zoning: Impacts and Future Directions. Northern Illinois University Law Review 22: 419-457.

CAST (Council for Agricultural Science and Technology), 2008. Conversion of Agriculture and Energy: Considerations in Biodiesel Production. Cast Commentary QTA2008-2. Ames, IA, USA.

CAST, 2006. Convergence of Agriculture and Energy: Implications for Research and Policy. Cast Commentary QTA 2006-3. Ames, IA, USA.

Directorate-General for Agricultural and Rural Development, European Commission. 2006. Brazil's Agriculture: A Survey. Monitoring Agri-trade Policy Rep. No. 02-06.

ESA (Economics and Statistics Administration), US Department of Commerce, 1947. 1945 Census of Agriculture. Washington, DC, USA.

ERS (Economic Research Service), USDA (United States Department of Agriculture), 2009. Farm Income and Costs: Farms Receiving Government Payments. Available at: http://www.ers.usda.gov/Briefing/FarmIncome/govtpaybyfarmtype.htm (updated 6 Feb. 2009).

ERS, USDA, 2005. Food Security Assessment. GFA-16. Washington, DC, USA.

Eitel, M.R., 2003. The Farm and Ranch Lands Protection Program: An Analysis of the Federal Policy on United States Farmland Loss. Drake Journal of Agricultural Law 8: 591-630.

Even, W.J., 2005. Green Payments: The Next Generation of U.S. Farm Programs? Drake Journal of Agricultural Law 10: 173-204.

FAO (Food and Agriculture Organization of the United Nations), 2010. The State of Food and Agriculture 2009. Rome, Italy.

FAO, 2009a. Summary of World Food & Agricultural Statistics. Rome, Italy.

FAO, 2009b. FAOStat. Key Statistics of Food and Agriculture External Trade. Available at: http://faostat.fao.org/site/342/default/aspx.

FAO, 2006a. The State of Food Insecurity in the World 2006. Rome, Italy.

FAO, 2006b. Statistics Division. Major Food and Agricultural Commodities and Producers. Available at: http://www.fao.org/es/ess/top/country.html.

FAO, 2006c. The State of Food and Agriculture 2006. FAO Agriculture Series No. 37. Rome, Italy.

FAO, 2002. World Agriculture: towards 2015/2030. Summary report. Rome, Italy.

FAO, 2001. Food Security and the Environment. Available at: http://www.fao.org. worldfoodsummit/english/fsheets/environment.pdf.

Foley J.A., Defries, R., Asner, G.P., Barford, C., Bonan, G., Carpenter, S.R., Chapin, F.S., Coe, M.T., Daily, G.C., Gibbs, H.K., Helkowski, J.H., Holloway, T., Howard, E.A., Kucharik, C.J., Monfreda, C., Patz, J.A., Prentice, I.C., Ramankutty, N. and Snyder P.K., 2005. Global Consequences of Land Use. Science 309: 570-574.

FSA (Farm Service Agency), USDA, 2010. Conservation Reserve Program Monthly Summary. January 2010. Available at: http://www.fsa.usda.gov/.

Glauber, J.W., 2010. Prospects for the U.S. Farm Economy in 2010. USDA 2010 Outlook Forum (18 February 2010). Available at: http://www.usda.gov/oce/forum/2010_Speeches/Speeches/GlauberJ.pdf.

GAO (Government Accountability Office), 2009. Biofuels: Potential Effects and Challenges of Required Increases in Production and Use (GAO-09-446). Washington, DC, USA.

GAO, 2007. Foreign Assistance: Various Challenges Impede the Efficiency and Effectiveness of U.S. Food Aid (GAO-07-560). Washington DC, 2007.

GAO, 2006. Agricultural Conservation: USDA Should Improve Its Process for Allocating Funds to States for the Environmental Quality Incentives Program (GAO-06-969). Washington, DC, USA.

Grossman, M.R., 2006. Agriculture and the Polluter Pays Principle: An Introduction. Oklahoma Law Review 59: 1-51.

Grossman, M.R., 2005. Commodity Programs, Cross Compliance and Conservation in United States Agricultural Legislation. Rivista di Diritto Agroalimentare e dell' Ambiente 2005: 39-59.

Grossman, M.R., 1992. Agricultural Land-use Law in the United States. In: Grossman, M.R. and Brussaard, W. (eds.). Agrarian Land Law in the Western World. Oxon, UK, pp. 171-195.

Grossman, M.R., 1985. Exercising Eminent Domain Against Protected Agricultural Lands: Taking a Second Look. Villanova Law Review 30: 701-766.

Grossman, M.R. and Fischer, T.G., 1983. Protecting the Right to Farm: Statutory Limits on Nuisance Actions Against the Farmer. Wisconsin Law Review 1983: 95-165.

Hansen, L., 2006. Wetland Status and Trends. In: Wiebe, K. and Gollehon, N. (eds.). Agricultural Resources and Environmental Indicators, chapter 2.3. ERS, USDA, EIB No. 16. Washington, DC, USA.

Heimlich, R.E. and Anderson, W.D., 2001. Development at the Urban Fringe and Beyond: Impacts on Agriculture and Rural Land. ERS, USDA, AER No. 803. Washington DC, USA.

IWG (Interagency Working Group on Food Security) and FSAC (Food Security Advisory Committee), 1999. U.S. Action Plan on Food Security: Solutions to Hunger. Washington, DC, USA.

Kayden, J.S., 2000. National Land-Use Planning in America: Something Whose Time Has Never Come. Washington University Journal of Law and Policy 3: 445-472.

Libby, L.W., 2002. Rural Land Use Problems and Policy Options: Overview from a U.S. Perspective. Paper prepared for Research Workshop on Land Use Problems and Conflicts, Orlando, 21-22 Feb. 2002.

Libby, L.W., 2000. Farmland as a Multi-Service Resource: Policy Trends and International Comparisons. Paper prepared for International Symposium on Agricultural Policies Under the New Round of WTO Agricultural Negotiations, Taipei, Taiwan, 5-8 Dec. 2002.

Lilliston, B. (ed.), 2007a. A Fair Farm Bill for Renewable Energy. Institute for Agriculture and Trade Policy, Minneapolis, MN, USA.

Lilliston, B. (ed.), 2007b. A Fair Farm Bill for the World's Hungry. Institute for Agriculture and Trade Policy. Minneapolis, MN, USA.

Lubowski, R.N., Vesterby, M. and Bucholtz, S., 2006a. Land Use. In: Wiebe, K. and Gollehon, N. (eds.), Agricultural Resources and Environmental Indicators, 2006 Edition, chapter 1.1. ERS, USDA, EIB No. 16. Washington, DC, USA.

Lubowski, R.N., Vesterby, M., Bucholtz, S., Baez, A. and Roberts, M.J., 2006b. Major Uses of Land in The United States. ERS, USDA, EIB No. 14. Washington, DC, USA.

Malcolm, S.A., Aillery, M. and Weinberg, M., 2009. Ethanol and a Changing Agricultural Landscape. ERS, USDA, ERR No. 86. Washington, DC, USA.

Murphy, S., Lilliston, B. and Lake, M.B., 2005. WTO Agreement on Agriculture: A Decade of Dumping. Institute for Agriculture and Trade Policy, Minneapolis, MN, USA.

NASS (National Agricultural Statistics Service), USDA, 2009. 2007 Census of Agriculture vol. 1. Washington, DC, USA.

NASS, USDA, 2004. 2002 Census of Agriculture. Washington, DC, USA.

NCGA (National Corn Growers Association). 2008. U.S. Corn Growers: Producing Food & Fuel. Available at: http://ncga.com/files/pdf/FoodandFuelPaper10-08.pdf.

NRCS (Natural Resources Conservation Service), USDA, 2010a. Environmental Quality Incentives Program. Available at: http://www.nrcs.usda.gov/programs/eqip/index.html#prog.

NRCS, USDA, 2010b. Conservation Stewardship Program. Available at: http://www.nrcs.usda.gov/programs/new_csp/csp.html.

NRCS, USDA, 2007. 2007 National Resources Inventory, Glossary of Key Terms. Available at: http://www.urcs.usda.gov/technical/NRI/2007/glossary.html.

NRCS, USDA, 2001. National Resources Inventory: Highlights. Available at: http://www.nrcs.usda.gov/technical/NRI/pubs/97highlights.pdf.

Nickerson, C. and Barnard, C., 2006. Farmland Protection Programs. In: Wiebe, K. and Gollehon, N. (eds.). Agricultural Resources and Environmental Indicators, chapter 5.6. ERS, USDA, EIB No. 16. Washington, DC, USA.

Nord, M., Andrews, M. and Carlson, S., 2009. Household Food Security in the United States, 2008. ERS, USDA, ERR No. 83. Washington, DC, USA.

OECD (Organisation for Economic Co-operation and Development), 2004. Agriculture and the Environment: Lessons Learned from a Decade of OECD Work. Available at: http://www.oecd.org/dataoecd/15/28/33913449.pdf.

Office of the Chief Economist, USDA, 2010. USDA Agricultural Projections to 2019. OCE-2010-1. Available at: http://www.ers.usda.gov/publications/oce101/.

Offutt, S., Somwaru, A. and Bohmann, M., 2002. What's at Stake in the Next Trade Round. Agricultural Outlook 297: 43-45 (Dec).

Paster, E., 2004. Preservation of Agricultural Lands Through Land Use Planning Tools and Techniques. Natural Resources Journal 44: 283-318.

Pew Initiative on Food and Biotechnology, 2004. Feeding the World: A Look at Biotechnology and World Hunger. Washington, DC, USA.

Pimentel, D., 2006. Soil Erosion: A Food and Environmental Threat. Environment, Development and Sustainability 8: 119-137.

Pinstrup-Anderson, P., Pandya-Lorch, R. and Rosegrant, M.W., 1999. World Food Prospects: Critical Issues for the Early Twenty-first Century. IFPRI, Food Policy Report. Washington, DC, USA.

Renewable Fuels Association, 2009. Growing Innovation: 2009 Ethanol Industry Outlook. Available at: http://www.ethanolrfa.org/pages/annual-industry-outlook.

Renewable Fuels Association, 2006. From Niche to Nation: Ethanol Industry Outlook 2006. Available at: http://www.ethanolrfa.org/pages/annual-industry-outlook.

Richardson, J.J., Jr., 2003. Downzoning, Fairness and Farmland Protection. Journal of Land Use and Environmental Law 19: 59-90.

Shapouri, S., Rosen, S., Meade, B. and Gale, F., 2009. Food Security Assessment, 2008-2009. ERS, USDA, GFA-20. Washington, DC, USA.

Shapouri, S. and Rosen, S., 2004. Fifty Years of U.S. Food Aid and Its Role in Reducing World Hunger. Amber Waves 2(4): 38-43 (Sept).

Scholl, J., 2009. Food Security is National Security. American Farmland Trust. Available at: http://blog.farmland.org/2009/12/food-security-national-security/.

Sullivan, P., Hellerstein, D., Hansen, L., Johansson, R., Koenig, S., Lubowski, R., McBride, W., McGranahan, D., Roberts, M., Vogel, S. and Bucholtz, S., 2004. The Conservation Reserve Program: Economic Implications for Rural America. ERS, USDA, AER No. 834. Washington, DC, USA.

Taylor, M.R., 2001. The Emerging Merger of Agricultural and Environmental Policy: Building a New Vision for the Future of American Agriculture. Virginia Environmental Law Journal 20: 169-190.

Tweeten, L., 1998. Food Security and Farmland Preservation. Drake Journal of Agricultural Law 3: 237-250.

United Nations-Energy, 2007. Sustainable Bioenergy: A Framework for Decision Makers. Available at: http://esa.un.org/un-energy/pdf/sustdev.Biofuels.FAO.pdf.

United Nations, Population Division, 2009. World Population Prospects: the 2008 Revision. Executive Summary. ST/ESA/SER.A/287/ES. New York, NY, USA.

United Nations, Population Division, 1999. The World at Six Billion. ESA/P/WP.154. New York, NY, USA.

US AID, 2009. Food Security. Fact Sheet. Available at: http://www.usaid.gov/press/factsheets/2009/fs091019_2.html.

US Census Bureau, 2009. Statistical Abstract of the United States: 2010. Available at: http://www.census.gov/statab/www/.

US Environmental Protection Agency, 2010. EPA Finalizes Regulations for the National Renewable Fuel Standard Program for 2010 and Beyond. EPA-420-F-10-007. Available at: http://www.epa.gov/oms/renewablefuels/420f10007.pdf.

USDA, 2009. News Release, Agriculture Secretary Vilsack Announces $1.17 Billion in Conservation Reserve Program Rental Payments. Release No. 0497.09.
USDA, 2006a. Conservation and the Environment. 2007 Farm Bill Theme Paper No. 2.
USDA, 2006b. Strengthening the Foundation for Future Growth in U.S. Agriculture. 2007 Farm Bill Theme Paper No. 5.
USDA and Council on Environmental Quality, 1981. National Agricultural Lands Study. Final Report. Washington DC, USA.
US Department of State, 2009. Global Hunger and Food Security Initiative: Consultation Document. Washington, DC, USA.
Wiebe, K., 2003. Linking Land Quality, Agricultural Productivity, and Food Security. ERS, USDA, AER No. 823. Washington, DC, USA.
Vesterby, M., 2001. Land Use. Agricultural Resources and Environmental Indicators, chapter 1. ERS, USDA, AH No. 722. Washington, DC, USA.
World Bank, 2010. Global Economic Prospects 2010: Crisis, Finance, and Growth. Washington, DC, USA.
Yacobucci, B.D., 2009. Biofuels Incentives: A Summary of Federal Programs. CRS, R40110.

Statutory and regulatory measures

European Union

Council Regulation No. 1257/1999, 1999 O.J. (L 160) 80, as amended (most parts repealed 1 January 2007).
Council Regulation No. 1698/2005, 2005 O.J. (L 277) 1 (effective 1 January 2007).

US statutes

US Code (USC). Individual sections, as cited. http://uscode.house.gov/.
Agricultural Foreign Investment Disclosure Act, 7 USC §§ 3501-3508; regulations, 7 CFR part 781.
Agricultural Management Assistance Program, 7 USC § 1524(b); regulations, 7 CFR part 1465.
Biomass Crop Assistance Program, 7 USC § 8111.
Clean Air Act, 42 USC §§ 7401-7671q.
Conservation Reserve Program, 16 USC §§ 3831-3835a; regulations, 7 CFR part 1410.
Conservation Security Program, 16 USC §§ 3838-3838c; regulations, 7 CFR part 1469.
Conservation Stewardship Program, 16 USC §§ 3838d-3838g; regulations, 7 CFR Part 1470.
Energy Policy Act of 2005, Public Law 109-58, 119 Stat. 594 (2005).
Energy Independence and Security Act of 2007 (EISA), Public Law 110-140, 121 Stat. 1492 (2007).
Environmental Quality Incentives Program, 16 USC §§ 3839aa-3839aa-9; regulations, 7 CFR part 1466.
Farmland Protection Policy Act, 7 USC §§ 4201-4209; regulations, 7 CFR part 658.
Farmland Protection Program, 16 USC §§ 3838h-3838j; regulations, 7 CFR part 1491.
Food, Conservation, and Energy Act of 2008, Pub. L. No. 110-234, 122 Stat. 923 (2008) (2008 Farm Bill).
Grassland Reserve Program, 16 USC §§ 3838n-3838q; regulations, 7 CFR part 1415.

Highly Erodible Land Conservation, 16 USC §§ 3811-3814; regulations, 7 CFR part 12
Soil and Water Conservation Assistance, 16 USC § 3830 note.
Wetland Conservation, 16 USC §§ 3821-3824; regulations, 7 CFR part 12.
Wetlands Reserve Program, 16 USC §§ 3837-3837e; regulations, 7 CFR part 1467.
Wildlife Habitat Incentives Program,16 USC § 3839bb-1; regulations, 7 CFR part 636.

US regulations

Code of Federal Regulations (CFR). Individual sections, as cited. Available at: http://www.gpoaccess.gov/cfr/index.html.
Environmental Protection Agency, Regulation of Fuels and Fuel Additives: Changes to Renewable Fuel Standard Program; Final Rule, 75 Federal Register 14,670-14,904 (26 March 2010), codified at 40 CFR part 80, subpart M.

State laws

Illinois, 2008. Agricultural Areas Conservation and Protection Act, 505 ILCS 5/1-20.3.
Oregon, 2007. Measure 49. Just compensation for land use regulation. ORS 195.300-.36.
Oregon, 2004. Measure 37: Compensation for loss of value due to land use regulation. ORS 197.352.

Court decisions

Bormann v. Board of Supervisors, 584 N.W.2d 309 (Iowa 1998), *cert. denied*, 525 US 1172 (1999).
Gacke v. Pork Xtra, LLC, 684 N.W.2d 168 (Iowa 2004).

Chapter 11
Intellectual property rights and food security: the international legal battle over patenting staple crops[511]

Melanie Wiber

1. Introduction

At the March 2006 meetings of the Conference of the Parties to the Convention on Biological Diversity (COP/CBD), the conflict between developing countries and industrial giants came to a head over the so-called Terminator Technology.[512] Terminator Technology is the colloquial name for genetic technology that restricts the use of patented plants by causing second-generation seeds to be sterile. Western nations such as Canada and the UK, while both signatories to the Convention on Biological Diversity (encouraged by the US, which is not a signatory) supported a move to end the moratorium on the use of the technology. Influenced by signatories such as Brazil and India and by many NGOs interested in food security, the CPO/CBD rejected the call for a 'case by case approval process' and instead endorsed a further moratorium. This debate and others like it are symptomatic of the growing disparity between North and South on issues of genetic modification (GM) of important human food crops and animals.

The World Trade Organization (WTO) Agreement on Trade Related Aspects of Intellectual Property Rights (TRIPS) exacerbated this debate, since the United States employed its own trade legislation[513] to force recalcitrant developing nations to sign up to TRIPS (DeBièvre, 2002: 6). The stick was loss of access to American markets; the carrot was developing nation concessions on market access for textiles and agricultural products (DeBièvre, 2002: 6). While the 'single undertaking' approach and the 'take-it-or-leave-it' logic of the negotiations left many developing states with little room for manoeuvre (Cullet and Raja, 2004: 106), many developing

[511] I thank the participants of Workshop I during the 25th Anniversary Celebrations, as well as Bernd van der Meulen, for many helpful comments on this paper.

[512] Conflicts over biotechnology date back to the Uruguay round (1986-1994) of the General Agreement on Tariffs and Trade (GATT) and from the beginning they have involved a North-South confrontation (Downes, 2005: 5). These conflicts 'helped to precipitate the collapse of the 1999 WTO Ministerial session in Seattle' (McAfee, 2003: 210), and are 'crystallizing some of today's major political and social controversies' (Müller, 2006b: 3-4).

[513] Section 301 of the US Trade Act was amended in 1984 to give the President the authority to withdraw trade benefits from another country or impose duties on any of its goods entering the US. In 1988, the '301 process' was enhanced with investigative powers to identify those countries that 'denied 'adequate and effective protection' of intellectual property rights' (see Drahos and Braithwaite, 2002: 88-89).

nations continue to fight against the exponential growth of intellectual property in food crop genetics.

Food security and poverty alleviation have been advanced as key arguments both for and against patenting genetic components of plants and animals that are significant human food sources. Food security is a complex issue (Anonymous, 1993; AAA, 1993; Hopkins, 1992; Wade, 2004). It requires not only sufficient production levels, but also that adequate supplies be delivered to the market, and that consumers are able to purchase enough high quality food to meet basic nutritional needs (Downes, 2004, 2005). Given that biotechnology is primarily applied to issues of production, the potential impact of genetically modified foods on food security is highly contentious (Stone, 2002). This paper explores the growing tensions between new developments in intellectual property rights in life and the issue of food security. It will argue that no other debate more clearly demarcates the agricultural and trade-related realities dividing North from South, realities that continue to threaten increased poverty and hunger for millions of the world's poor.

The genie is out of the bottle; genetic modifications are part of the reproductive pool of many globally important crops and animals. It seems very unlikely that we can put the genie back in the bottle, even if we could get universal agreement on that course of action. And genetically modified germplasm appears to be moving around in nature at a rate that no one had predicted (McAfee 2003: 208), although this too is contested. So what then continues to fuel this debate?

2. The great moral debate

An interesting feature of what Stone (2002) calls the Genetic Modification Wars is the moral tone taken by protagonists both for and against protecting intellectual property rights in agricultural genetic resources. It is important to note that this moral tone also flavours the wider debates about the actual effects of globalisation, with trade liberalisation advocates claiming the inequality gap is narrowing as prosperity increases, while those critical of liberalisation argue that world poverty has been increasing, not declining (Wade, 2004). Each side accuses the other of disseminating 'the big lie' (Wade, 2004: 568). The interventions of Civil Society Organisations (CSOs), Non-Governmental Organisations (NGOs), social activists, trade ministers, WTO bureaucrats and corporate representatives all sound equally principled *and*, oddly enough, equally strident. And when it comes to the role of genetically modified crops in mitigating world poverty, the voices grow even more strident.[514] On the one side are claims that genetically modifying plants and

[514] As Müller (2006b) notes, accusations of 'frankenfoods' that imply a science out of control, are countered by accusations of 'crimes against humanity' for holding up technology that could feed the hungry of the world.

animals offers a more *efficient and predictable* mechanism for improving varieties, by directly controlling both yield and required inputs such as pesticides or antibiotics (McAfee, 2003: 205). A second pro-GM argument is that, to the extent that a developing nation is willing to protect exclusive intellectual property rights in new GM plants, they will attract more investment and experience greater economic growth (Davalos *et al.*, 2003), particularly through the potential for an industrial-style transformation of their agriculture (Stone, 2002). Anti-GM activists, however, claim that patents on crop germplasm will destroy centuries-old farming practices such as seed saving, seed sharing and selective breeding that have enriched our common agricultural heritage (Downes, 2004) and may even precipitate a global ecological crisis (Müller, 2006b). Furthermore, the high cost of buying seed will act to further impoverish the global poor and potentially force them off their lands (Downes, 2004). Both sides in this highly polarised debate claim that they know the best moral choices for all parties involved.

For example, both sides claim to know how to put a stop to 'piracy' (Drahos and Braithwaite, 2002). The piracy claims as applied to intangible property are of two, interrelated types (Downes, 2004: 367). Industrialised countries claim that pirates in developing nations are stealing the innovations and inventions of the best minds in America. The US International Trade Commission, for example, estimated in 1999 that intellectual property piracy of all types was costing US corporations between 100 and 300 billion dollars per year (Downes, 2004: 367; see also Drahos and Braithwaite, 2002). Anti-TRIPS activists, on the other hand, have come up with their own calculations of the cost of piracy. Some have estimated that the US alone has bled over 50 billion dollars per annum from developing nations as a result of stolen agricultural germplasm, with an additional 32 billion dollars per annum taken by the pharmaceutical industry (Downes, 2004: 367). Patents are either the panacea or the *modus operandi* of property theft.

3. Moral signposts

The moral tone to this debate is interesting for two reasons: first, intellectual property law in general, and patent law in particular, are silent on the moral nature of the innovation – useful yes, moral no (Boylan and Brown, 2001; Seide and Stephens, 2002; Wilson, 2002). Thus, for example, the 1952 US Patent Act (Section 201) explicitly allows for patents on innovations 'presumed to be unknown' in the US – whether or not already patented elsewhere – a denial of 'prior art' that some say fuels US piracy (Downes, 2004: 13). Second, from an anthropological perspective, claims to the moral high ground are often a sign of significant political instability, which in turn call for critical analysis of the resulting power struggles (Wolf, 1972). Without that critical analysis, we may be swayed by the moral arguments without situating them in an understanding of who gains and who loses given a particular moral stance or of how arguments based on morality often enhance power differentials (see also Wolf, 2001). Unfortunately, as Wolf also notes, we often 'confront the problem of understanding power at a time when the very signposts of understanding are growing confused and irrelevant' (Wolf, 1972: 258). This observation certainly applies to the Genetic Modification Wars.

What are the confusing and irrelevant signposts of understanding on the biotechnology debate? They relate to issues such as defining the public good or the global commons, distinguishing the private from the public sector, conceptualising property systems and their incentive structures, querying the role of 'prior art' in new inventions, evaluating competing knowledge systems, recognising which trade restrictions are 'unfair trade practice', and judging when justifiable economic rewards veer off into cartelism and monopolistic capitalism (Wiber, 2009). Each of these signposts is confusing and irrelevant for the simple reason that there seems to be no solid epistemological ground to resolve the debate. I only have space for a brief discussion of some of these, but even a short survey will illustrate the problem we face in arriving at any definitive conclusions.

4. Commons versus private property

One argument that is advanced when debating the role of GM plants in food security relates to a larger, long-term debate about the highly porous public/private divide (Weintraub, 1997). The question raised here is which of the global stock of genetic resources properly belongs to the global commons and which to the private sector. As 'intangible property', genetic resources are of a peculiar nature (Bettig, 1996). Bettig argues, for example, that the 'peculiar nature' of intellectual property places it within the realm of 'public goods' (the product cannot be used up by any one consumer) and that these commodities are therefore 'prone to market failure' (Bettig, 1996: 2). Capitalist firms, however, have developed a variety of mechanisms to overcome this 'public goods as commodities syndrome', including copyright, patents, trademarks, compulsory licensing, packaging, encryption, price discrimination and so on. Given previous experience with music and video in their home market, some American firms in particular were well positioned to take advantage of the information explosion in biotechnology (see Drahos and Braithwaite, 2002). But this advantage has relied on some convoluted rhetorical devices.

For example, the 1983 Food and Agricultural Organization (FAO) International Undertaking for Plant Genetic Resources (IU) recognised free access to all plant genetic resources, 'based on the universally accepted principle that plant genetic resources are a heritage of mankind and consequently should be available without restriction' (quoted in Downes, 2004: 13). However, within a few short years, this IU was to be modified as a result of protests that the 'fruits of free enterprise' should not be lumped together under this 'heritage of mankind' designation (Downes, 2004: 13). In other words, while deploring open access in general, the industrial North has framed the 'common heritage of humankind' such that genetic 'raw materials' *must* continue to constitute a global open-access resource[515] (McAfee, 2003: 211). To maintain this *de facto* state of open access, they have negotiated requirements under the Convention on Biological Diversity (CBD) such that signatory states must grant access 'on reasonable terms' to the biodiversity found in their territories (Davalos *et al.*, 2003). For the 'fruits of free enterprise', however, the North has argued that the genetic components should be granted private property status, largely on the basis of the 'value added' to them by scientific manipulation (McAfee, 2003: 212). But McAfee notes that this argument is only successful for the North if the value-added nature of pre-genomics plant breeding does not deserve private property rights, a contradiction resolved by consigning it to a kind of: 'scientific dark age...as if plant breeders...and farmers...have been toiling in ignorance and superstition' (McAfee, 2003: 212). So are all plant improvements deserving of private property rights, or are none?

In another convoluted argument, biotechnology firms have also treated the output of public scientific research institutions as an open-access resource. For example, the Consultative Group on International Agricultural Research (CGIAR) holds in trust one of the world's largest collections of plant genetic resources as part of the public domain. They have often been unable to prevent the patenting of genetic material sourced from these holdings, as the protracted battle over the US Proctor patent on the yellow bean illustrates (Magnus, 2002)[516]. Meanwhile, biotech firms resist any demands that they should be required to engage in transfer and benefit sharing agreements with public scientists or institutions. And state or parastatal agricultural research regimes of 50 years' standing are being privatised under restructuring pressure from the World Bank, leaving public crop seed production and distribution severely disrupted (Zerbe, 2001). Here, the logic appears to be that of economic growth through monopoly rents (Harvey, 2001), such that something which is shared among many is of little economic value, while that which is gathered up into corporate hands and protected under private property rights can generate wealth. Behind the rhetorical devices then is a real issue of who should benefit from the privatisation of public genetic resources.

Some have argued that the difference between private and public biotech research may be key to addressing food security; protecting the public research sector may allow for more 'pro-poor' biotech research (Stone, 2002). But Stone notes that the

[515] It is interesting that concerns about the tragedy of the commons do not appear to figure in these arguments for an open-access global commons in genetic resources, despite what some see as a 'gold rush style' race to the patent office (Drahos and Braithwaite, 2002).

[516] In 1999 Proctor received a US utility patent and a Plant Variety Protection Certificate for a yellow variety of bean. This was successfully challenged by the Center for International Tropical Agriculture, the Mexican government and others based on the argument that Proctor's bean is genetically identical to plants previously found widely cultivated in Mexico and lodged in CGIAR genetic seed banks (Magnus, 2002: 269). According to a 2005 ETC Group report (see www.etcgroup.org), however, Proctor has successfully manipulated the appeals process of the US patent system to extend its monopoly on this bean variety for over 6 years, or nearly one third of the 20-year patent term.

debates about the impact of genetic modification and intellectual property rights on food security have obscured these differences and have tarred all genetic modification with the same brush. Other pitfalls may lie in the path of pro-poor public research. This is evident in the arguments in favour of 'two-track' legislation to facilitate access to potentially important genetic resources originating in countries of the South (Davalos *et al.*, 2003). In the Philippines, for example, recent legislation allows public research such as the 'pure science' conducted by universities to be 'fast tracked', with less stringent genetic access agreements, presumably since there is a public benefit. On the other hand, commercial agreements will be more strictly monitored (Davalos *et al.*, 2003). But there are two problems related to this approach. First, as Müller (2006b: 7) notes, there are significant questions about 'whether the forms of innovation, promotion, and regulation [are] sufficiently trustworthy to defend the public interest' in such research. And second, given the above-mentioned tendency for biotechnology firms to use public knowledge in their patent applications (Drahos and Braithwaite, 2002), the two-track system of genetic resource access may only result in biotech firms appropriating pro-poor research for future, unforeseen gains (Grajal, 1999).

5. The role of property

In this public/private divide, property plays an important role: private property, common property, cultural property. Unfortunately, the instrumental role that many on various sides of the debate intend that property should play is more a matter of hope than fact. Furthermore, the concept of property is tossed about without any theoretical rigor (Von Benda-Beckmann *et al.*, 2006), with little distinction made between the separate analytical layers at which property manifests itself: in *ideologies*, in *legal institutions*, in actual *social relationships*, and in *social practices* (of production, exchange and reproduction) (Von Benda-Beckmann and Von Benda-Beckmann, 1999). None of these layers can be reduced one to the other. So, for example, while six agro-chemical companies (Bayer, CropScience, Monsanto, Syngenta, Avanta and DuPont) hold the vast majority of patents on GM seeds (Müller, 2006b: 6), it is not at all clear at this point whether all state legal regimes will respect these patents, or whether all states will be in a position to enforce patent protection. While placing intellectual property rights within the WTO will mean that those members found in non-compliance[517] will be open to sanctions under the WTO's Dispute Settlement Understanding (Downes, 2004), there may be little such countries can do to prevent non-compliance at the grassroots level (DeBièvre, 2002). Their infrastructure for enforcement may simply be inadequate to the task, and the attempt to upgrade this infrastructure will probably only take

[517] Non-compliance here would include failing to develop national legislation in line with international standards, and might also include failing to 'create domestic enforcement institutions [that] guarantee IPR-holders access to domestic courts' (DeBièvre, 2002: 8). As DeBièvre points out, enforcement is thus decentralised in that individual states must ensure compliance, but also in that private actors must be granted access to WTO rights protection.

resources away from health, education and nutritional systems that might have a greater pro-poor impact (Labonte and Schrecker, 2004).

The difficulty of policing GM organisms should not be downplayed. There are frequent reports, for example, of GM tomato seeds being smuggled into the Middle East or GM soy seed into Africa. Stone (2002: 619) reports that GM cotton was grown extensively in India on an illegal basis before its formal approval in March 2002. Intellectual property is said to be 'intangible' – and this is illustrated at one level by the innocuous appearance of GM seeds – highly portable, easily disguised, and impossible to mark with copyright symbols (McAfee, 2003). Thus, the proprietors of genetic patents are spending billions of dollars in research to identify their patented material in seeds and in processed goods, which is one of the reasons why the Terminator Technology is of interest to them[518]. The genie may be out of the bottle in more ways than one, and the Terminator Technology may allow proprietors to confine the rewards of GM seeds to those owners recognised under intellectual property protection.

In the meantime, there is a great deal of conflict over which body of law should regulate genetic properties. The WTO developed TRIPS as a way of getting around the interminable wrangling that characterised the World Intellectual Property Organization (WIPO) (DeBièvre, 2002). TRIPS is touted as a way to create a global 'level playing field' for intellectual property rights; but there are also voices in favour of *sui generis* (of its own kind) regulation that conforms to the plant breeders' rights (PBR) model proposed by the CBD (Downes, 2004: 370). Unfortunately, neither of these models (*sui generis* or TRIPS) deals adequately with the *plurality* of property ideologies and legal institutions at work in most contemporary states, and that significantly affect actual productive relationships. These institutions are often rooted in different sources of legitimacy, including the official legal system of the state with all its internal contradictions, traditional or customary law, multiple bodies of religious law, plus rapidly developing international law. As just one example of the many potential pitfalls of the actual situation of legal pluralism, consider how many commentators on both sides of the debate simply and uncritically treat the ownership of genetic resources as securely lodged at the state level (Davalos *et al.*, 2003; Grajal, 1999). But this legal and administrative convenience is in fact highly contested.

This contestation is illustrated by the equally complex disagreement among the diverse members of the anti-GM camp about the nature of geographical Centers of Origin, and the ability to award property rights fairly and accurately to those

[518] On the other hand, critics of the patents point out that very little money is being invested in investigating the long-term ecological, biological and human health impacts of genetic modification (see Müller, 2006a,b).

regions that are said to be the source of origin for a plant variety. As the UN Food and Agricultural Organization (FAO) has commented (quoted in Downes, 2004: 368):

> 'any region in the world is dependent on genetic material which originated in other regions for over 50 per cent of its basic food production, and, for several regions of the world, such dependency is close to 100%.'

Corn is an excellent example, as it has deep historical roots in a number of regions of Central and South America (Brush, 1998) and GM varieties have reached into some of the most isolated of these regions, threatening the purity of traditional landraces. But corn is also a global crop in the sense that it is now cultivated in most of the major agricultural regions of the world. Rice is no less a complex case, as are beans. The patenting of genetic material of such important food crops has generated not only unequal contests between cultural groups and Western corporations, but also among the different cultural groups that occupy geographical source areas. There has been conflict over which group can properly claim rights in key plants of interest. Speaking of 'cultural property' requires delineating cultural boundaries with some precision, which is never an easy prospect (Brush, 1998; Brush, 1993; Orlove and Brush, 1996). To take just one example, there have been questions about the historical construct of 'the Maya', and also who among 'the Maya' could be said to hold cultural property rights in traditional medicinal plants (Nigh, 2002).

And this raises some very vexing questions of property systems and their incentives structures, a topic I have addressed with respect to intellectual property elsewhere (Wiber, 2005). First, there is the question to whom these rights should be awarded and on what basis: as cultural property to the entire group (however defined), as the collective property of a set of practitioners, or as individual property. Second, there is the problem of what kind of rights to recognise, as in the debate as to whether or not the right of alienation should be included. And third and most taxing is the problem of moral fairness of any distribution of rights and resulting rewards. Consider an international corporation, for example, that seeks to acquire rights to potentially useful plant genetics, while at the same time owning plant varieties developed at their research farms, the laboratories that do genetic testing, the genetic patents and process patents that emerge from the lab research, the databases that place the results in a broader genetic context, the patents for herbicides and pesticides used in the production systems for those plants, most commercial seed outlets, and even a scientific publishing house. This highly concentrated 'ownership' of rights in particular kinds of objects, processes and outcomes requires new levels of analysis to identify the consequences for others. For example, many argue that we need to protect academic freedom, public interest in crop seed biodiversity, or any other of a long list of potential values threatened by such concentration. It seems unlikely that we can address this kind of concentration through intellectual property agreements that attempt to meet the obligations to reward source regions under the Convention on Biological Diversity. Such attempts have often been derailed by the resulting political struggles over potential financial gains and who should control them; the state, the source region, or a particular cultural group from that region. As Davalos *et al.*, (2003) point out, the range of groups claiming rights in genetic resources is a policy nightmare, and one that has only been heightened by the promise of significant monetary rewards.

6. Battlefields of knowledge

If we turn to science, do we find a value-neutral place to interrogate the utility of GM foods in fighting world hunger? Surely we ought to be able to establish the truth of claims by the bioengineering sector on this point? Unfortunately, there are no clear signposts in science either. The competing knowledge systems – whether folk or formal – are locked in debates not only between each camp, but also within each camp. McAfee (2003: 209) notes, for example, that the biotechnology proponents speak of genetic use restriction technologies (V and T-GURTS), recombinant DNA technology, messenger RNAs, and germ-line cells as if genes were 'a discrete, causal unit of heredity', such that bioengineers can map them, move them around and thereby 'reprogram' life. In this rhetoric, organisms are simply an advanced form of the 'Cartesian machine' (McAfee, 2003)[519]. But other geneticists critique this 'trope' of the gene. Donna Haraway is critical of the role of science in particularising the complete to make the parts subject to different rules than the whole; a process based on Western reductionist scientific knowledge production. In genetic research, the scientific labour of particularising the complete can then become the basis for a property claim (Haraway, 2000).

Genes, such critics argue, are *not* individual units that can be moved around at will; they are part of a larger and complex working whole that is not easily broken down into discrete parts. For this reason, genetic research has been highly inefficient and plagued by 'a degree of randomness' that routinely produces unintended outcomes (McAfee, 2003: 206), including vertical and horizontal gene flow,[520] unexpected and deleterious plant characteristics, ecological rigidity, and fuelling of the 'pesticide treadmill' (McAfee, 2003: 208). Living organisms have agency in the sense that once released into the environment, they escape human control (Müller, 2006b: 4), and behave in highly unpredictable ways. And as Müller (Müller, 2006b: 10) notes, what we know about the actual nature of genetic modification, its successes and failures, is constrained by the 'silencing' of those who work in

[519] This Cartesian notion of biological organisms has taken root in legal reasoning, as I have argued elsewhere (Wiber, 2009). See also Müller (2006a).

[520] Vertical gene transfer is through sexual reproduction (i.e. pollen drift) while horizontal transfer is through microbial vectors (as when bacteria and virus transfer genetic material to higher plants and animals) (McAfee, 2003: 208).

large biotech laboratories.[521] A more nuanced understanding of the pitfalls and limitations, thus, may be hard to get from such 'scientist-entrepreneurs' (McAfee, 2003: 204).

Nevertheless, one might hope that the evidence exists with which to establish dispassionately whether or not there have been or could be successful cases of more nutritious, more productive bioengineered plants that might improve food security or improve incomes through increased sales of commercial crops. But perhaps precisely because of the real complexity that plagues genetic manipulation, the evidence appears rather contradictory. Altieri (2002: 620) reports that yields have not increased with existing commercial bioengineered crops, as 'soybean yields tend to be lower, cotton yields are unchanged, and maize yields are higher only under high pest pressure'. McAfee (2003: 205) points to the way that GM cotton crops unexpectedly proved susceptible to higher than normal seasonal temperatures in Texas and Missouri, a fact that cost Monsanto a great deal of money as farmers sued for lost income. Other researchers, however, examine canola production in Canada and argue that yields *are* higher, but that this higher production is not translating into an increase in farm income (Müller, 2006a). This mixed outcome may be explained by the fact that at least 73% of GM crops actually being cultivated are those designed to tolerate proprietary herbicides, which has only served to 'substantially increase the sale of chemicals' (McAfee, 2003: 212-213). Oxfam has argued that very little bioengineering has been designed to meet the needs of the rural poor or to enhance their productivity (cited in Downes, 2004: 369). Such pro-poor bioengineering developments are unlikely to generate the large profits attached to existing products, and thus such initiatives will be abandoned to the under-funded public research sector.[522] Yet others argue that the replacement of local crop diversity with mono-cropped GM varieties will further the post-Green Revolution trend of loss of genetic diversity and farmer resilience (Downes, 2004). And as McAfee (2003) reminds us, the Green Revolution may be a good guide for the potential impact of GM seed in actual farming systems, as the productive quality of seed is often not the main problem. Other inputs such as marginal lands, insufficient manure or water supply, and lack of access to markets may be more determinant factors. In other words, science practitioners also struggle to weigh the many divergent considerations that must be factored into an assessment of the impact of bioengineering on food security.

[521] Müller's arguments about those who are silent in the genetic debates must be tempered by a consideration of the role such laboratory scientists play as expert witnesses in IP court cases, a role I will return to below.

[522] Downes (2004: 373) reports, for example, that in 1998 CGIAR spent 25 million dollars on genetic research while the corporate giant Monsanto alone invested 1.26 billion dollars.

7. Sorting out the stakeholders

The question of policy solutions to the debate does not appear to lie in negotiations among the multiple agents involved at national and international levels. One feels the need for a universal acronyms translator to keep track of them all, including: WTO, WIPO, UPOV (International Union for the Protection of New Varieties of Plants), EPO (European Patent Office), OECD (Organization for Economic Cooperation and Development), IIPA (International Intellectual Property Alliance), ETC Group (Action Group on Erosion, Technology and Concentration; formerly RAFI), ITDG (Intermediate Technology Development Group) and others. While on the surface, these groups would seem to sort neatly into pro- and anti-genetic modification, in fact, the political manoeuvring between and among them is Byzantine.

As Müller (2006b: 8) notes, for example, 'it is impossible to speak of one public of GMO protestors'. Associations of consumers have different concerns than do farmers' organisations, while anti-globalisation action groups, organic farmers and ecological action groups have pursued different and often unrelated agendas in the GM debate. Consumers wonder if GM foods encourage allergies in their children or promote higher rates of cancer. Organic farmers, trying to carve a successful niche out of a very troubled agricultural industry, want to restrict the distribution of GM germplasm in the environment to protect their investments in this niche market. Non-organic farmers may want GM varieties in the hopes of higher yields, but protest the high cost imposed by GM contract farming. Those concerned with the potential ecological consequences of GM germplasm in the environment, may want no GM release at all (Crouch, 1998-2006). And some environmentalists are concerned at the way that trade legislation via TRIPS invalidates environmental legislation within the nation state (Cullet and Raja, 2004). This has made for heated debates and some open conflict in farmers' organisations and between and among farmers, consumers and environmental activists. Raising the question of GM foods in a farmers' market setting, as I have done, can trigger many anti-GM comments, but can stimulate defensive responses as well.

Consider too the relationship between the WTO and the CPO/CBD, characterised on the one hand by open conflict, and on the other by ongoing co-operation and alternative terminological wrangling. As the debate over biotechnology has developed, the CBD has often been viewed as the best international mechanism to constrain TRIPS, particularly as UPOV is generally seen as too closely allied with the WTO objective of exclusive property rights in plant varieties.[523] But at the core of the Biological Diversity convention is the

[523] Downes (2004: 375) notes that UPOV requires a 20-year period of exclusive rights to plants, with 'the rights of farmers to retain and use protected seeds left to the discretion of national governments'.

> 'notion that the economic value of natural genetic resources, realized through the development of new uses in biotechnology, would provide economic incentives for the sustainable use of these resources' (Davalos *et al.*, 2003: 1512).

So while developing nations see the CBD as a way to protect their genetic resources and to distribute more fairly the benefits that arise from their commercial development, the biotechnology-rich Northern states tend to interpret it as yet another device to facilitate genetic access (Davalos *et al.*, 2003). The main players in the North read the CBD text in a way that prioritises access to genetic resources, while the South reads the CBD text as conserving biodiversity[524] and establishing a fair and sustainable use of its components (Davalos *et al.*, 2003: 1514).

Meanwhile, international organisations such as the CPO/CBD, UPOV and others, seem particularly vulnerable to political pressure. The debate over the wording of the Terminator sanction is a case in point. The ETC Group reported in April 2003, for example, that an expert panel recently convened by the CBD in Montreal to consider the impact of the Terminator seed technology for small farmers, indigenous peoples and local communities was presented with a highly critical paper that had been prepared by UPOV. UPOV has undergone significant change in recent years, as more and more members have been attracted from developing states in response to the alternative to TRIPS that UPOV plant protection offers. But increased membership from developing countries has not translated into a uniformly pro-poor stance in UPOV. The above anti-Terminator paper, for example, became the focus of an intense period of wrangling between UPOV and US patent officials as UPOV had taken the position that the Terminator built-in biological control over seeds would far exceed the legal monopoly granted through conventional intellectual property mechanisms (Mulvany, 2003). As long as the UPOV paper was critical, the US represented UPOV as 'not competent' to comment on the possible intellectual property implications of Terminator seeds. When the report was significantly modified by UPOV to eliminate their criticism, however, their competency was no longer in question. The report now is a 'sanitized and shorter 'position paper' that carries none of the criticisms of the original report' (Mulvany, 2003).

In a similar move, another international body (in this case the Food and Agricultural Organization Committee on Agriculture) was subject to attempts on the part of the US, Australia and New Zealand to dismantle the Panel for Eminent Experts and Ethicists (Mulvaney, 2003). This panel, which brings together an international collection of scholars, scientists and ethicists to consider the implications of food

[524] Grajal (1999) argues that biodiversity protection measures need to focus more on direct threats such as deforestation, overfishing, land tenure conflict, and habitat conversion. But with CBD he notes that many governments are now tougher on museums and ethnobotanists than on hydropower developments and logging companies.

and agricultural policies and practices, was attacked for taking the position that the Terminator Technology was generally unethical.[525] The resulting disagreement in the FAO boardrooms also unfolded along North/South lines.

Some anti-GM activists see the reasons for US aggression in international GM diplomacy as simply and easily explained by recent changes in the volume of goods that the US exports in which are lodged IP rights. Downes (2005: 6) reports that:

> 'In 1947...intellectual property comprised almost 10% of all US exports. By 1986 intellectual property formed 37% of US exports. In 1994, intellectual property accounted for over half of the United States exports.'

To protect their international trade in technology rich products and to protect their favoured status in the global division of labour where Northern countries generate innovations and the South constitutes their primary market, the United States has been a prime mover in the international privatisation of genetic resources and in protecting intellectual property rights (Downes, 2005). Others suggest that the American aggression in negotiations is also linked to recent growing criticism of the biotech industry and of intellectual property developments from some surprising sources, including the Royal Society in London and Alan Greenspan, former chairman of the US Federal Reserve (Mulvany, 2003). In either case, bargaining both over how IP is to be protected and over how IP rights are to be enforced has taken place in 'an environment characterised by a highly asymmetrical distribution of power' (DeBièvre, 2002: 3-4). As DeBièvre points out, this enduring imbalance in power raises serious questions about the legitimacy and sustainability of both the content of WTO agreements and their judicial enforcement (DeBièvre, 2002: 3-4).

Such realpolitik at the international level of diplomacy may explain why the many accords, agreements, conventions and international laws that have proliferated exponentially in recent years under the auspices of the above regulatory organisations have generally been unhelpful in resolving the debate over the role of GM in food security. Aside from TRIPS, a brief list of these agreements might include: NAFTA (North American Free Trade Agreement), GATT (General Agreement on Tariffs and Trade), the Convention on Biological Diversity Biosafety Protocol, Leipzig Global Plan of Action, Decision 391 of the Andean Pact countries, various plant variety protections (PVP), and the FAO International Seed Treaty, to name but a few. These international agreements and conventions are largely unhelpful in settling the bioengineering debate because they are so often self-contradictory with respect to basic principles (Cullet and Raja, 2004). For example,

[525] It has been suggested that bioethics, 'based as it is on some of the yardsticks of classical Western philosophy' is insufficient to deal with the 'thick' problems associated with the many social and cultural contexts in which biotechnology plays a role (Pálsson and Hardardóttir, 2002). While Pálsson and Hardardóttir refer to the Human Genome Project, genetically modified foods have many of the same moral and ethical complexities.

many are at one and the same time trying to protect biological diversity, generate economic growth, regulate access to genetic resources, and facilitate biotechnology agreements between parties of vastly unequal power in such a way as to generate more equitable distribution of rewards and benefits (Davalos *et al.*, 2003; DeBièvre, 2002). It should come as no surprise that the concrete mechanisms to obtain these varied outcomes are difficult to arrive at. It should also be no surprise that the interpretation of these articles and agreements is increasingly split along a North/South fracture line.

For example, while Article 27.3B of the TRIPS agreement has been under review since 1999 (Downes, 2004), the contrasting parties take quite different views of what this review might encompass. Article 27.3B requires that all WTO members protect plant varieties either through patents or some variety of plant breeders' rights. The countries of the South see this review process as an opportunity to address significant problems with the basic tenets of Article 27.3B and particularly the dangers to farmers' rights and of narrowing the global genetic base. The main state actors of the North, on the other hand, have generally viewed the review process as one that will negotiate timing and national implementation schedules of the existing agreement.

One should wade into these waters with significant trepidation; the phrase: 'where angels fear to tread' comes to mind. Nevertheless, from the anthropological perspective and from the perspective of law and governance, the issue of GMOs and food security certainly cries out for more critical analysis. Where to begin?

Buried in all this hyperbole, are a few facts that rarely come to light. First, both the pro- and anti-GM camps have operated in a 'breathless rhetoric of speed' (Sunder Rajan quoted in Müller, 2006b: 6). There appears to be no time to assess the impact of existing GM before pushing more onto the market. This rhetorical device of time uses two arguments: 800 million people are going hungry, and individual countries will lose out in the race to 'conquer markets and control patents' if they don't move quickly (Sunder Rajan quoted in Müller, 2006b: 6.). As Müller notes, the consequence is that little long-term research is being undertaken to assess potential impacts on the environment or farm systems (Sunder Rajan quoted in Müller, 2006b: 7). Second, the known outcomes of biotechnology are mixed, including as they do 'new therapeutics, diagnostics, reproductive technologies, and engineered crops' (Seide and Stephens, 2002: 71), as well as limitations on farmers' rights, outrageous patent claims (Basmati rice, Mexican beans, the neem tree, turmeric), copious amounts of chill litigation, bioprospecting and some biopiracy (Magnus, 2002; Mulvany, 2002; Nigh, 2002; Wiber, 2009). It also appears that despite the Malthusian arguments offered by the biotech industry, biotechnology has little to do with addressing poverty and food insecurity; both of these have little to do with production levels and have much more to do with distributive issues (Downes, 2004; Stone, 2002). Transgenic crops do not appear to increase yields (Altieri, 2002; McAfee, 2003), and there is very little transgenic research going on into predominantly subsistence crops such as chick pea, mung bean, sorghum, cassava, and pigeon pea (Altieri, 2002). Many farmers in the South see a significant risk to their normal farm practices of saving and sharing seed under new national regulations that have been emerging as a result of international pressure (Arnejo, 2003). And there is a significant pattern of international pressure being brought to bear on the issue of standards for global protection of intellectual property (Drahos and Braithwaite, 2002). Perhaps the trail of lesser known facts may prove the best signposts to critical analysis, as they point the way to the twin questions of the deployment of power, and who benefits?

8. Questions of power

As one traces out the moves and countermoves in this debate and the numerous agents with quite different power bases who are involved, the increasing geopolitical complexity confronting us today becomes glaringly apparent.

Genetic modification in major crops has promoted changes in trade patterns, for example, that have a far greater effect on food security for the world's poorer nations than the effect of biotechnology in their own farm operations. For example, after the US developed genetically engineered sweeteners from corn, sugar imports from the Philippines and from the Caribbean declined precipitously, costing those nations respectively 600 and 400 million dollars between 1980 and 1987 (Downes, 2004: 369). Severe ecological consequences then unfolded in the Philippines as former sugar plantation lands were transformed for tiger shrimp production, with salt water pumped into excavated ponds; some located as far as four miles inland from the marine sources of salt water (Murray Rudd, personal communication). Biotechnology trials on cocoa, palm oil and vanilla may soon threaten farmers in similar ways in Ghana, Cameroon, Ivory Coast and Zanzibar (Downes, 2004).

Concentration of ownership into the hands of a few (and largely US) corporate actors is also an important factor. As Downes (2004: 374) notes, the impact of TRIPS has not been trade liberalisation but the granting of corporate monopolies over life forms. One example is Brazil, where after the introduction of plant variety protection in 1997, the US firm Monsanto went from 0% involvement to 60% control of the local maize seed market within one year (Downes, 2004.). As mentioned above, Monsanto's tactics to protect their intellectual property rights in GM seed must be considered in the context of their growing domination of agricultural seed distribution and other agricultural inputs such as herbicides. Monsanto and similar corporate actors have been very aggressive in buying up domestic and international farm seed producers (Center for Food Safety, 2005), and in patenting not only new genetically-modified germplasm but also the laboratory techniques and the various proteins that are used to produce them. Monsanto introduced GM varieties for canola, cotton, potatoes, and soy beginning in 1996, and GM varieties now dominate in each of these important crops in area of acreage planted (Center for Food Safety, 2005: 8). And because Monsanto also produces the dominant herbicides used in agriculture, they benefit from their introduction of GM strains that tolerate these herbicides; since the introduction of Roundup Ready varieties, Roundup herbicide application in the United States has increased by over 63 million kilograms (Center for Food Safety, 2005: 8). This dominance in the seed and herbicide marketplace is why American and international farmers' organisations filed a private, antitrust action against Monsanto, claiming that Monsanto and its co-conspirators (DuPont, Dow Chemical, Novartis, AstraZeneca and others) used patents and licensing arrangements to fix prices and restrain trade in both the GM corn and soybean seed markets (Drahos and Braithwaite, 2002: 164)[526].

It would also appear that pro-GM claims that patent protection would result in technology transfer to developing nations have been overstated. The World Bank World Development Report 1998-99, for example, stated that very little GM research and development is taking place in the South (quoted in Downes, 2004: 46; see also World Bank, 1999-2000). The report for 1999-2000 continues to deplore this state of affairs and to note that important genetic resources flow out of developing countries without compensation (World Bank, 1999-2000: 103). On the other hand, farmer *deskilling* in areas where GM contract farming is routinely carried out is a well-documented phenomenon (Stone, 2002: 619). Farmers' production patterns and farm practices are dictated by the requirements and sensitivities of the new GM varieties, and by the nature of legal contracts with the seed providers. Long-term knowledge about traditional varieties and their micro-environmental advantages and production requirements are disappearing along with the varieties themselves, as this knowledge is no longer being passed on to successive generations. The 'vital role of skill in sustainable small-holder agricultural production' is downplayed at the same time that much of this skill is being lost to future farmers (Stone, 2002).

Much of this state of affairs can be traced to the outcome of social processes at quite a different (macro) level. At the level of multilateral trade negotiations, many have commented on the democracy deficit in the way that bilateral or multilateral treaties are negotiated. Those who will be most affected by the outcome of these negotiations, including producers, consumers, and even scientists, are simply not involved. Drahos and others, for example, have been sharply critical of the way that TRIPS was negotiated in the first place (quoted in Downes, 2004: 6). Drahos argues that three conditions should have been obtained if TRIPS were to have been negotiated through democratic bargaining: first, all the relevant interests would need to be at the table (condition of representation); second, all parties should have full information about the consequences of various courses

[526] In 2005, the US Court of Appeals for the 8th Circuit upheld a lower court's October 2003 decision to deny class-action status in this antitrust case against Monsanto and other companies.

of action (condition of full information); third, one party should not coerce the others (condition of non-domination). Drahos and others have argued that none of these conditions prevailed during the TRIPS negotiations, with most nations from the South being under-informed, downright misled, or actively coerced in the negotiations.

This power quotient at the macro level is also visible in the way that the more powerful trading nations have used bilateral or multilateral trade agreements to oblige developing countries to meet or exceed Western intellectual property protection (Downes, 2004: 375; See also Drahos and Braithwaite, 2002). For example, while the European Union has generally resisted American efforts to introduce GM crops into European farming systems, the EU Free Trade Agreement with Mexico (modelled on the North American Free Trade Agreement or NAFTA) requires Mexico to uphold the provisions of UPOV 1991 and to accord 'the same level of protection' to intellectual property rights as they would receive in their country of provenance (Downes, 2004: 375). And as Downes (2004) notes, as career bureaucrats usually negotiate treaties and the wording is usually confidential until the treaties are actually signed, 'consultation with either parliaments or public opinion' is negligible.

One can also trace the role of power in the way that IP and its protection have entered into the legal domain. Recent disputes over GM plants, including corn, soy, alfalfa and canola, have involved courts in a number of states (the United States, Mexico, Canada, Germany), and courts in other jurisdictions have carefully watched the resulting decisions. This in part explains the tremendous resources being thrown into litigation by both sides in the debate. In these disputes, the cases have often boiled down to which kinds of rights will be privileged: farmers' rights or intellectual property rights (Wiber, 2009). Farmers' rights were defined by the FAO in 1989 as 'rights arising from the past, present, and future contributions of farmers in conserving, improving, and making available plant genetic resources, particularly those in the centers of origin/diversity' (quoted in Downes, 2004: 14). In practice, however, this farmer contribution to plant genetic resource conservation and enhancement has not led to recognition in the form of property rights, a significant problem in preserving important landraces (Cullet and Raja, 2004). Farmers' rights and farmers' autonomy (i.e. the freedom to make independent decisions about the best use of his/her own private resources) are often confused in the literature, although they are related. Unfortunately, more recent international agreements have left it up to the discretion of national governments to implement protection for farmers' rights and/or farmers' autonomy and to recognise when either of those were at risk (Cullet and Raja, 2004). As the rights of farmers to save, use and exchange seed are not protected under TRIPS or UPOV models, many legal jurisdictions where IP and farmers' rights have been contested have defined farmers' rights narrowly (Downes, 2004: 43-45; Müller 2006a, Wiber, 2009). The result has been a significant attack on farmers' rights and on farmers' autonomy.

Two cases from Canada, described in detail in Müller (2006a), illustrate some of the resulting legal contradictions. In one case, Monsanto claimed that wherever their GM canola gene appeared, the seed and the plant that grew from it were their property and they could claim it and were entitled to royalty payments. That such plants were owned by a farmer and found on land also owned by that farmer was no defence to patent infringement. Expert testimony from Monsanto crop scientists was key to throwing doubt on Percy Schmeiser's argument that the GM canola seed had 'blown onto' his fields (or that pollen from neighbouring GM canola fields had) and that he was therefore an 'innocent bystander'[527] (see also Wiber, 2009). The Supreme Court of Canada found that Percy Schmeiser was infringing Monsanto's patent rights as he had GM canola plants growing in his fields and had saved the seed from the resulting crop (Müller, 2006a). In the other case Müller (2006b) discusses, however, organic farmers from Saskatchewan agreed that Monsanto owned such GM plants wherever they showed up, and should thus take responsibility for them when they contaminated farmers' organic fields[528]. However, in their legal struggles, the organic farmers found that Monsanto would not agree to any liability obligation. Monsanto argued that they might clean up such invasive plants as a voluntary measure, but that they were not responsible for and would not claim ownership of the plants when it did not suit them to do so (Müller, 2006a). Monsanto's argument in this second case was that the Canadian authorities had approved the release of GM canola varieties into the environment and that 'There was no way to put the genie back into the bottle. GM canola is now part of the environment.' (quoted in Müller, 2006b: 94). Monsanto threw the obligation back onto the Canadian regulatory agencies that had subsequently approved standards for certifiable organic crops, after GM varieties were loose in the environment. From Monsanto's perspective, it was up to the organic farmers to find a way to adhere to these standards. These Canadian cases have been decided in such a way as to effectively allow Monsanto to have their cake and eat it too, in terms of claiming or not claiming their GM progeny.

As in other countries, Canadian farmers are subsequently obligated to pay large financial costs as the penalty for normal farm system operations such as saving seed or lending each other equipment that might harbour seeds. This fact more than any other has cast doubt on the pro-GM arguments regarding food security. If farmers in the North are unable to legally protect their autonomy and their rights to own, trade and sell seed once seed patents are recognised under their state laws, there is very little chance that farmers in the South will fare any better. It also seems clear that the opportunity to generate income from intellectual property protection in the South will depend on a major restructuring of land

[527] See *Monsanto Canada Inc. v. Schmeiser* 2004 Supreme Court Canada 34, available online at http://decisions.fct-cf.gc.ca/en/2001/2001fct256/2001fct256.html, accessed December 28, 2007.
[528] See Hoffman v. Monsanto Canada Inc., Judgment 11 May 2005, SKQB 225. See also http://www.monsanto.ca/monsanto/layout/news/07/05-03-07.asp accessed December 28, 2007.

and resource ownership. Poor farmers with small-holdings cannot afford to buy expensive seed every year, nor are they of much interest to corporate players who seek contract farmers. However, if debts and rising production costs force small-holders to liquidate their real estate assets, and the subsequent concentration of those assets puts them into the hands of larger farm operations, then the major players in the agro-industrial complex will have greater opportunities for financial returns on their IP investments. This may be the only route to higher productivity through GM plants, but the resulting food crops will only come at the expense of higher production costs on these larger farm enterprises. This will mean that the resulting crops are unlikely to be available in local markets and within the purchasing power of poorer members of Southern nations.

9. Conclusions

It has been argued that the timing for the wider introduction of GM innovations in global farming systems could not have been worse (Stone, 2002). While the introduction of GM varieties into United States farm operations and food markets generated an astonishingly low level of public debate, mad cow disease in Europe and in Canada, and the later avian influenza scares, as well as other food safety concerns resulted in a very different scenario in the rest of the world. Stone (2002: 611) has argued that the response of the major GM stakeholders was to usher in 'a Golden Age of misinformation', matched or exceeded by anti-GM stakeholders. Monsanto and six other biotechnology firms responded to the anti-GM campaign in Europe by forming a public relations consortium called the Council for Biotechnology Information[529]. To counter food safety concerns, the CBI played what Stone (2002: 611) calls 'the Malthus card'. Acceptance of GM products was to be ensured through linking GM closely to questions of feeding the world's hungry. Several years after Stone published his article, the CBI website is still dominated by headlines about the pro-poor benefits of higher yields for GM crops, the global warming benefits of GM biofuels, voluntary GM seed donations to the world's poorest farmers, and the health benefits of GM crops. One story focuses on the approval of a GM rice variety in China, and on how both 'green' (requiring less fertiliser) and 'gold' (modified to stimulate production of vitamin A) varieties of rice might be expected to 'improve food production' to 'feed 1.3 billion people'. Although the story admits that China's population is one of growing affluence, the story goes on to tout the 'welfare gains' of GM adoption. 'The economic gains from GMO adoption are substantial,' said an August 2003 report on the economic impacts of genetically enhanced crops in China. In the most optimistic scenario, where China commercialises both *Bt* cotton and GM rice, the welfare gains amount to an additional annual income of about US$5 billion in 2010. This amounts to about $3.50 per person. This is not a small amount in a country, where according to the World Bank, 18% of the population had to survive on less than $1 per day in 1998.

[529] Available at: www.whybiotech.com.

The report cited comes from the proceedings of the 2003 International Conference of Agricultural Economists, and is one of the few academic citations referred to in the article. But a quick reading of the source itself shows that the authors are quoted somewhat out of context, as they specified that their analysis was based on 'tentative experimental data for the rice sector' and that their conclusions were aimed at the relative return on biotechnology investments (Huang *et al.*, 2003).

Stone and others have been equally critical of the anti-GM movement. The use of rhetorical devices to stigmatise all genetic research in human food crops has contributed to the misinformation campaign. And the end result is that we have very few signposts to a reasonable assessment of the debate. While we might welcome the opportunity for anthropologists and other scholars of the grass roots to examine 'the social life of genetically modified seeds' in real farming systems in many different parts of the world (Stone, 2002: 619), it seems to me that we need to pay equal attention to study 'up the food chain' to examine the social life of genetically modified seeds in the boardrooms where international trade agreements are hammered out, and where the North/South divide finds expression in significant power imbalances. From this perspective, it would appear that opportunities to genetically modify major staple crops has only enhanced the power of the North to extract maximum trade gains from the global food market. So long as this is the case, the impact of genetic modified crops on food security will likely be a negative one.

References

Altieri, M., 2002. Comment on 'Both Sides Now'. Current Anthropology 43: 619-620.

Anonymous, 1993. American Anthropological Association Position Statement on African Famine. Surviving Famine and Providing Food Security in Africa. Anthropology Today 9: 25.

Arnejo, A.D.M., 2003. The community registry as an expression of farmers' rights: experiences in collective action against the Plant Variety Protection Act of the Philippines. CAPRi-IPGRI International Workshop on Property Rights, Collective Action and Local Conservation of Genetic Resources, Rome, Italy.

AAA (American Anthropological Association), 1993. AAA Position Statement on African Famine. Surviving Famine and Providing Food Security in Africa. Anthropology Today 9: 25.

World Bank, 2000. Entering the 21st Century. The World Development Report 1999-2000 World Bank, Washington, DC, USA.

Bettig, R.V., 1996. Copyrighting Culture: the political economy of intellectual property. Westview Press, Boulder, CO, USA.

Boylan, M. and Brown, K.E., 2001. Genetic Engineering. Science and Ethics on the New Frontier. Prentice Hall, Upper Saddle River, NJ, USA.

Brush, S.B., 1998. Bio-cooperation and the benefits of crop genetic resources: the case of Mexican maize. World Development 26: 755-766.

Brush, St. B., 1993. Indigenous knowledge of biological resources and intellectual property rights: the role of anthropology. American Anthropologist 95: 653-671.

Center for Food Safety, 2005. Monsanto vs. U.S. Farmers. A Report of the Center for Food Safety. Center for Food Safety, Washington, DC, USA.

Crouch, M.L., 1998-2006. How the terminator (gene) terminates. In Occasional Paper: The Edmonds Institute, Washington, DC, USA.

Cullet, P. and Raja, J., 2004. Intellectual Property Rights and Biodiversity Management: the Case of India. Global Environmental Politics 4: 97-114.

Dávalos, L.M., Sears, R.R., Raygorodetsky, G., Simmons, B.L. Cross, H., Grant, T., Barnes, T., Putzel, L. and Porzecanski, A.L., 2003. Regulating access to genetic resources under the Convention on Biological Diversity: an analysis of selected case studies. Biodiversity and Conservation 12: 1511-1524.

DeBièvre, D., 2002. The WTO Agreement on Intellectual Property: Global Protection, Decentralized Enforcement, and Local Exceptions. MPG 2000 + Workshop 'Politik und Recht unter den Bedingungen der Globalisierung und Dezentralisierung'. Max Planck Institut fur Gesellschaftsforschung. Köln, Germany.

Downes, G., 2004. TRIPs and food security. Implications of the WTO's TRIPs Agreement for food security in the developing world. British Food Journal 106: 366-379.

Downes, G., 2005.Implications of TRIPs for food security in the Majority World. Comhlamh Organization, Dublin, Ireland.

Drahos, P. and Braithwaite, J., 2002. Information Feudalism. Who Owns the Knowledge Economy? The New Press, New York, NY, USA.

Grajal, A., 1999. Biodiversity and the nation state: regulating access to genetic resources limits biodiversity research in developing countries. Conservation Biology 13: 1-6.

Haraway, D., 2000. Deanimations: Maps and Portraits of Life Itself. In A. Brah and Coombes, A.E. (eds.). Hybridity and its Discontents. Politics, Science, Culture. Routledge, New York, NY, USA.

Harvey, D., 2001. The art of rent: globalization and the commodification of culture. In: Harvey, D. (ed.). Spaces of Capital. Towards a Critical Geography. Routledge, New York, NY, USA, pp. 394-411.

Hopkins, R.F., 1992. Reform in the international food aid regime: the role of consensual knowledge. International Organization 46: 225-264.

Huang, J., Hu, R., Van Meijl, H. and Van Tongeren, F., 2003. Economic Impacts of Genetically Modified Crops in China. 25th International Conference of Agricultural Economists, Durban, South Africa, 2003. International Association of Agricultural Economics.

Labonte, R. and Schrecker, T., 2004. Committed to health for all? How the G7/G8 rate. Social Science and Medicine 59: 1661-1676.

Magnus, D., 2002. Intellectual property and agricultural biotechnology. Bioprospecting or biopiracy? In: Magnus, D., Caplan, A. and McGee, G. (eds.). Who Owns Life? Amhurst, Prometheus, New York, NY, USA, pp. 265-276.

McAfee, K., 2003 Neoliberalism on the molecular scale. Economic and genetic reductionism in biotechnology battles. Geoforum 34: 203-219.

Müller, B., 2006a. Infringing and Trespassing Plants: patented seeds at dispute in Canada's courts. Focaal 48: 83-98.

Müller, B., 2006b. Introduction. GMOs - Global objects of contention. Focaal 48: 1-16.

Mulvany, P., 2002. Sustaining Agricultural Diversity and the integrity and free flow of Genetic Resources for food for agriculture. In Forum for Food Sovereignty.

Mulvany, P., 2003. Who Calls the Shots at UPOV? US Government and Multinational Seed Industry Force UPOV to Abandon Critique of Terminator. ETC Group, Ottawa, Canada.

Nigh, R., 2002. Maya Medicine in the Biological Gaze. Bioprospecting Research as Herbal Fetishism. Current Anthropology 43: 451-477.

Orlove, B.S. and Brush, S.B., 1996. Anthropology and the conservation of biodiversity. Annual Review of Anthropology 25: 329-352.

Pálsson, G. and Hardardóttir, K.E., 2002. For whom the cell tolls: Debates about biomedicine. Current Anthropology 43: 271-301.

Seide, R.K. and Stephens, C.L., 2002. Ethical Issues and Application of Patent Laws in Biotechnology. In: Magnus, D., Caplan, A. and McGee, G. (eds.). Who Owns Life? Prometheus, New York, NY, USA, pp. 59-73.

Stone, G.D., 2002. Both Sides Now. Fallacies in the genetic-modification wars, implications for developing countries, and anthropological perspectives. Current Anthropology 43: 611-630.

Von Benda-Beckmann, F. and Von Benda-Beckmann, K., 1999. A Functional Analysis of Property Rights, with Special Reference to Indonesia. In: Van Meijl, T. and Von Benda-Beckmann, F. (eds.). Property Rights and Economic Development. Kegan Paul International, New York, NY, USA, pp. 15-56.

Von Benda-Beckmann, F., Von Benda-Beckmann, K. and Wiber, M., 2006. The Properties of Property. In: Von Benda-Beckmann, F. and Von Benda-Beckmann, K. and Wiber, M. (eds.). Changing Properties of Property. Berghahn, New York, NY, USA, pp. 1-39.

Wade, R.H., 2004. Is Globalization Reducing Poverty and Inequality? World Development 32: 567-589.

Weintraub, J., 1997. The theory and politics of the public/private distinction. In: Weintraub, J. and Kumar, K. (eds.). Public and Private in Thought and Practice. Perspectives on a Grand Dichotomy. University of Chicago Press, Chicago, USA, pp. 1-42.

Wiber, M., 2009. What Innocent Bystanders? The Impact of Law and Economics Reasoning on Rural Property Rights. Anthropologica 51: 29-38.

Wiber, M.G., 2005. The Voracious Appetites of Public versus Private Property: A View of Intellectual Property and Biodiversity from Legal Pluralism. In CAPRi Working Paper 40.

Wilson, J., 2002. Patenting Organisms. Intellectual Property Law Meets Biology. In: Magnus, D., Caplan, A. and McGee, G. (eds.). Who Owns Life? Prometheus Books, New York, NY, USA, pp. 25-58.

Wolf, E., 1972. American Anthropologists and American Society. In: Hymes, D. (ed.). Reinventing Anthropology. Vintage Books, New York, NY, USA, pp. 251-263.

Wolf, E., 2001. Pathways of Power. Building an Anthropology of the Modern World. University of California Press, Berkeley, CA, USA.

World Bank, 2000. Entering the 21st Century. World Development Report 1999/2000. Oxford University Press, New York, NY, USA.

Zerbe, N., 2001. Seeds of hope, seeds of despair: towards a political economy of the seed industry in southern Africa. Third World Quarterly 22: 657-673.

Chapter 12

Food security as water security: the multi-level governance of virtual water

Dik Roth and Jeroen Warner[530]

1. Introduction

No food without water. The food we eat requires water for its production. According to the International Water Management Institute IWMI, 'we "eat" between 2,000 and 5,000 litres of water per day' (IWMI, 2007). However, water is increasingly scarce in many regions of the world. The Water Decade (1980s) brought water to many, but many more still went without. Globally, irrigated agriculture – an important source of food and other crops – accounts for more than 70% of freshwater use (Molden, 2007). Agriculture faces the enormous challenge of satisfying the growing demands for food, feed and increasingly also fuel, boosted by growing giant economies like India and China. At the same time, rapid processes of change like urbanisation and industrialisation generate new water demands outside agriculture that cannot be satisfied from unused water sources. Therefore, it is argued, this growing competition for water between various users and uses will necessitate water transfers out of agriculture into other sectors and uses (Molden, 2007). These processes can be expected to be full of political tension and conflict (Mollinga, 2008).

Do we have the scenario for a water crisis, a food crisis or even water wars here? Not so, optimistic scholars claim, if we were to embrace the concept of 'virtual water', 'the water required to produce water-intensive commodities such as grain' (Allan, 2002; see also Allan, 1997, 1998). The basic idea is simple: the production of any agricultural or industrial product requires smaller or larger volumes of water (its 'water footprint'; Hoekstra, 2003). The lion's share of available water resource consumption – up to 90% – goes to food production, leaving a meagre 10% for drinking and other uses (Allan, 1998). Resulting water scarcity can be solved and economic efficiency of water use increased by adapting regional and global food trade to existing variations in water scarcity. In this view, water embedded in food supplements the water balance in (semi-)arid countries so that, on balance, these countries are not water-short.

Thus, the concept of virtual water links water scarcity to global, regional and other food trade flows, and therefore to issues of food security and food sovereignty. The World Water Council defines food security as 'the capability of a nation to provide

[530] Author names are in alphabetical order; authors have contributed equally to this publication.

access to everyone in the country to adequate, nutritious and safe food now and in the future' (World Water Council, 2004). 'Food sovereignty' denotes people's right to define and control issues pertaining to agriculture, food production, trade and consumption in a way that supports their rights to sustainably produced safe and healthy food (Windfuhr and Jonsén, 2005).

In view of the growing international concerns over climate change-induced weather extremes and their impact on water resources, food production and food security, virtual water is set to rise on the international political agenda as the smart solution to future challenges. We will argue here that, in its transformation from scientific analytical concept to policy prescription, it becomes more explicitly normative in its content and complex in its real-life consequences, and hence also more problematic. Implementing a virtual water policy entails conscious social and political choices based on problem perceptions and definitions, conceptualisations of solutions, and assumptions and ideological values pertaining to the new and 'better' conditions to which such solutions should lead. These choices are, importantly, made on the basis of certain assumptions about resource governance, governability, and legitimate state control; about the degree to which real-life societies can be manipulated into the blueprints of government policies. More specifically, as a market-based form of food and natural resource governance, it is likely to have considerable impact upon local, regional or national food security and food sovereignty by influencing the capacity of food producers to access and control crucial resources like land and water, and allocate these resources to specific forms of food production on which their livelihoods are based.

In earlier work we have critically discussed virtual water and its assumed 'politically silent' character (Warner and Johnson, 2007), and related it to specific food and (virtual) water policies in Egypt and India (Roth and Warner, 2008). In this chapter we explore its assumed political silence from yet another angle: by focusing on the socio-legal and governance dimensions of the virtual water concept. Socio-legal approaches to property rights in the domain of natural resources and analytical approaches to 'governance' in the domains of food and natural resources are two key themes of the Law and Governance Group of WUR. The group's analytical rather than prescriptive approach to food and natural resources governance raises questions that are seldom asked in the virtual water literature.

One rather crucial question concerns the degree of social, legal and political control that can be exerted over the key natural resources in agricultural production, land and water. Even if we were to take the role of 'the invisible hand' in virtual water trade for granted (which we prefer not to), other crucial but too rarely asked questions on resource governance remain unanswered.[531] While water

[531] From Germany, Laube and Youkhana (2006) and Horlemann and Neubert (2007) broach similar territory in questioning the virtual water concept as a 'solution' to scarcity problems; see further down.

accountants count the benefits of economically and ecologically efficient virtual water markets, they tend to leave the other side of the virtual water equation unexplored: what will happen to the land and water resources 'freed' by virtual water trade from water-guzzling forms of food production? Most authors on virtual water seem to assume that, either automatically (the Invisible Hand) or with proper control, intervention and prediction, water security for all can be guaranteed. But is it really so straightforward to set the controls for optimum reallocation of resources like land and water for alternative uses? Who actually decides what is produced where and under which conditions? What is the role of existing notions of property rights to resources like land and water, and can approaches that take into account the local existence of complex and competing definitions of property rights be reconciled with views that take such state control for granted? We will argue here that only by leaving these questions basically unaddressed, can the image of virtual water as a politically silent solution be upheld.

This chapter consists of the following sections: after this introduction, we will give a short explanation of our approach to governance to deal with the questions raised above. In the next section we sketch the background of the emergence of the virtual water concept and engage with it. We shall discuss its possible contributions and pitfalls, also in the light of recent developments on the food and resource markets. After that, we discuss the assumptions of 'governability', more specifically pertaining to the control over land and water for food production. In the last section we present a short conclusion, the main message of which will be that normative approaches to governance and naive assumptions of governability of land and water resources are untenable. Seen in this light, the belief in virtual water as a 'solution' is untenable as well. As the concept is moving away from the desks of the water accountants to the world of policy-making and implementation in the 'real' world, a crucial paradox manifests itself: the more seriously virtual water policies are pursued to solve water scarcity problems in a politically silent (and economically invisible) way, the more political noise this will create as a consequence of the very interventions needed to implement virtual water policies.

2. Engaging with 'governance'

The concept of 'governance' (re-)emerged in the 1990s and has since experienced a meteoric rise in the policy literature. However, there appears to be little consensus on the precise meaning of the concept and its implications for both analysis and application (Nuijten *et al.*, 2004; Peters, 1998; 2000). Much confusion about the concept is caused by mixing up the analytical (what 'is') and the normative (what 'should be') (see Nuijten *et al.*, 2004; Von Benda-Beckmann *et al.*, 2009). Adding to the confusion is the direct association – at least in the eyes of some – of 'governance' with the explicitly normative concept of 'good governance'. The impact of such implicit interpretations of governance has been conspicuous in development: while DFID and the World Bank, for instance, go by relatively neutral definitions, USAID more normatively likens governance to democratic decision-making (Grindle, 2007: 556-557). This vagueness serves a purpose: it invites a wide variety of actors to subscribe to it and formulate a diversity of approaches and (sometimes contradictory) policy recommendations, reflecting various degrees of instrumentality, normativity or analysis associated with it. This may be the daily bread and butter of politicians - for analytical purposes, however, it proves problematic.

The oft-used catchphrase 'from government to governance' (Rhodes, 1996) has further added to the confusion, suggesting that 'before' all governing was top-down, while 'now' there is – or at least should be – a more or less harmonious distribution of responsibilities between public, private and civil society actors to tackle complex contemporary policy challenges. It also assumes that this latter state of affairs is clearly preferable, although it needs perfecting. The state, formerly dominant in resource management, has suffered from overload, and is now sharing responsibilities with the private and civil-society sectors. Hence the ways in which steering in society takes place is seen as shifting from 'government' to 'governance', in the process becoming more transparent, democratic, efficient and accountable. One of the most confusing aspects of such evolutionist thinking is that it is not only based on untenable assumptions about processes and conditions of steering now and in the future, but also on equally untenable assumptions about processes and conditions in the past.

Opinions differ on whether the concept has something new to contribute at all. Thus, Biswas states that 'at least in the water sector, there is the risk that old wine is being recycled in a new bottle with a label of governance'.[532] Von Benda-Beckmann *et al.* (2009) however remark that 'old wine in new bottles' is too simple. First, they argue, the politics behind this attention to governance have had a real impact on both thinking and policies, leading to e.g. decentralisation, structural reforms, and interventions for 'good governance'. These changes need our analytical attention and deeper understanding. Second, there are real changes in the scope and scale of processes of governance, number and types of actors involved, networks through which they are connected, and normativities on which contestations and legitimations are based. Compounding these developments, an 'institutional void' (Hajer, 2003) seems to be opening up 'where there are no accepted rules and norms according to which politics is to be conducted and policy measures to be agreed upon' (Hajer, 2003: 175). This raises new questions about legitimacy, legal responsibility, and the seat of legitimate authority in governance processes (see Von Benda-Beckmann *et al.*, 2009).

An emerging 'global environmental governance' literature (e.g. Karns and Mingst, 2004; Wöstl *et al.*, 2008) suggests that institutions with a global span can be tweaked

[532] Available at: http://www.thirdworldcentre.org/governance.html. Accessed April 2010.

to further goals that will make the globe a better place for all, e.g. in water saving and poverty reduction. These goals, and the means to attain them, appear as an uncontested issue of institutional (re)design. For Mehta *et al.* (1999: 18) governance 'has become something of a catch-all to describe the ways in which the activities of a multitude of actors, including governments, NGOs and international organisations, increasingly overlap .... It describes a complex tapestry of competing authority claims' (Mehta *et al.*, 1999: 18). These authors zoom in on an oft-neglected aspect of normative and instrumental approaches to governance: governance processes are inherently *political* processes, involving actors and institutions contesting existing or proposed structures and instruments for controlling society, competing for legitimate authority, power and control over resources.

The existence of such different approaches to governance raises questions about what governance does, what it can do and what it should do. In order to avoid the type of confusion discussed above and to be able to pursue the analytical agenda set by von Benda-Beckmann *et al.* (2009), we take an analytical, multi-level perspective as represented by Nuijten *et al.* (2004: 123) who define governance in terms of 'processes of steering, ordering, ruling and control', as well as their results. Contrary to normative approaches, in this approach no *a-priori* assumptions are made about the governing actors, those 'to be governed', their relationships and steering capacities, directions and outcomes, and the existence of shared goals.

The present contribution applies such an analytical perspective of governance to virtual water as a form of multi-level (global, state, societal) governance. We shall mainly scrutinise the two following key elements of governance attached to virtual water as a policy prescription: first, the assumption that water scarcity issues can (and should) be governed through the market; second, the idea that land and water under existing uses and property regimes can be 'freed up' unproblematically and in a politically silent manner, to be transferred to other uses.

These questions both problematise, in Nuijten *et al.*'s (2004) terms, the processes of steering and control (resource governance as a neutral 'toolbox') and their intended results - who is likely to get what out of a virtual water-policy, where, why, and how?

3. Virtual water: an emerging debate

3.1 Averting the water crisis? From water wars to virtual water peace

The narrative of water crisis and even water wars has proved to be a popular one. Gleick (1993) for example, proclaimed water to be 'in crisis', a sentiment echoed by the World Water Council.[533] This scarcity crisis was assumed to inevitably lead to violent conflict. Influential articles and books appeared claiming that 'water wars' were imminent (Starr, 1991; Bulloch and Darwish, 1993; Shiva, 2002). A deadly mix of global mismanagement, population pressure, resource capture and scarcity, these authors believed, would mean that water would soon run out. Water, always considered 'low politics', would be dominating the future. Kaplan (1994) called for US intervention to avoid resource wars and anarchy. Growing attention to climate change and its consequences for humanity again propelled the fear of 'water wars' up the political agenda. Thus, Ban Ki Moon, Secretary General of the United Nations, sees climate change as 'likely to become a major driver of war and conflict' (Buckland, 2007: 1).

However, while some authors still zoom in on the bad news of a water crisis (e.g. Barlow, 2007), others come up with better news: The threat of crisis can be averted by better policy and more efficient water use. Thus, the Comprehensive Assessment of Water Management in Agriculture (Molden, 2007) answers the question whether there will be enough resources (land, water, human capacity) to produce food over the coming 50 years or whether we will run out of water as follows: 'It is possible to produce the food – but it is probable that today's food production and environmental trends, if continued, will lead to crises in many parts of the world' (Molden, 2007: 1).

There was more good news: contrary to what is suggested by the spectre of water wars, there is no clear causal relationship between resource scarcity and war. Putting into perspective the neo-Malthusian (e.g. Homer-Dixon, 1999) scenario of resource crisis leading to war, Wolf (1998) had already noted that water wars have been extremely rare in history while, in situations of growing scarcity, the number of co-operative river agreements is even steadily growing. From the NGO community Buckland (2007), discussing the assumed relationship between climate change and war for food, water, and the refugee problem, stresses that there is no strong evidence of a causal relationship between water crisis and the emergence of water wars.[534]

Thus, in reaction to this, gradually a more optimistic 'water peace' camp made itself heard, which saw the post-Cold War period as an opportunity for reform towards environmentally sound, participatory, integrated, co-operative basin water management. This view was supported by the concept of virtual water.

[533] Available at: www.worldwatercouncil.org. The BBC news homepage wrote on 15 March 2000: 'a war over water? Government ministers and experts will be gathering in The Hague at the Second World Water Forum this weekend to try to avoid it' (available at: http://news.bbc.co.uk/2/hi/middle_east/677547.stm).
[534] For the political role of water in the Middle East, see Selby (2005). This author rejects both the water war thesis and the 'liberal functionalist' stance (that water nurtures co-operation). He rightly pleads for a political-economy approach to water and water scarcity.

If economic efficiency of water use can be increased by adapting food trade to existing variability in water scarcity, a net water-saving effect can be reached by concentrating production of water-intensive food crops and products with low water productivity in water-abundant countries, while turning water-scarce countries into food importers and producers of less water-intensive crops. While agricultural production in the former tends to be based on rain-fed agriculture, in the latter it requires complex interventions in the 'real' water cycle (irrigation systems, dams and other large infrastructure) that often contribute to depletion of scarce water resources and ecological degradation (Chapagain *et al.*, 2005).[535]

This view, moreover, gives us a completely different perspective of water management. Water managers and analysts always used to conceive of water as a physical good that is hard to transport over long distance and can only be controlled with heavy infrastructure; and, as a vital good, often subsidised (Bakker *et al.*, 2006). As a result, the state was the natural water manager, whose 'hydraulic mission' it was to bring water to people mainly through large infrastructural projects (see Molle *et al.*, 2009). This supply-side bias of 'water resources development', however, is now reaching its limits.

If virtual water is recognised as an option, water can be moved around much more easily. The advantages would be many: it can also be handled by private actors, does not deplete the local resource base, and does not have to come for free. Virtual water allows a scope beyond the national political economy. Food is largely produced in temperate zones and sold to drier zones[536], so that the world food market is, in fact, an efficient mechanism for reallocating scarce water resources. Growing food where water is abundant and selling it where it is scarce eases the pressure on resources in arid countries and prevents 'water wars' from breaking out (Allan, 2002). The 'iron fist' or 'velvet glove' (Jessop, 2003) can be replaced by the 'invisible hand' of the world market to redistribute water. Gradually the concept was taken up by academics and policy-makers alike.

Virtual water gained a global platform when 'water prince' Willem-Alexander of the Netherlands mentioned it in his closing address to the Third World Water Forum in Japan, March 2003.[537] It has since started on a remarkable march through development policy, agricultural research and planning institutions (Rosegrant *et al.*, 2002; UNESCO, 2006; World Water Council, 2004) to become a dominant mindset in the water sector, triggering several research projects to calculate where

[535] The concept can also be applied to other (non-food; non-agricultural) goods and services (Hoekstra and Hung, 2003) and resources (e.g. land, labour).

[536] Although large agricultural exporters such as Australia and large swathes of the US would not qualify as 'temperate'. In the US, the gigantic Ogallalla aquifer is rapidly depleting due to virtual water exports while irrigated areas in California are drying up fast as well (Aldaya *et al.*, 2008).

[537] Available at: e.g. www.adb.org/documents/periodicals/adb_review/2003/vol35_1/vision.asp. For earlier use see: http://www.tradeobservatory.org/headlines.cfm?refID=17590.

in the world water is, or can be, most efficiently used and transported (Hoekstra, 2003). These calculations have yielded lots of quantitative information on trade flows and their virtual water contents.

Virtual water was primarily meant to visualise the quantity of water needed for food (and non-food) production. However, water use for agriculture is under growing pressure of reallocation to non-agricultural sectors and uses.[538] Main competitors in the arena of water transfers are irrigators on the one hand, and urban, domestic and industrial users on the other. The latter tend to be winning out, as water use for irrigation is regarded as the least beneficial and productive in market terms. Reallocation is difficult, and the limited capacity of the state to fulfil all water demands is socially and politically sensitive. 'Closure' of river basins and the tensions caused by competing claims of upstream and downstream administrative regions or states add to this political sensitivity. As noted, the most pessimistic water analysts believe that transboundary water scarcity can even cause water wars (see Kumar *et al.*, 2005; Warner, 2003).

Bringing together issues related to food (feed, biofuel) production, consumption and trade, the virtual water metaphor has stimulated scientific efforts to estimate and map virtual water flows between (and within) nations and regions in the world. This has yielded the water accounts and 'footprints' (Hoekstra, 2003) that are prominently present in the virtual water literature. Thus, the exact relationship between water use and food (feed, biofuel) production can be made visible. But the debate is more than just an abstract academic exercise. It has also entered the policy domain, where recipes ('solutions') for dealing with water scarcity are being produced and environmental conservation and food strategies based on it are recommended as good policy for water-scarce areas (Warner and Johnson, 2007; Wichelns, 2004). Hoekstra and Hung (2003: 46) stress that the next step is 'to go beyond 'explanation'' and use the concept as a basis for interventions that enhance global water use efficiency. In a nutshell, the concept's main practical ambitions are, first, to solve water scarcity problems at various scales and levels of governance by making optimal use of comparative advantage and differences in resource availability; second, to prevent water conflicts in a 'politically silent' manner (Allan, 1998; Hoekstra, 2003).[539]

One way of dealing with water scarcity is, for instance, through the import of staple foods like wheat and rice. Putting into practice the virtual water concept,

---

[538] In 2000, an estimated 67% of the world's total freshwater withdrawal went to agriculture. By 2025, water requirements in agriculture will have increased by 1.2 times, in industry by 1.5 times, and for domestic consumption by 1.8 times (UNEP, 2002).

[539] Discussing virtual water trade between India and Bangladesh and its prospects in food policy, Parveen and Faisal (2004) alert us to virtual water trade becoming linked to sensitive political issues like international boundary conflicts (with negative consequences for food security); another illustration that virtual water is far from politically 'silent'.

'countries are, in effect, using grain to balance their water books' (Brown, 2006: 55). Food imports to a water-scarce country can release scarce water for economically more productive (e.g. industrial) uses (De Fraiture *et al.*, 2004). To give a basic indication: the amount of virtual water in international trade is estimated at 1.6 trillion cubic metres annually (Chapagain *et al.*, 2005; UNESCO, 2006)[540]: 16% of all water use is now for export (Hoekstra and Hung, 2003). As virtual water enters the domain of policy-making, it becomes entangled with the socially and politically sensitive issue of food policy and, hence, with questions of food self-sufficiency, food security, and food sovereignty (see also below).[541] It is here that we leave the dream castle of neutral virtual water policies.

In the virtual water debate, water is considered an economic good (e.g. Hoekstra and Hung, 2003). Much virtual water literature reflects this economic bias, engaging mainly in the calculation of crop virtual water contents, trade-related water flows, and estimated volumes of savings, which have a temporal and a spatial dimension: the food export from water-abundant to water-scarce countries and the virtual water storage capacities of food in periods of water scarcity respectively (Chapagain *et al.*, 2005). Some authors distinguish several levels of intervention at which these efficiencies can be reached: the local level, where pricing structures will stimulate water-saving technology and create awareness of scarcity; the catchment or river basin level, where available water will be re-allocated to uses with the highest marginal benefits; and the global level, where comparative (dis-)advantage will stimulate import or export of virtual water (Chapagain *et al.*, 2005).

3.2 Virtual water: a fresh approach to water scarcity and food production

The use of the concept has its merits and pitfalls. Let us first look at its possible contributions.[542] First, it makes explicit a dimension of food (and other goods) that often remains (largely) hidden: the role of water in its production and the scarcity and opportunity costs of this resource (Wichelns, 2004; 2005; 2010). It is an instrument to take into account the important factor of soil moisture or 'green water' (Allan, 1998; Aldaya *et al.*, 2008). The relationship between water scarcity and food production makes it possible, in principle, to relate production, consumption, and trade policy to food self-sufficiency, food security and food sovereignty in a broader resource scarcity and sustainability perspective.

[540] About one fifth of total world trade; 80% of virtual water flows is embedded in agricultural -, 20% in industrial products. Produced with an actual 1.2 trillion cubic metres per year, a 352 billion m^3 saving is claimed, especially in crop production. The 'water footprint', which can be determined internally or externally, for individuals or nationally, expresses water use in relation to consumption (Chapagain *et al.*, 2005; UNESCO 2006: 392).

[541] For virtual water policy in relation to food production in Egypt and the Indian state of Punjab, see Roth and Warner (2007).

[542] For a fuller discussion, see Roth and Warner (2007).

Second, it globalises discussions on water scarcity, ecological sustainability, food production and consumption. Virtual water is an important 'tele-connection' in the global water system, bringing together biophysical, institutional and governance dimensions of water scarcity (Crasswell, 2005). Therefore, it is a promising entry point for analyzing the global, regional and local dimensions of issues concerning food and natural resources in light of the growing interdependencies of a global market.

Third, virtual water stimulates reflection on processes of socio-economic change in relation to changing food consumption patterns leading to growing and changing demands, production needs, pressures on current production systems and the environment (Hoekstra, 2003).[543] These changes have serious consequences for food production and trade, and natural resource use (water, land, forest). They also affect social relationships, networks and power relations. The rapid economic growth of giants like India and China and the emergence of large middle classes that exert a growing demand and develop new consumption habits (e.g. more meat) are a major challenge. Attention to virtual water contributes to the debate if it is embedded in a broader framework. Meat consumption, for instance, is not only water-intensive, but may also boost the cutting of existing forests for soy cultivation (see Chapter 14 in this volume).

Fourth, the concept may increase awareness of water scarcity among consumers, public officials and producers (Wichelns, 2004, 2005). This may stimulate the development of water-saving policies, 'more crop per drop' technologies, water pricing and allocation that reflects scarcity. Water allocation between sectors in ways that create high returns and employment levels are crucial for economic and political stability (Allan, 2002). According to Hoekstra (2003) it is indispensable for developing rational national policies and promoting water savings 'through enhancing food security by appropriate agreements and increasing reciprocity in agricultural food products trade' (Hoekstra, 2003: 21; Wichelns, 2004, 2005). It may also be helpful in valuating water-related processes and trends. In China, for instance, 'real' water is now increasingly transferred from the water-abundant South to the water-scarce North, while North China exports much water-intensive food to the South (Ma *et al.*, 2005). Intriguingly, while some authors see virtual water as a possible alternative to real water transfers, others see the latter as an alternative to virtual water (Kumar and Singh, 2005).

Finally, virtual water is useful in understanding what water-related food policies would actually mean for demand and supply. Though starting from an environmental doom perspective, Brown (2006) gives an interesting picture of China turning virtual water theory into practice. China's grain production has been dropping, in part due to water shortages, from 392 million tonnes in 1998 to 358

[543] Thus, every individual turning to meat consumption enlarges his water footprint by 4,000 litres a day!

million tonnes in 2005, a drop that 'exceeds the annual Canadian wheat harvest' (Brown, 2006: 45). According to Brown, the world grain markets will ultimately be the battlefield on which the 'water wars' will be fought.

3.3 Virtual water as another metaphor for the real world: need for caution

Let us now highlight some criticism of the concept. Apart from discarding the label as wrong-headed and irrelevant,[544] Merrett (2003a,b) feels that what he considers the false and misleading assumptions underlying the concept puts analysts on the wrong track: it focuses on the water contents while the most problematic dimensions of virtual water is its free-market bias, the issue of political control, the lack of attention to negative impacts of food imports on domestic agriculture and rural livelihoods, foreign exchange needs, and the dependence on import subsidies. Chapagain *et al.* (2005) add the stimulating effect on urbanisation and new environmental and other externalities in exporting countries.

The most cited examples in the virtual water literature are not always the most convincing ones. To what extent can Egypt's food policy be analyzed as based on a conscious policy decision to implement 'virtual water policy'? Arguably, ecological and demographic processes left little room for alternatives. Egypt does not 'save' much water through its imports; starvation of its population is the only alternative to virtual water import. It is now further developing a water-saving programme, but has no strategic interest in greatly reducing its claim on water (De Fraiture *et al.*, 2004; Wichelns, 2005). Furthermore, the issue cannot be seen in isolation from Egypt's geopolitical status in the Middle East. As a major regional political ally of the United States, it controls resources not commanded by countries like Ethiopia (Warner, 2003).

Allan's approach leans heavily on a classical political-economy analysis in terms of comparative advantage (Allan, 2003). However, as Wichelns (2005: 430) remarks, comparative advantage is more than just resource endowments. Even the production of water-intensive goods may give a comparative advantage to a water-scarce country or region.[545] Therefore, virtual water does not suffice to determine optimal production and trade strategies for a country. Kumar and Singh (2005) show that there is no correlation between relative water availability

[544] Merrett discards the concept as redundant: why not just speak of water requirements? Why are food imports presented as water imports? Furthermore, there is nothing 'virtual' in water needed for food production: virtual water is real water (2003b).
[545] Wichelns (2005) discusses wheat production in Saudi Arabia and Sudan, and grass and ethanol in the USA.

in a country and its virtual water trade.[546] Much trade takes place between either water-abundant countries or for reasons unrelated to water scarcity. According to De Fraiture *et al.* (2004) in 2025 more than 60% of cereal trade will be unrelated to water. An important additional argument is that water tends not to be the decisive factor in countries' crop production and trading strategies (De Fraiture *et al.*, 2004; Wichelns, 2005, 2010).[547]

The water focus tends to background other production factors (land, labour, genetic resources). Moreover, it leaves largely unanswered the question what is going to happen with the water that is 'saved' by virtual water imports, whether and how it can be used more productively.[548] De Fraiture *et al.* mention the Asian paddy areas as an example where 'saving' is hardly useful or even possible (Chapagain *et al.*, 2005; De Fraiture *et al.*, 2004; IWMI, 2007; Wichelns, 2010).[549] Food crops might even be replaced by more water-intensive (food or non-food) crops because of their higher market value. Lichtenthäler (2002), for example, shows how Yemeni farmers, out-competed by cheap imports and food aid, have shifted to *qat* (a stimulant consumed by chewing) production. Wichelns (2004) stresses the role of comparative advantages (which is not just resource endowments but also includes a comparison of production techniques and opportunity costs). Kumar and Singh (2006), therefore, seem right in concluding that a discussion of food security and food policy should not be based on a water-resources perspective only, but also take into account factors like the availability of arable land and size of landholding.

The concept is excessively optimistic about the role of the global food market in solving water scarcity problems. It seems that virtual water trade takes place on a level playing field, that there are only winners and no losers. We think that much more caution is needed here. Though presented as an unproblematic solution designed on the basis of rational calculation of comparative advantages and efficiencies, there is much disagreement on globalisation of trade, increasing interdependence, the market and the inequalities it creates or reproduces (see Hoekstra and Hung, 2003; Warner, 2003). Many developing countries are

[546] Countries with high water availability like Japan, Portugal and Indonesia have high virtual water imports, while Afghanistan, Malawi, India, Thailand and Denmark are water-stressed but export much virtual water. Many water-rich countries have few arable land resources available for using their water resources in crop production. Virtual water flows often go from water-poor but land-rich countries to water-rich but land-poor countries (Kumar and Singh, 2005: 765, 785). See also the recent debates about 'land-grabbing' discussed below.

[547] Political and economic considerations like production increase for food self-sufficiency, poverty reduction and food security may be more important and constrain policy options for virtual water trade, especially for poor countries sensitive to price fluctuations on the food market (De Fraiture, 2004; Kumar and Singh, 2005; World Water Council, 2004). A contrary trend pushes countries away from food self-suffiency, to meet the growing water demand from other sectors (Kumar and Singh, 2005).

[548] Note that not all 'saved' water can be regarded as real savings. In monsoon regions, for instance, water capture and storage are difficult, and alternative uses may be hard to find (IWMI, 2007).

[549] How much water is saved also depends on the definition used: the volume used by the exporter or the volume the importer would have used in the absence of export? (De Fraiture, 2004; Hoekstra, 2003).

economically too weak and vulnerable to opt for the kind of world market dependence associated with virtual water. In addition, issues of food self-sufficiency in relation to other (geo-)political factors may be more important in determining national policies than water scarcity issues (De Fraiture *et al.*, 2004). Gupta and Deshpande (2004: 1222), for example, state for India that:

> '... given the large size of the Indian population, the role of agriculture both directly and indirectly in the Indian economy and the desire to become and be seen as a developed nation, food security for the country is something on which no Indian may be willing to compromise. The begging-bowl image of the sixties cannot be accepted. Therefore, a planned import of food grains to feed the burgeoning population of the country in the coming years is not considered any further'.

Recent global developments seem to bear out these critical remarks. The rise of biofuels (and, until recently, their EU endorsement) has primarily fuelled debates about allocation of scarce resources to their production. Rising food prices in 2008 were associated with pressures on the world market caused by the growing demand for (and speculation with) food, feed and biofuel crops, as well as knock-on effects from other commodity markets, triggered by failed harvests in China. Such trends may make policies based on virtual water even less attractive for countries like India and China. Another recent development related to the 2008 food price increase and the growing distrust in regional and global food markets is the surge in foreign private and state investments from food-importing countries in agricultural land in several countries in mainly Africa and Asia (Von Braun and Meinzen-Dick, 2009; Smaller and Man, 2009).[550]

Smaller and Mann (2009) add an important but underexposed dimension to this issue: the water component of the recent investments in land. The authors state that 'what are often described now as land grabs are really water grabs' (2009: 3). According to the authors, water is one of the main long-term factors driving this trend of foreign investments in land. Investments are focused in countries with a high irrigation potential like Sudan and Mozambique. These are opportunities for countries like Saudi Arabia, which has recently decided to phase out its wheat production by 2016 (Smaller and Mann, 2009). For our discussion it is interesting to note here that this country (like many others), driven by distrust of the market rather than believing in it, seeks investments in agriculture abroad rather than relying on the virtual water market. This trend from local farming towards 'long-distance farming' primarily aims to satisfy food and other needs, leading to the

[550] Von Braun and Meinzen-Dick (2009) discern two types of countries involved here: food-importing countries with land and water constraints but much capital (e.g. Gulf states), and countries with large populations and food security concerns (e.g. China, South Korea, India). Their target are primarily countries with lower production costs.

increased commoditisation of land and water 'subject to globalised rights of access' (Smaller and Mann, 2009: 7). While proponents see benefits for the poor (jobs, infrastructure, poverty reduction), the critics fear that the unequal power relations involved in these processes may in the end damage local livelihoods of inhabitants dependent on local resources like land and water. Especially where land and water are used under customary tenure arrangements, conflicts may occur leading to social and political instability (Von Braun and Meinzen-Dick, 2009).

Windfuhr and Jonsén's (2005) food sovereignty policy framework aims at 'refocusing the control of food production and consumption within democratic processes rooted in localised food systems' (2005: preface). Virtual water hardly deals with crucial questions about food security, the right to food, food self-sufficiency and food sovereignty, though its impact on them may be considerable. Food security can be reached through strategies based on self-sufficiency, imports or mixes of these.[551] Food sovereignty, stressing people's rights to define and control food-related issues (see above), goes beyond food security. The aggregate concept of 'food self-sufficiency' stands apart from both food sovereignty and food security, in the sense that a country may be self-sufficient in food production while part of its citizenry suffers from chronic malnutrition or even acute hunger. There is a considerable gap in terms of starting points between the virtual water approach, and food security and sovereignty approaches. While the former has solving a water scarcity problem as its main objective and sees global food markets primarily as a policy instrument to reach it, the latter regard food security as their main objective for which food sovereignty is a precondition and the formulation of a basic human right to food a major instrument (Windfuhr and Jonsén, 2005; World Water Council, 2004).

In a war or crisis environment, or war-prone neighbourhood, states have sought food self-sufficiency. The current economic crisis has impelled the rediscovery of export bans and inter-country barter deals (Burger *et al.*, unpublished data). The Middle East is a prime example. Residents of 'rough neighbourhoods' will try and avoid dependence on outside sources. Thus physical water trade initiatives (such as Turkey's various Peace Pipeline schemes) have not seen enthusiastic uptake, and water-poor countries such as Saudi Arabia decided to deplete water resources to grow cereals and tomatoes in the desert rather than buy them from its unruly neighbours Iran and Iraq. The same mechanism takes place at the state-society interface: as Lichtenthäler notes, sub-state actors like Yemeni tribes, in turn, seek self-sufficiency to avoid dependence on corrupt rulers. Tribesmen are fully aware that food – like 'real' water in oases – can be used as a source of control (Lichtenthäler, 2002: 12-13).

[551] Food grants are another option (Wichelns, 2004).

In conclusion, we should approach with caution the claim that virtual water is part of a solution to water scarcity by saving scarce water in water-stressed areas in a politically silent way. The concept is particularly problematic when introduced in a policy context. Most problematic is the overly optimistic faith in the market as a water and food governance mechanism (Pierre and Peters, 2000) and the assumed politically silent character of decisions concerning food imports taken in such a framework. This is illustrated by recent developments in the markets for staple foods, in recent trends in the control over land and water resources, and by the recent food price hike. But the return of a strong belief in state intervention has its own practical difficulties. We will elaborate on this latter point in the next section.

4. Virtual water governance: some uncomfortable questions

Above we have dealt with one side of the virtual water equation: the trade in agricultural products. This section focuses on the other side: the assumption that water (and agricultural land) can be unproblematically reallocated from wasteful food production to other uses and users. Virtual water analysts do their accounting on a nation-state level, suggesting that you can tweak and fine-tune agricultural production (and virtual flows of water) according to the lowest ecological costs and most efficient use of available resources. However, studies from a number of overlapping fields like social and legal anthropology, political science and political ecology show that deciding what is produced where, by whom and under what conditions is not always as simple as that, and often cannot be determined in a top-down fashion by government policies. Such issues are hardly dealt with in the virtual water literature, primarily because of its disciplinary nature (water engineering, resource economics) and macro-orientation (Youkhana and Laube, 2009). In the politically silent space of scientific reports, the neat virtual water figures and accounts look impressive and appealing to policy-makers. However, the macro structures and figures can easily distract from real-life people and their economic activities, forms of political organisation, institutions, cultural practices and social networks. It is these that determine the degree to which virtual water policies might work out in specific localities of resource use and be influenced by affected people.

Here we enter the domain of governance of resources between the national and the local levels. Virtual water as a policy solution takes for granted the feasibility of reallocating water and land 'saved' from specific forms of production with the aid of a virtual-water policy to other functions, uses and users. However, in our view 'implementing' this savings side requires increased and deeper intervention in existing practices of resource use and allocation, in definitions of legitimate rights to land and water, and in jurisdictions regarded as authoritative in these domains. Again, one should be prepared here for political noise rather than silence.

Above we have discussed the importance of making a distinction between normative or prescriptive and analytical approaches to governance, and stressed the need for taking an analytical perspective. Normative and instrumental views of governance are often based on implicit assumptions about who should 'do' the governing, who should be allowed around the table when decisions are made, and which instruments can be used to steer human behaviour in society. Whether it is the state, the market or civil society that is supposed to realise certain desirables, efforts to control and steer are often based on social and legal engineering approaches that reduce complex problems to relatively simple, linear processes within manageable proportions. This makes them digestible for steering instruments like laws and policies, after simple solutions have been devised (Cleaver and Franks, 2008; Mollinga, 2008). Narratives with a high appeal for policy-makers – e.g. virtual water as a solution to water scarcity – further reduce the scope for reflection and engagement with nasty questions. But can the use of key natural resources like water and land be governed in this way? How realistic is this assumption?

The state has coercive, economic and communicative instruments at its disposal to influence the behaviour of its citizens. For the 'hydraulic state', development of water resources was tantamount to survival. There are many examples of large-scale forms of social, legal and economic engineering in states with a 'hydraulic mission' (Molle *et al.*, 2009). Major approaches have been, and still are: bringing water to people (supply management; water resources development, e.g. by diverting rivers, river-linking or water transfers); or bringing people to water (mass resettlement). However, increasing worries about scarcity and river basin closure (however spurious; see Molle, 2008; Williams, 2001) put serious limits on supply-based approaches (unless based on massive water transfer or river linking projects as in China or India), while growing populations limit options for large-scale resettlement. Contemporary governance strategies, therefore, also seek to affect the allocation, distribution and efficiency of use (more crop per drop) of water through technical, legal (property rights) and economic (price incentives and controls; quotas) instruments.

Virtual water has been presented as a new addition to the palette of policy instruments a state has at its disposal. Thanks to virtual water 'an increased range of options become available to the strategic-level decision-maker.'[552] A key 3rd World Water Forum session bore the slogan 'from unconscious behaviour to conscious decisions', and advocated the promotion of virtual water as a solution for arid countries.[553] But what might this involve? Above we have mentioned some of the benefits of virtual water as an instrument of communicative governance. It visualises supposed misallocations of water and other resources. The related water footprint (Hoekstra, 2003) seeks to raise public awareness and incite a change of

[552] Available at: www.awiru.co.za/pdf/trutonanthony2.pdf.

[553] Available at: http://www.waterfootprint.org/Reports/Virtual_WaterTrade_and_geopolitics.pdf.

behaviour in consumption, dietary habits and lifestyles. As we have seen, virtual water is primarily seen as an instrument of economic governance through the market. This is shown by the UNESCO/IHE Virtual Water Programme, which shows the ambitions of providing governments with information and tools to reach water saving by engaging in virtual water trade.[554] However, the policy-induced change of function foreseen for land and water resources in order to 'cash' the water benefits of virtual water trade nationally or regionally would probably be more or less contested, and lead to approaches based on combinations of finding negotiated solutions with more coercive methods. Below we will further discuss the lack of attention to social and political contestation in such processes.

4.1 Virtual water and the new neglect of local contexts of land and water use

In recent years we see a growing awareness of the importance of analyzing water scarcity and other water-related issues in the specific context of use, management and governance. Though water problems have been increasingly presented as 'global' and part of a global water and climate crisis, this is true only to some extent. In its actual physical properties, uses, and use and control-related histories, socio-cultural and political arrangements, water is, by definition, highly contextual. Water problems and solutions always relate to specific socio-political, economic, cultural and agro-ecological contexts as well as histories. Policies in the context of which 'solutions' to water problems are framed will have to take these contexts into account (see Donahue and Johnson, 1998; Mollinga, 2008).

However, this is not always recognised by those who embrace 'governance' as a solution to a host of water problems. Contrary to this attention to water use and control contexts, normative interpretations and biased uses of 'water governance' are often at the heart of pleas for 'global water governance', involving specific policies and solutions that seem to move away from such specific contexts.[555] Thus, Pahl-Wöstl *et al.* (2008: 421) state that 'many anthropologists and others continue to argue that one needs to understand local rights, needs, and stakeholders in order effectively to address governance issues'. On the basis of their 'adaptive governance' framework, the authors stress that 'some kind of formal global co-ordination is required in tandem with more decentralised network and market-based approaches' (Pahl-Wöstl *et al.*, 2008: 432). In such an approach, network governance and 'the market' rather than understanding contexts of water use and control seem to hold the key to problem-solving. This takes empirical analysis

[554] Available at: www.unesco-ihe.org/Project-activities/Project-database/Virtual-Water-Trade-Research-Programme.

[555] Seriously taking into account 'the local' (local histories, forms of embeddedness of e.g. land and water use, local dynamics of development interventions or state policies) is a real nightmare for policy-makers and implementers, but also for many economists, social, or administrative scientists bent on feeding the policy world with new recipes for interventions.

of water governance further away from an approach that 'steers a way between extreme localism and universalising positivism' (Cleaver and Franks, 2008: 172).

In our view a concept like virtual water requires due attention to its possible impact on all levels of resource use and governance, from global to local, rather than to national-global interactions only. Such attention has been remarkably lacking until this day. Critical views of the concept are too often limited to resource and trade economics (e.g. Wichelns, 2010). Youkhana and Laube (2006: 13; see Youkhana and Laube, 2009), however, go well beyond this, criticising the concept from a livelihood perspective. They conclude from their research in the West-African Volta region that application of the concept in real-life water use settings is problematic because it glosses over cultural, socio-economic, market and political-institutional factors. According to the authors, these very neglected factors are the major drivers of economic development and policy processes. Hence, they doubt whether 'the widespread application of hydrological rationality in the design of economic policies' is feasible. In addition, they point to the possible dangers, as such policies may have 'potentially disastrous effects on the livelihoods of rural populations as well as the political stability of developing states' (Youkhana and Laube, 2006: 13, 2009). Horlemann und Neubert (2007) reach similar conclusions, among others on the potentially negative impacts on the economies and livelihoods of regions with poor populations.

Thus, these authors contribute a new and analytically valuable element to the virtual water debate. However, they do not raise another crucial element of water and land governance: the complex character of the ways in which rights to natural resources like land and water are defined in various systems at various levels of governance. Food (feed, fuel) production takes place in a variety of agro-ecological, socio-cultural and political contexts in which people use land and water resources in their attempts to subsist on, or derive an income from, agriculture (see Donahue and Johnson, 1998). Implementing a virtual water policy could have serious implications for the livelihoods of food-producing people (rights to resources, crop production, food (in-)security, dependence on the food market, migration).

However, the actual use of land and water on which the livelihoods of rural populations is based will not be the exclusive product of top-down policies, regulations and governance structures, as actor-oriented sociology (e.g. Long and Long, 1992) and subaltern studies (e.g. Scott, 1985) show.[556] The use and control of natural resources will, at least partly, be shaped by local actors, their social and legal relationships to resources like land and water, and their perceptions of ecological, social, economic and political conditions and processes (see also De Haan and Zoomers, 2005). Confronted with virtual water policies, they will play a

[556] Scott (1985) shows that the marginalised are not powerless, but have ways of avoiding and resisting control: the 'weapons of the weak'. The relevant literature on the water sector is also impressive.

co-determining role in the extent to which such policies are accepted, internalised, circumvented, changed, ignored, protested or rejected.

4.2 Complex rights to natural resources

A major ambition of virtual water policy is redirection of water to other uses. The assumption is that populations can be unproblematically engineered into new agricultural practices (or be moved out of agriculture), relations to the market, and property rights regimes to land and water resources. However, in practice, much depends on the complex and often plural character of social and legal institutions and frameworks (law, property rights) that co-determine the (legitimate) access to natural resources in interaction with human behaviour. Especially legal anthropologists have made an important contribution to the analysis of natural resource conflicts by pointing out that the existence of multiple, often contradictory, definitions of rights to natural resources like land and water can be an important cause of such conflicts.

The existence of multiple legal systems is called legal pluralism or legal complexity (Von Benda-Beckmann *et al.*, 2006; for water rights see e.g. Boelens *et al.*, 2010; Bruns and Meinzen-Dick, 2000; Meinzen-Dick and Pradhan, 2002; Roth *et al.*, 2005).[557] Such approaches take into account the existence of a variety of socially and politically legitimate values, norms, principles and laws deriving from various sources of validity, legitimacy and authorisation in the same socio-political space (Von Benda-Beckmann *et al.*, 2006). The scale and scope of such legal systems or domains may vary considerably, and may involve, for instance, local socio-political groupings or entities (e.g. customary systems or forms of 'local law' that emerge out of the interactions between various bodies of norms or laws), the legal or norms systems of the state, of religious groupings, bodies of rules associated with specific development interventions or reforms (e.g. donor-driven land and water reforms) or those emerging from the domains of international or transnational principles, norms and laws.

Approaches that take into account legal complexity regard 'the legal' as a domain of political struggle and contestation. Rather than seeing law as a neutral problem-solver and instrument of governance, such approaches analyze it as a socio-political resource used by competing social actors to legitimise their claims in struggles about control over natural resources. This shows what many resource struggles actually are: political processes that may involve contestation, negotiation and violence at various levels and between a variety of state and non-state actors. Often, such struggles are not exclusively about the resources but also about much broader political agendas of authority, legitimisation and control. Thus, for instance,

[557] For a good overview article about land tenure in Africa and the role of legal complexity, see Sjaastad and Cousins (2009).

Hellum and Derman (2005) have shown how the national water reform policy in Zimbabwe, taking shape in the complex multi-level force fields of international norms about water use, national policy ambitions, and local norms pertaining to water use, is deeply influenced by the politically inspired 'fast-track' land reform programme implemented by the Mugabe regime to buy societal support for its ambition to remain in power.

Therefore, the focus on the mobilisation of competing sources of authorisation and legitimation of claims makes explicit the linkages to broader struggles about political power and authority in society, in relation to issues of resource control: while actors are actively seeking sources of authority for legitimisation of their resource claims, these sources of legitimising authority boost their political legitimacy and authority in society by being recognised as such (Sikor and Lund, 2009). These are inherently political processes that tend to be obscured by, for instance, instrumental approaches to governance like those on which virtual water policy is based. A more scientifically and empirically grounded approach to these 'water-saving' dimensions of virtual water policy would at least require serious attention to the social, legal and political processes around water transfers from existing uses under specific definitions of land and water rights into future uses under new definitions of such rights, usually in contexts of increasing competition and politicisation of water issues. This would at least require paying attention to legal complexity, issues of fairness, equity, and legitimacy, and recognition of the need to find inclusive negotiated solutions rather than 'simple administrative or market transfers' (Meinzen-Dick and Pradhan, 2005). It will be clear that attention to specific local use and control contexts of water use, in relation to its wider use, policy, legal and other governance dimensions, is indispensable. Defining the water crisis as 'global' and assuming (supra)national legal control, while neglecting such local contexts of resource control is a recipe for political trouble.

5. Conclusion: coming down to earth

In this chapter we have discussed virtual water as a concept used in the analysis of trade in food (and other agricultural products) as well as in attempts to solve water scarcity problems through virtual water trade. In our analysis of water governance through policies based on the virtual water concept we avoid normative and instrumental approaches to governance. These tend to be coloured by all kinds of assumptions, beliefs, and implicit values and norms (e.g. 'from government to governance'; 'governance without government') that depoliticise the concept, making it less useful for the scientific analysis of political-institutional processes.

Used as a concept for the analysis of the water content of food and other crops, it can be useful to create societal awareness about a variety of environmental and other problems like consumerism, changing food demand patterns, over-use of water resources due to supply orientation in water bureaucracies, or biases in the allocation of land and water resources to certain crops. However, when virtual water becomes a policy concept, we are more sceptical about the prospects of its use on various grounds. A major point of criticism we raised in this chapter concerns the belief in and acceptance of 'the market' as a governance mechanism for virtual water. Evidence that nation states base their food production and trade policies and strategies on water scarcity is very weak. Recent developments on the food and resource markets – growing global demands for food, feed and fuel crops; large-scale land leases across national boundaries ('land-grabbing') in reaction to the recent food crisis – show that a concern for water resources seems to play second fiddle to (geo-)political and food security considerations.

This chapter highlighted the importance of paying due attention to the nuts and bolts of the 'water-saving' side of the virtual water story. Normative approaches to governance, including those propagating virtual water as a policy instrument, define water issues as global and abstract away from local water use and control contexts. Such approaches propose reliance on combinations of national and global forms of governance, through e.g. laws or the market, to govern water resources locally. Only a few empirical studies that explore the scope for implementation of virtual water policy in specific livelihood contexts express their doubts about the capacity of virtual water policy to determine the behaviour of land and water users. Actor-oriented sociology, subaltern studies and approaches from anthropology of law that take into account legal complexity call into question simple assumptions about the relationships between the socio-legal engineering instruments of governance and human behaviour.

In conclusion, while appreciating the analytical value of the virtual water concept and the debates it has started, we see the ambivalence of virtual water as both a theoretical and a policy concept as rather problematic. It creates tensions and contradictions between the norms and prescriptions buzzing around in the policy world, and the social and political realities of real-life resource use and governance. More awareness is needed of the normative and ideological assumptions (e.g. about scarcity, the market) behind the impressive figures and formulas of water budgeting that form the scientific legitimation of virtual water as policy prescription. Virtual water policy entails political decision-making about the (re-)allocation of resources, with important consequences for the livelihoods of both rural and urban people. As an area of political contestation, virtual water and its consequences in agricultural policy and natural resources governance unavoidably concern issues of law, rights, authority and power, inclusion and exclusion, equity and legitimacy. To reach its global water-saving ambitions, 'political silence' will not do the trick, as the smooth operation of market forces attached to the concept tends to be a chimera. Virtual water policies will have to be more interventionist rather than less, with important consequences for resource governance.

The more interventionist, and thus politically outspoken ('louder') strand, has a strong tendency to overestimate the smooth operation of governance forces. However, to really steer food production and trade, a kind of authoritarianism and force may be needed that will not sit easily with national constituencies. While virtual water has proved a highly useful signal, it has also brought with it all kinds of 'noise'. Exposing the political mechanisms of virtual water usefully highlights this noise. After all, the political silence mainly benefits elites that wish to delay more sensible water management. When confronted with the real world, what is currently hailed as a politically silent solution can be expected to be shouted down by the deafening noise of extremely sensitive and fractious political processes.

References

Aldaya, M.M., Hoekstra A.Y. and Allan J.A., 2008. Strategic importance of green water in international crop trade, UNESCO-IHE Institute for Water Education, Value of Water Research Report Series No. 25. Delft, the Netherlands in collaboration with University of Twente, Enschede, the Netherlands, and Delft University of Technology, Delft, the Netherlands.

Allan, J.A., 1997. 'Virtual water', a long term solution for water short Middle Eastern economies? SOAS occasional paper no. 3.

Allan, J.A., 1998. Virtual water: a strategic resource. Global solutions to regional deficits. Ground water 36: 545-6.

Allan, J.A., 2002. Water resources in semi-arid regions: real deficits and economically invisible and politically silent solutions. In: Turton, A. and Henwood, R., (eds.). Hydropolitics in the developing world. A southern African perspective. African Water Issues Research Unit, Centre for International Political Studies (CIPS), University of Pretoria, Pretoria, South Africa, pp. 23-33.

Bakker, K., 2007. The commons versus the commodity: 'Alter'-globalization, anti-privatization and the human right to water in the global South. Antipode 39: 430-55.

Barlow, M., 2007. Blue Covenant. The global water crisis and the coming battle for the right to water. New Press, London, UK.

Boelens, R., Getches, D.H. and Guevara-Gil, A. (eds.), 2010. Out of the mainstream. Water rights, politics and identity. Earthscan, London, UK.

Brown, L.R., 2006. Rescuing a planet under stress and a civilization in trouble. W.W. Norton & Co, New York, NY, USA.

Bruns, B.R. and Meinzen-Dick, R. (eds.), 2000. Negotiating Water Rights. Intermediate Technology Publications/International Food Policy Research Institute (IFPRI), London, UK.

Buckland, B., 2007. A Climate of War. Stopping the Securitisation. International Peace Bureau, Geneva, Switzerland.

Bulloch, J. and Darwish, A., 1993. Water Wars: Coming Conflicts in the Middle East. Victor Gollancz, London, UK.

Chapagain, A.K., Hoekstra, A.Y. and Savenije, H.H.G., 2005. Saving water through global trade. Unesco-IHE Institute for Water Education, Delft, the Netherlands.

Cleaver, F. and Franks, T., 2008. Distilling or diluting? Negotiating the water research - policy interface. Water Alternatives 1: 157-76.

Craswell, E.T., 2005. Water and poverty in Southeast Asia. The research agenda from a global perspective. Asian Journal of Agriculture and Development 1: 1-11.

De Fraiture, C., Cai, X., Amarasinghe, U., Rosegrant, M. and Molden, D., 2004. Does international cereal trade save water? The impact of virtual water trade on global water use. Rep. 4, International Water Management Institute, Colombo, Sri Lanka.

De Fraiture, C., Giordano, M. and Liao, Y., 2008. Biofuels and implications for agricultural water use: blue impacts of green energy. Water Policy 10: 67-81.

De Haan, L. and Zoomers, A. 2005. Exploring the frontier of livelihood research. Development and Change 36: 27-47.

Donahue, J.M. and Johnston, B.R. (eds.), 1998. Water, culture, and power. Local struggles in a global context. Island Press, Washington, DC, USA.

Gleick, H.P. (ed.), 1993. Water in Crisis. Oxford University Press, New York, NY, USA.

Grindle, M.S., 2007. Good enough governance revisited. Development Policy Review 25: 553-74.

Gupta, S.K. and Deshpande, R.D., 2004. Water for India in 2050: First-order assessment of available options. Current Science 86: 1216-1224.

Hajer, M., 2003. Policy Without Polity? Policy Analysis and the Institutional Void. Policy Sciences 36: 175-195.

Hellum, A. and Derman, B., 2005. Negotiating water rights in the context of a new political legal landscape in Zimbabwe. In: Von Benda-Beckmann, F., Von Benda-Beckman, K. and Griffiths, A. (eds.). Mobile people, mobile law. Expanding legal relations in a contracting world. Ashgate, Aldershot, UK, pp. 177-198.

Hoekstra, A.Y., 2003. Virtual water: an introduction. In: Hoekstra, A.Y. (ed.). Virtual water trade. Proceedings of the international expert meeting on virtual water trade, IHE, Delft, the Netherlands, pp. 13-23.

Hoekstra, A.Y. and Hung, P.Q. 2003. Virtual water trade: a quantification of virtual water flows between nations in relation to international crop trade. In: Hoekstra, A.Y. (ed.). Virtual water trade. Proceedings of the international expert meeting on virtual water trade, IHE, Delft, the Netherlands, pp. 25-47.

Homer-Dixon, T.F., 1999. Environment, Scarcity, and Violence. Princeton University Press, Princeton, NJ, USA.

Horlemann, L. and Neubert, S., 2006. Virtueller Wasserhandel: Ein realistisches Konzept zur Lösung der Wasserkrise? Deutsches Institut für entwicklungspolitik/German Development Institute (DIE), Bonn, Germany.

IWMI (International Water Management Institute), 2007. Does food trade save water? The potential role of food trade in water scarcity mitigation. International Water Management institute, Battaramulla, Sri Lanka.

Jessop, B., 2003. Governance and Metagovernance: On Reflexivity, Requisite Variety, and Requisite Irony. In: Bang, H.P. (ed.). Governance as Social and Political Communication. Manchester University Press, Manchester, UK, pp. 142-172.

Kaplan, R.D., 1994. The Coming Anarchy. How scarcity, crime, overpopulation, tribalism, and disease are rapidly destroying the social fabric of our planet. Atlantic Monthly 273: 44-76.

Karns, M.P. and Mingst, K.A., 2004. International Organizations: The Politics and Processes of Global Governance. Lynne Rienner, Boulder, CO, USA.

Kumar, M.D. and Singh, O.P., 2005. Virtual water in global food and water policy making: is there a need for rethinking? Water Resources Management 19: 759-89.

Kumar, R., Singh, R.D. and Sharma, K.D., 2005. Water resources of India. Current Science 89: 794-811.

Lichtenthäler, G., 2002. Political ecology and the role of water: environment, society and economy in northern Yemen. Ashgate Publishing Ltd, Aldershot, UK.

Long, N. and Long, A., (eds.), 1992. Battlefields of knowledge. The interlocking of theory and practice in social research and development. Routledge, London, UK.

Ma, J., Hoekstra, A.Y., Wang, H., Chapagain, A.K. and Wang, D., 2005. Virtual versus real water transfers within China. Philosophical transactions of the Royal Society of London - B 361: 835-842.

Mehta, L., Leach, M., Newell, P., Scoones, I., Sivaramakrishnan, K. and Way, S.A., 1999. Exploring Understandings of Institutions and Uncertainty: New Directions in Natural Resource Management. Environment Group, Institute of Development Studies, University of Sussex, Brighton, UK.

Meinzen-Dick, R.S. and Pradhan, R., 2001. Implications of legal pluralism for natural resource management. IDS Bulletin 32: 10-7.

Meinzen-Dick, R. and Pradhan, R., 2005. Recognizing multiple water uses in intersectoral water transfers. In: Shivakoti, G.P., Vermillion, D.L., Lam, W.F., Ostrom, E., Pradhan, U. and Yoder, R. (eds.). Asian Irrigation in Transition. Responding to Challenges. IWMI/Sage Publications, New Delhi, India, pp. 178-205.

Merrett, S., 2003a. Virtual water and the Kyoto consensus. A water forum contribution. Water International 28: 540-2.

Merrett, S. 2003b. Virtual water and Occam's razor. Occasional Paper No 62, SOAS Water Issues Study Group, School of Oriental and African Studies/King's College London, University of London, London, UK.

Molden, D. (ed.), 2007. Water for Food, Water for Life. A comprehensive Assessment of Water Management in Agriculture. IWMI / Earthscan, London, UK.

Molle, F., 2008. Why enough is never enough: the societal determinants of river basin closure. International Journal of Water Resources Development 24: 217-226.

Molle, F., Mollinga, P.P. and Wester, P., 2009. Hydraulic bureaucracies: Flows of water, flows of power. Water Alternatives 2: 328-349.

Mollinga, P.P., 2008. Water, politics and development: Framing a political sociology of water resources management. Water Alternatives 1: 7-23.

Neubert, S., 2001. Wasser- und Ernährungssicherheit. Problemlagen und Reformoptionen. Politik und Zeitgeschichte 48-49: 13-22.

Nuijten, M, Anders, G, Van Gastel, J., Van der Haar, G., Van Nijnatten, C. and Warner, J., 2004. Governance in action. Some theoretical and practical reflections on a key concept. In: Kalb, B., Pansters, W. and Siebers, H. (eds.). Globalization and development; themes and concepts in current research. Kluwer Academic Publishers, Dordrecht, the Netherlands, pp. 103-30.

Pahl-Wöstl, C., Gupta, J. and Petry, D., 2008. Governance and the global water system: A theoretical exploration. Global Governance 14: 419-435.

Parveen, S and Faisal, I.M., 2004. Trading virtual water between Bangladesh and India: a politico-economic dilemma. Water Policy 6: 549-558.

Peters, R.G., 1998. Governance without Government? Journal of Public Administration Research and Theory 8: 223-243.

Pierre J. and Peters, B.G., 2000. Governance, Politics and the State. Macmillan Press, Houndmills, Basingstoke, Hampshire and London, UK.

Rhodes, R.A.W., 1996. The new governance: governing without government. Political Studies 44: 652-667.

Rosegrant, M.W., Cai, X. and Cline, S.A., 2002. Global water outlook to 2025. Averting an impending crisis, International Food Policy Research Institute (IFPRI) and International Water Management Institute (IWMI), Washington, DC, USA/Colombo, Sri Lanka.

Roth, D., Boelens, R. and Zwarteveen, M. (eds.), 2005. Liquid Relations. Contested Water Rights and Legal Complexity. Rutgers University Press, Piscataway, NJ, USA.

Roth, D. and Warner, J., 2008. Virtual Water, Virtuous Impact? The Unsteady State of Virtual Water. Agriculture and Human Values 25: 257-270.

Scott, J.C., 1985. Weapons of the weak: everyday forms of peasant resistance. Yale University Press, New Haven, CO, USA.

Selby, J., 2005. The geopolitics of water in the Middle East: fantasies and realities. Third World Quarterly 26: 329-349.

Shiva, V., 2002. Water Wars. Privatization, Pollution and Profit. South End Press, Boston, MA, USA.

Sikor, Th. and Lund, C., 2009. Access and property: A question of power and authority. Development and Change 40: 1-22.

Sjaastad, E. and Cousins, B., 2009. Formalisation of land rights in the South: An overview. Land Use Policy 26: 1-9.

Smaller, C. and Mann, H., 2009. A thirst for distant lands: Foreign investment in agricultural land and water International Institute for Sustainable Development. Winnipeg, Canada.

Starr, J.R., 1991. Water Wars. Foreign Policy 82: 17-36.

UNEP (United Nations Environment Programme), 2002. Vital water graphics. An overview of the state of the world's fresh and marine waters, United Nations Environment Programme (UNEP), Nairobi, Kenya.

UNESCO, 2006. Water: a shared responsibility. The United Nations World Water Development Report. United Nations Educational Scientific and Cultural Organization (UNESCO) and Berghahn Books, New York, NY, USA.

Von Benda-Beckmann, F., Von Benda-Beckmann, K. and Eckert, J., 2009. Rules of law and laws of ruling: law and governance between past and future. In: Von Benda-Beckmann, F. Von Benda-Beckmann, K. and Eckert, J. (eds.). Rules of law and laws of ruling. On the governance of law, Ashgate, Farnham, UK, pp. 1-30.

Vond Benda-Beckmann, F., Von Benda-Beckmann, K. and Wiber, M.G., 2006. The Properties of Property. In: In: Von Benda-Beckmann, F. Von Benda-Beckmann, K. and Wiber, M.G. (eds.). Changing Properties of Property. Berghahn Books, New York, NY, USA, pp. 1-39.

Von Braun J. and Meinzen-Dick, R. 2009. 'Land grabbing' by foreign investors in developing countries: Risks and opportunities. International Food Policy Research Institute, Washington, DC, USA.

Warner, J., 2003. Virtual water, virtual benefits? Scarcity, distribution, security and conflict reconsidered. In: Hoekstra, A.Y. (ed.). Virtual water trade. Proceedings of the international expert meeting on virtual water trade, IHE, Delft, the Netherlands, pp. 125-35.

Warner, J.F. and Johnson, C.L., 2007. 'Virtual Water', Real People. Useful Concept or Prescriptive Tool? Water International 32: 63-77.

Wichelns, D., 2004. The policy relevance of virtual water can be enhanced by considering comparative advantages. Agricultural Water Management 66: 49-63.

Wichelns, D., 2005. The virtual water metaphor enhances policy discussions regarding scarce resources. Water International 30: 428-37.

Wichelns, D., 2010. Virtual water: a helpful perspective but not a helpful policy criterion. Water Resources Management, DOI 10.1007/s11269-009-9547-6.

Williams, P., 2001. Turkey's H2O Diplomacy in the Middle East. Security Dialogue 32: 27-40.

Windfuhr, M. and Jonsén, J., 2005. Food sovereignty. Towards democracy in localized food systems. ITDG Publishing, Warwickshire, UK.

Wolf, A.T., 1998. Conflict and Cooperation along International Waterways. Water Policy 1: 251-65.

World Water Council (ed.), 2004. E-Conference Synthesis: Virtual Water Trade - Conscious Choices, World Water Council. 4th World Water Forum, Marseille, France.

Youkhana, E. and Laube, W., 2006. Cultural, socio-economic and political constraints for virtual water trade: perspectives from the Volta Basin, West Africa, University of Bonn, Centre for Development Research, Department of Political and Cultural Change. Research Group Natural Resources and Social Dynamics, Bonn, Germany.

Youkhana, E. and Laube W., 2009. Virtual water trade: a realistic policy option for the countries of the Volta Basin in West Africa? Water Policy 11: 569-81.

Chapter 13
Food security in a Bolivian indigenous territory: exchange relations of subsistence farming reproduced in indigenous governance

Michiel Köhne

1. Introduction

Petro Fernandez[558] lives in San Lorenzo, a small village in the indigenous lowlands of Lomerío, Bolivia. He wants to clear a new field because the one that he has used for the last five years is running out of steam. At the end of the dry season in September he wants to select a piece of land in the woods, clear the bushes, fell the trees and burn the scrap. Than he plans to fence off the land with poles and barbed wire: against animals and to show that it is his. The amount of work to be done is impossible for a man alone, even considering that in this village the mention of a man also implicates the work that his wife and children can do. But not even the help of his father and brothers, and their wives and children, will suffice.

Juan Soquere, who is *Cacique Mayor*, the highest authority in this village, wishes to resolve a conflict between two families over a bridge that needs to be rebuilt for easy access to the family fields. He can ask them to listen to his reasoning, but that is all he can do by himself. If they do not wish to co-operate, he cannot fine them, put them in jail, or simply order the bridge to be built. There is no way he could enforce such decisions by himself. Neither can he rebuild the bridge himself or assign it to one of the lesser caciques. Their job is an honorary one, and for their livelihood they need to work their fields. So they cannot afford to lose too much time on their honory jobs. Nor are there any civil servants that he can order to do this task.

In order to farm its land a family in San Lorenzo depends on the helping hands of neighbouring families, just as village authorities need the full participation of the villagers, their labour to carry out his decisions and their co-operation to solve conflicts and to decide on arrangements for the future and to enforce settlements. The co-operation at the level of families, at village level, and at the territorial level is crucial to food security in this indigenous area, because for

[558] All names of persons in the text have been altered.

their food Lomerians rely fully on the local agrarian production and the way this is governed by local authorities. Household members eat what they manage to produce on their subsistence fields, what they manage to barter, and what village members bestow on them in times of need, or they do not eat.

2. Entitlements shaped by reciprocity and brokerage

To understand food security in the setting as exemplified by the interdependencies described above, one needs to look further than simply at local food availability. Food security is embedded in local practices of food production and food distribution. It is about who manages to get access to, and control over, food in this setting of interdependencies. Sen makes the now popular distinction between endowments and entitlements to show how even when there is plenty of food individuals or groups can still die of hunger because they lack effective command of the available food (Sen, 1981).

Leach *et al.* (1999) add the notion that endowments are neither simply 'there', but are also gained by people. They explain the relationship between the two by focussing on how endowments become entitlements in a context of institutions. Nevertheless, the distinction remains elusive because it cannot explain why for example a right to land would be an endowment. For such a right can only exist in any meaningful way because of its embeddedness in institutions. According to Leach *et al.* (1999) 'the distinction between them depends on empirical context and on time, within a cyclical process. What are entitlements at one time may, in turn, represent endowments at another time period, from which a new set of entitlements may be derived.' If that is the case I wonder whether the use of two concepts to explain the importance of institutions for the usefulness of material goods for social actors is not in the end obscure because of its unclear delineation of material goods.

I would rather make a distinction between material resources, such as land and labour on the one hand and access to these resources on the other. People derive this access from cultural practices which give locally relevant meaning as to how actors can use their agency in order to derive benefits from things. Ribot and Peluso call this access 'the ability to benefit from things – including material objects, persons, institutions, and symbols' (Ribot and Peluso, 2003). Access is constituted by a broad set of factors, a 'range of powers – embodied in and exercised through various mechanisms, processes, and social relations, of which property is just one'.

To Leach *et al.* (1999) formal rights are endowments which are not conceptually distinguished from other endowments such as labour and skills. Starting from an access perspective a formal right can be analysed as a specific form of discourse which together with other discourses and interactions is used by actors to give a certain meaning to the use of things, including an attribution to its level of legitimacy and thus to the effectiveness of the power involved.

The process in which access is established is one of contestation in a web of power relations. An analysis of access entails the identification and mapping of the mechanisms by which access is gained, maintained, and controlled. 'Because access patterns change over time, they must be understood as processes' (Ribot and Peluso, 2003). They discuss rights-based access and other structural and relational mechanisms of access such as: access to technology, to capital, to markets, to labour and labour opportunities, to knowledge, to authority and access through social identity and via the negotiation of other social relations. These are heuristic categories, none of which is distinct or complete. 'Each form of access may enable, conflict with, or complement other access mechanisms and result in complex social patterns of benefit distribution.'

In the present study access through social relations of dependence and exchange is especially important to explain mechanisms of access to capital, labour, knowledge, which in turn shapes local access to food security. Interdependencies between villagers in the field of agricultural subsistence production significantly influence the nature of exchange relationships. Short labour-intensive periods in five-year agricultural production cycles make exchange of labour a necessity. Whereas money is often not available as a means to settle debts and pay for labour, trust-based reciprocity provides a basis for this.

The point of this chapter is to show how exchange relationships of reciprocity at the level of agricultural subsistence production permeate village governance. From there it gets mixed with less reciprocal forms of exchange to further shape higher level governance in the area in which the territorial indigenous organisation Cicol, NGOs, and state institutions at the municipal level and up are the main actors. I will demonstrate that at the discursive level reciprocity is carried over into the legitimating discourses in the public sphere. At the material level, conditions of an agricultural production that relies on the articulation of networks of relationships are replicated in village and territorial governance. Here actors can only obtain power 'to produce' by using their networks of relationships which connect these local authorities to resources such as labour, money and knowledge: from villagers, development organisations, donors or state institutions.

I understand village and territorial governance to be a combination of organising practices and legitimating discourses. These organising practices consider the administration, regulation, jurisdiction and service provision that are generally considered part of the 'public domain' (Nuijten, 2004). Organising practices should not be understood as the result of a common understanding or normative agreement, but as the result of a process of organising in a force field around specific kinds of problems and resources (Nuijten, 2005). This process of organising has been riddled with struggle, amongst Lomerían actors with conflicting interests (McDaniel, 2002), but mostly with other interest groups, largely concerning issues of access to natural resources and of political representation (McDaniel, 2003).

The legitimating discourse contains many different contesting voices, including both legitimating stories from the powerful and from the non-powerful who agree to operate within a certain web of power relations. This acceptance may be positive as well as negative (Bakker, 2001). In the process of legitimation practices are also relevant as they too can legitimise or delegitimise governance practices.

The concept of brokerage shows very well how both the organising practices and the legitimating discourses that make up the governance in a specific field, may be linked to relations of dependency. A broker is generally seen as a mediator where different social forms intersect with one another (Bierschenk *et al.*, 2002). A more universal approach to the problem of intersecting life worlds has been developed by Long in his work on the concept of interface, 'an intermediary position between different groups and individuals involved in planned intervention' (Long, 2001).

Bierschenk *et al.* (2002) focus on a specific type of brokerage that they call development brokerage. A development broker is an actor that serves as intermediary between 'donors' and potential 'beneficiaries' of development aid. This form of brokerage is specifically directed toward the acquisition of development revenues. Bierschenk *et al.* (2002) want to emphasise how development brokers and classic village authorities interact. I use the concept not to understand the relationship between development brokers and classic village authorities, but to explain the role of brokerage in village and territorial governance. I want to emphasise how actors in village and territorial governance occupy intermediary positions between different groups, both with their own life-worlds, practices and discourses, a divide which is used as a source of power.

My contention that the governance power of brokers is based on relationships of interdependency does not mean that they are passively dependent actors who are simply used by actors on either side of the divide. On the contrary, a broker actively uses his position as an intermediary to manipulate relevant knowledge in one field into relevant knowledge for the other field. At the same time he takes care not to lose his position. He is 'the man who makes things happen', legitimating his actions as being 'devoid of personal interest, motivated only by the well-being of the community' (Bierschenk *et al.*, 2002).

The broker tries to make himself indispensable using two strategies. He translates that which makes sense on one side of the divide into that which makes sense on the other side. This strategic management of information is called management of meaning by Bierschenk *et al.* (2002). And the broker emphasises his unique knowledge of and power over what happens on the other side while at the same time claiming to be one of us, rather than one of them. Bierschenk *et al.* call this: valorising his own activity (Bierschenk *et al.*, 2002).

The interdependencies of brokerage are close to those on which clientelism is based. However, I prefer the more neutral concept of brokerage which can explain the power coming from these interdependencies and distinguish this from power based on primary sources of power, rather than lump them together. In a case like Lomerío there is no extreme inequality, but rather a quite egalitarian order. The differences between the richest and the poorest are significant but not large enough to be a sufficient basis to exert power. This draws attention to relationships as a source of power.

To understand brokerage, imagine for example a Lomerían villager who has worked some time for an NGO. Perhaps he travels to town, learns to understand the discourse and the needs of development organisations, builds up some personal connections with a few co-workers, experiences a project from their point of view. When he comes back to his village he will have a network in and knowledge about both worlds. He will thus be able to translate village needs into development discourse and to specifically address a specific NGO and activate his personal contacts, while at the same time he will gain trust in the village, being one of them, and have the special power to deal with this foreign world.

Brokerage is a practice in which an actor, who is situated as intermediary between different social groups, uses this position as a source of power. This power is based on so-called 'secondary resources', resources that are derived from networks of relationships with actors who control 'primary resources' such as labour, money and knowledge (Bierschenk *et al.*, 2002). Although brokerage power depends on relationships rather than on primary resources, it is these primary resources that turn interdependencies into a source of power.

In this chapter I will concentrate on my field findings that show the importance of interdependencies as a basis for economic life in Lomerío. Earlier studies have elaborated on the importance of the exchange relationships within villages (Rozo López, 2000). I will extend their perspective to an analysis of both present-day village and territorial governance of access to food and the role in this of the main indigenous authorities and outside actors such as state, municipal and NGO institutions.

In Section 3 I will introduce the case study. In Section 4 I will describe the subsistence farming as it takes place in Lomerío and show how it relies on co-operation between the villagers. Section 5 describes the role of village and territorial authorities in agricultural production and analyses their role as brokers. Section 6 offers a further analysis and Section 7 my conclusion.

3. Lomerío, a Chiquitano territory

My fieldwork in the Bolivian lowlands has been directed toward the indigenous territory of Lomerío, which is the home to a Chiquitane group whose local organisation is dubbed 'Cicol'. The fieldwork on which the findings in this chapter are based took place during three months in the summer of 2005, mainly in the village of San Lorenzo.

Lomerío is the 260 thousand hectare Chiquitane territory in the Bolivian lowlands north of Santa Cruz. The denomination 'Chiquitane' is used for the descendants of around fifty ethnic groups forged into one at the beginning of the 18th century. Jesuits provided them with shelter in their mission posts against Spanish and Portuguese slave-hunters, and enabled this more sedentary lifestyle necessary to keep safe from slave hunting by providing cattle and iron tools to work the land. After the Spanish chased away the Jesuits in 1767 many became enslaved on local haciendas and from 1880 until 1945 in rubber plantations in the Bení, in the North-West of Bolivia. Others fled into even more inaccessible areas such as Lomerío. Here, the oldest villages, including San Lorenzo, were established around 1900 (Rozo López, 2000).

The *Decreto de Empadronamiento* made debts hereditary in such a way that landlords were secured of free labour generation after generation. This system of debt slavery was officially abolished in 1937, but in Lomerío it only came to an end when the 1953 *Reforma Agraria* was finally implemented ten years after its introduction. In the 1960s residents of San Lorenzo and surrounding villages organised themselves against the slave-keepers and started a *sindicato agrario* to become independent of the haciendas. Labour was no longer an indigenous resource which was freely available. In the seventies a new free indigenous resource was booming: wood. This resulted in the ransacking by logging companies of almost all precious woods from the area, and the roads built to take the wood out opened the area for an influx of trade and development and religious organisations.

In Lomerío, production is almost entirely subsistence agricultural farming. Less than 1% of the land is used for cropping or fenced off for cattle. The rest is used for free ranging, hunting, and fishing and for the gathering of wood and non-wood products. Lomerío has become home to about 7000 people spread over 28 hamlets of 10 to 50 houses. The largest village is San Antonio with about 200 houses.

3.1 San Lorenzo, a village

San Lorenzo is the second biggest village in Lomerío and has always been an attractive settlement mainly due to its reliable water source. A little stream runs past the village, which is in full flow in the wet season, but becomes a series of puddles in the dry season. This water source has recently been upgraded with the addition of a small artificial lake along with the instalment of a diesel pump, water piping and four communal taps in the village. The water proves a useful resource and together with strong political unity this has made San Lorenzo an attractive location for development projects. In this way San Lorenzo has become a comparatively wealthy village, only second to the largest village San Antonio, home to the Catholic church which has been the source for a lot of funding and consequently improved living standards. Once started as a one-family settlement, San Lorenzo has gradually grown into a 50-house village and is still growing. Its communal land property amounts to 4722 hectares.

3.2 Cicol, the territorial organisation

To the villagers Cicol is very much the successor of the *syndicatos* of the sixties and the *cooperación* with the 'Fundacion Barbiera' of the seventies. The *syndicatos* and the *cooperación* are both organisations of co-operation between different families that have been initiated by development organisations. Both forms of co-operation stopped functioning as soon as this external support ended. The NGO APCOB (*Apoyo para el Campesino-Indígena del Oriente Boliviano*) was established in 1978 in order to help the lowland indigenous peoples to organise themselves into federations and assisted in the establishment of the CIDOB (*Confederación Indígena del Oriente Boliviano*). CIDOB was supposed to represent all lowland indigenous peoples and, in order to achieve this, Apcob tried to convince these peoples to organise themselves at a level in between CIDOB and the village level.

Jürgen Riester, director of the CIDOB and a German anthropologist who had been researching the Chiquitanes among other lowland peoples, came to the villages of Lomerío in 1982 to teach them about the CIDOB initiative and about the forms that other indigenous peoples had chosen to organise themselves. The Guaraníes de Izozo are presented as an example. They are one of the very few lowland peoples that traditionally organise themselves at the level of people under a representative board. The leader of the Izozo at that time was also one of the initiators of the CIDOB. Representatives from Lomerío were invited to come to Santa Cruz, which is a two day walk and half a day in a bus, to participate in workshops with other peoples. Notwithstanding strong advice against co-operation with APCOB by the Catholic priest, who tried to scare the Lomeríans by telling them that the APCOB people were Communists, thus attributing devilish malice to them. To those who seem less inclined to keep clear of APCOB the church also threatens to deny future development benefits.

There was clearly a world of difference between the development goals of Riester and those of the priest. Riester wanted to empower the Lomeríans through the recovery of their own culture after so many years of adaptation and flight. The priest on the other hand wished to bring an end to the backwardness of the people as soon as possible by his teachings of the modern perceptions of the Catholic

Church. The church was especially strong in villages where they supplied a lot of development projects, thus diminishing the need for the grassroots organisation that APCOB planned to help with.

The result of this was that only six people went to the workshop, three of whom were from San Lorenzo. On their return they managed to convince seven villages to start a provisional board of Lomerío, the *Consejo provisional de Lomerío*. In 1982 when Cicol was installed, it was initially identified as a co-operation of farmers, and indigeneity was still an insult. It took 15 years and a dramatic change in development discourse and politics until finally in 1997 Cicol was officially re-established as an indigenous organisation.

Initially Cicol was mainly welcomed to protect Lomerío against the exploitation of precious woods at a time when lorries were driving to and fro. Lomeríans perceived this exploitation as thieving of 'their wood', although part of it was based on concessions by the state. APCOB convinced them that their rights to this wood superseded the rights of the loggers and tried to convince the state of their views by writing forest management plans and the installation of a sawmill which even resulted in the production of FSC-certified wood. Connected to this fight for recognition Lomeríans participated in national indigenous protests against the government in a quest to receive territorial land rights under the new land reform act, the *Ley INRA*.

4. Family farming

Almost the entire workforce of Lomerían families is dedicated to the production of their daily food. This is largely directed at their '*chaco*', the household plot. The villages grant rights to every household to clear the wood from a piece of land to grow vegetables and fruits. Almost every household also owns a few cows. The size of a *chaco* and the number of cows one can keep depends for the most part on one's workforce.

The size of the household *chaco* is largely determined by the availability of labour within the household because of the labour that is required to clear the field and fence it off. To clear the land, they fell the trees and bushes to let them dry, and then burn them. This fertilises the land which can then be used for five years. After that a new *chaco* is cleared and the old one is left to regenerate, which takes about twenty years. This *chaco* is always fenced off, to keep free ranging cattle out, and is usually about one hectare. *Chaco*-making is concentrated at the end of the dry season, in September.

Typical *chaco* products are: corn, rice, beans, onions, potatoes, beans and some trees with pineapple, papaya and mango. After three to five years other crops follow: yucca, bananas, sugar cane (Schwarz, 1994). Only a very small portion of the harvest is sold, 10% of the corn and less than 5% of the rice (Godoy and

McDaniel, 1998). The same is true for the oranges coming from the many privately owned orange trees in the village. They are usually grown around the family houses where they offer protection against sun and wind. The quality of the fruit is limited nevertheless and it does not pay to transport them to the Santa Cruz market, where they would have to compete with fruits from more fertile and less dry areas. Large bags are nevertheless sent to Santa Cruz as exchange gifts to family members who live in town.

In order to have some income from farming a family buys a cow as soon as they can afford to. These cows will feed themselves free-ranging on the pampas, in the woods, on the old *chacos*, on the village green and on the football field and they can drink from the artificial lake and the little brook. Cows are one of the very few products that can be sold for money. They are a sound investment. Free-ranging cattle in particular require very little labour. They are primarily kept as a way to spread the risk and thus as source of security in times of need, when there is a failed harvest or when a productive member of the family falls ill or dies.

In order to earn a steadier income from cattle, one needs to take it one step further and make a *portrero*, that is, to clear and fence off pastureland. Old *chacos* are often a starting point for this, but one hectare will only feed one cow, so a lot of extra work is involved. Nevertheless, cattle in *porteros* are more productive because of the protection it offers against: predators, wandering off, cattle theft, poisonous plants, and falls from rocks when trying to get to the water. The costs incurred are a huge investment in labour to clear the ground and to bring daily drinking water in the dry season.

The latter is also costly in terms of money: namely, the purchase of barbed wire. People sometimes prefer to make their *chacos* next to each other so that they need less barbed wire to protect their crops from, mostly domestic, animals. Nevertheless, the fences also express rights of access and use of the land and its fruits, so a great deal of confidence is needed to do this. This is one of the reasons why adjoining parcels are often assigned according to family ties which enables savings on fences.

4.1 Exchange among families

As we have seen, labour is an important resource in almost all agricultural production, both to start the *chaco* that provides daily food for the family, and to fence off a *portrero* to ensure a little income. Two main types of labour exchange are used to deal with the short labour-intensive periods that every household encounters every once in a few years: the *minga* and the *cooperacíon*.

A *minga* starts with its convocation. The male head of the household invites the other families to help during one morning which he pays for by providing the workers with lunch and *chicha*, the local low-alcohol brew based on corn. Attending a *minga* once invited is regarded as the duty of a *comunario*, which is what you are called once you have been fully accepted as a village member.

Chicha is used as a means of payment and plays a central role in village life. It has been an important drink for centuries, as was already noted by one of the Jesuit padres in Mission times (Knogler, 1979). Its importance does not lie in its exchange as a commodity, but in its contribution to the festive rituals around the exchange. At the festivities both the gift and the acceptance of drinks, all in its correct order, are equally important (Rozo López, 2000). In some villages this *chicha*-based reciprocity has become compromised by the activities of the evangelical church which forbids their converts the consumption of alcoholic beverages. This clashes with the traditional use of drinking festivities to settle debts.

Trust is very important for the *minga*: if not enough people comply with a host's invitation or do not work hard enough he will end up bearing all the costs for the production of food and *chicha*, but with the work unfinished. The personal invitation to attend a *minga* or other irregular social events is considered very important to ensure the reciprocal bonds which include the obligation to attend. San Lorenzo has installed a P.A. system which nowadays replaces these personal invitations, which cuts the cumbersome duty of visiting all fifty families. Nevertheless, the elderly dislike the impersonality of the bullhorn and complain that nobody has come to invite them.

Sanctions for not co-operating well enough are of a reciprocal nature. If many families are of the opinion that you have not lived up to your *comunario* obligations this is easily sanctioned by not showing up when you invite people for a *minga*. Another way of sanctioning is openly joking about and criticising the culprit or even formally deciding to limit the co-operation at his *mingas*.

The *cooperación* is very much like the *minga*, but encompasses a smaller group of people within the village. Five or six villagers decide to co-operate, and thus become *socios* on work that they cannot do alone, for example to make a *chaco* or build a house. The person for whom the others are working is required to provide the workers with food and *chicha dulce*, *chicha* without alcohol, but is also obliged to participate in the work when the next person in the group asks his socios to work for him. Only after a full circle of work has been completed will the *cooperación* celebrate this with a *chicha* party. The *cooperación* is often used for long-term collaboration and depends very much on the mutual trust between the *socios*.

These forms of co-operation are necessary to cater for peaks in labour need. There is more to it though. They also strengthen the social cohesion by establishing networks of mutual exchange and interdependency. These networks are also reflected in the mutual help that families offer each other in times of need thus providing social security (Benda-Beckmann, 1988).

Although reciprocity is central in these practices of exchange, the *minga*, the *cooperación* the *chicha* parties and the associated discourses, the equivalence of the exchange value is not unproblematic. This equivalence needs to be negotiated against the backdrop of a reality in which disparities, even slight ones, are relevant. This includes disparities in wealth, authority, knowledge, labour power and other mechanisms of access to resources, including access through social identity and via other social relationships.

Inequality is, for example, reflected in the food that one is able to serve at a minga. But it can also be found in the need to build credit with the wealthier families that you might need in times of want, by working very hard in their mingas and by not demanding too much from their work in your minga. It is also shown in the clear improvement of one's position in village decision-making that is related to the brokerage power one gets through Cicol jobs.

Moreover, some families in San Lorenzo can afford financially not to participate in other people's *mingas* because they can simply pay for the labour they need rather than call for a *minga*. This is frowned upon by others who find it the obligation of the better-off to provide *mingas*, as a means of building social cohesion (Rozo López, 2000). This shows that the *minga* is not only a tool for production. It is also a means to reinforce interrelations between families by addressing the interdependency of the *comunarios* and emphasising the importance of reciprocity.

Of course, co-operation based on reciprocity is especially important to those families who cannot afford the money to pay for labour. Today, money is gradually becoming more available to some of the families in San Lorenzo. This coincides with a trend where the richer families pay for the required labour, while poorer families make use of the *cooperación* rather than the *minga*. The last seems only to function well in villages where almost all families rely on reciprocity and where monetary wealth does not give some the possibility to opt out of this system.

Overall, the disparities in wealth in Lomerío are minimal. This is due to the fact that production largely depends on personal labour due to the fact that land is equally available to all. Nevertheless, an important influence of development projects is the creation of temporary jobs and relationships with the outside world, which as a result raises inequalities in wealth. Project resources that are brought to the village are usually set up in such a way that they benefit the families more or less equally. This is due to open deliberations on these projects in village assemblies.

5. Lomerío's governance of food production

5.1 *Cacique mayor* and Cicol board as central brokers

In the governance of access to food security in Lomerío both the village *Cacique Mayor* and the territorial board of Cicol occupy a central position connecting the other main actors in the field of food production and food security. These others are: the male and female heads of the families, some village authorities, a number of NGOs, INRA and the municipal government.

At village level there are a large number of authorities, who are elected on a yearly basis by the villagers. There is a general board, the *Cabildo* headed by the *Cacique Mayor*, next to specialised authorities for almost anything: schooling, health, vaccination, women interests, religion, sport, etc. There are also three authorities related to a supra village level. There is the *Corregidor*, who is supposed to uphold Bolivian law on behalf of the central government, but claims to work primarily in accord with the villagers' wishes like the rest of the authorities. He is elected by the villagers from seven villages and organises the administration and the application of local law (Benda-Beckmann *et al.*, 1997; Köhne-Hoegen, 2006). There is a member of the *Comité de Vigilancia* of the municipality, the organ that is to oversee municipal budgets. And there is the *Agente Kantonal*, who, as part of the municipal government, is supposed to act as an intermediary between the municipality and a number of villages. The *Agente Kantonal* is gaining influence in parallel with the growing importance of the municipality in territorial governance.

At the village level the *Cacique Mayor* acts as the central co-ordinator between the wishes of the different male and female heads of the families and the plethora of village authorities. As soon as anything comes up that cannot be dealt with by the co-operation of one or two families or by one of the authorities, they come and talk to the *Cacique Mayor*. He then puts the issue on the agenda at the weekly village assembly for discussion and to gain support for a solution. If at these meetings it is decided that help from outsiders is necessary, the *Cacique Mayor* is the one to contact these outsiders.

At the territorial level Cicol plays a central role. This organisation was founded in 1982 and is based in a small Lomerían village, Puquio. The most important areas in which it promotes the co-operation of the villages of Lomerío are the autonomy of indigenous government and the protection of natural resources against exploitation by non-Lomerians. In order to achieve this, Cicol aims to secure the recognition of a territorial land title and has secured the recognition of Lomerío as a municipality. It also played a central role in a large project on the exploitation of wood which has now been abandoned and in small projects around agricultural co-operation, education, health and development projects in general. Just like the *Cacique Mayor*, the Cicol directors assume a central role in the agenda-setting at the weekly open assemblies with the Lomerians and in the required contacts with outsiders.

The main external governance of the internal affairs of Lomerío, i.e. from NGOs or state institutions, is in the form of projects. Rules and policies coming from the state hardly ever become relevant. State surveillance is zero and village and territorial governance come first. The only time when the state becomes relevant is at election time or when crimes are committed that the village authorities cannot deal with themselves (Köhne-Hoegen, 2006).

Projects on the other hand, usually funded by NGOs, but sometimes co-ordinated with state institutions are extremely important. They are set up to provide all kind of services to the Lomerían people that they cannot afford themselves, such as healthcare, schooling, quality housing, infrastructural works, etc. The importance of these projects for the acquisition of these resources makes the wheeling and dealing with NGOs a major technique in governance practice. Both the *Cacique Mayor* and the Cicol board need to portray the Lomerians in such a way that it fits with the latest fashion in development policies. In this portrayal the Lomerians not only need to appear needy enough to fit the proper policy goals. The central brokers also need to emphasise that a project run by them will lead to the success that an NGO requires to survive, thus fulfilling the institutional demands of the development industry.

5.2 Municipality and state less important

Municipalities have only become relevant for rural areas in Bolivia since the *Ley de Participación Popular* of 1994, which shifted the management of the rural areas from the central government to the municipalities. Later they even became relevant in Lomerío when in 2000 Cicol's request was granted and Lomerío got the status of municipality with a municipal government based in San Antonio.

Although Cicol expected to be able to control the municipality by winning the elections, it turned out that they had underestimated the power of the central political parties who managed to win the elections with promises. During the first four years of the municipality, it turned out to be rather powerless. The funds of the central political parties quickly dried up after the elections had been secured and the municipality was not well enough, if at all, rooted in Lomerío society to base its power on its brokerage position. NGOs have been bypassing the municipality and went straight to Cicol whom they considered the legitimate representation of the Lomerian people.

In spite of the locally elected municipal representation NGOs thus kept approaching the Lomerians through Cicol. It took until the next elections for the party supported by Cicol to win. Since then NGOs have started to acknowledge a two-headed representation of the Lomerians, making their support dependent on the joint approval of both Cicol and the municipality.

At the level of the central state the land reform policies are the most relevant for Lomerío, because they are the basis for the right to indigenous territorial titles. The state institution directly responsible for these policies is the *Instituto Nacional de Reforma Agraria* (INRA), which is the National Institute of Agrarian Reform. Its dealings with Lomerío mainly affect the relationship between the Chiquitanes and outsiders, which is not the focus of this chapter. Nevertheless, indirectly it also affects the governance of Lomerían internal relations (Köhne, unpublished data).

5.3 Village governance of food security

Village life in San Lorenzo is rich in public decision-making. In the village general assemblies are held every week and more often if the occasion so requires, which is at least a few times a month. It is the responsibility of the villagers to attend these meetings and decisions taken in your absence are assumed to be with your consent. Nevertheless, non-attendance of those directly involved is taken into account at assemblies and is often reason to postpone decision-making. The villagers and all their authorities keep in touch by a P.A. system that allows people to inform the village and is in use almost every morning at around 6 before people leave to work their *chaco*. Messages vary from summons to assemblies, church mass, soccer matches, *trabajo publico* (community works), parties and funerals to messages on the availability of bread, beef, etc.

All village authority positions are unpaid jobs for which you can get elected and which are seen as village duties. In line with this obligatory nature, all authorities are supposed to carry out the wishes of the general village assembly, rather than their own considerations. The effects of almost all decisions made by village authorities depend almost entirely on the co-operation of the other villagers. The *Cacique Mayor* can, for example, decide that a certain piece of land needs to be cleared, but if many villagers do not agree, they will simply refuse to show up. If a majority does not show up, the fine for not showing up at scheduled community work, will not be collected.

Village governance deals with a range of issues around agricultural production. Most important is the regulation of natural resource management, such as the regulation of access to land by the allocation to individual families of rights of use on communally owned land, to build a house, or to use as *chaco* or *portrero*. But access to land includes the opening of roads and construction of bridges to facilitate access to the *chacos*. Village governance also includes: planning of *chaco* burning, regulation of water use, and projects such as for non-traditional irrigated cropping.

A right to land was secured for the first time when, in the 1960s, pressure from the Catholic church brought about an implementation of the 1953 land reform, the *Reforma Agraria*. This provided most villages in Lomerío with a communal title to village land and to variable amounts of its surrounding lands. Since then, every village has developed its own norms and practices with respect to the regulation of access to, and use of, these lands. In spite of obvious differences in details, there is a lot of similarity on the general rules.

In general, therefore, every *comunario* has a right to an amount of land related to the working capacity of the household. This includes three plots of land in the village in which you are born: a small one to build your house, a *chaco* of around one hectare per male adult and *portero* land at a rate of one hectare per cow. Household size can range from one, a single, divorced or widowed man or woman, to more than twenty members, when up to four generations form a household.

You become a *comunario* by being born in the village, or by marrying a villager. Otherwise you need the permission of the village assembly. To be accepted it is important that the assembly can trust you to comply with your duties as a *comunario*. Although it is possible for both Chiquitanos and non-Chiquitanos to become a *comunario*, the trust required to be accepted will be harder to gain for a non-Chiquitano and the help that you need to set up village life with a house and a *chaco*, etc., will be less readily provided.

Apart from their own land, villagers also work their communal property. Every Saturday morning all men of the village are summoned to take their tools to a spot designated by the *Cacique Mayor* for the work on some communal project for the *trabajo publico*. This varies from the weeding of the *Central Plaza*, the central grass square of the village, and the painting of the community centre to the construction and upkeep of access ways and bridges to *chacos*, to the building of classrooms, a newly planned church and the newly-finished waterworks.

Other communal works concern the co-operation in development projects. At the time when I was there an irrigation project had just started. *Chacos* are typically non irrigated lands, which therefore only grow certain crops. However, through Cicol the village obtained an Italian horticulture project for which a 2 hectare area bordering on the artificial lake is cleared and fenced off and made suitable for crops that need a daily dose of water to survive in the dry climate, such as: tomatoes, lettuce, carrots, cabbage, onions. An engine is to be installed by the project to pump water from the lake to the highest point of the field.

The role of the village authorities in this project is to ensure that the project will be seen by the supporting NGO as a success in all its stages: at the time of its securing, in the discussions with the development organisation and in its execution. This might entail the securing of the attendance of the villagers at teaching sessions and the securing of their labour to clear the field, to feed the plants while waiting for the engine to arrive, to keep the villagers motivated for the project in the face of setbacks, and in the presentation of the positive effects of the project on village life.

Village governance may also include the provision of social security (Benda-Beckmann, 1988). When a male head of a San Lorenzo family died, at the end of the burial the very next day the Corregidor, standing at the grave, had called on the solidarity of all the villagers with the widow and her eight children six of whom were still living at home. The next day there was a special village assembly where it was decided to help the family of the deceased with both a short-term and long-term plan. For now everyone was summoned to help with food and other resources and a letter would be written to Cicol by the *Corregidor* on behalf of all the villagers requesting their help. In this way the family would be helped until, at the end of the dry period, a new chaco is created for her to grow her own food and thus take care of her family, until her children are old enough to clear a new chaco.

5.3.1 Brokerage

The importance of brokerage relationships in the governance of access to food security can be illustrated by looking at the role of the *Cacique Mayor*. The *Cacique Mayor* used to be an important actor in the governance of San Lorenzo, because he was the one who mobilised the villagers to take care of their own interests. In the years between the liberalisation from slavery in the 1960s and the start of Cicol in the 1980s, a good governor was one that could do this by showing off his own devotion to this cause. You needed muscles and a big mouth. Competition between current, former and future caciques existed, but was mitigated by the fact that every man shared both the opportunity and the burden of this task. There were differences in economic wealth, but they were much smaller than nowadays and therefore less decisive in village decision-making. The major source of differences at the time was longevity of settlement. Long-time settlers had more to say based on tradition and acquired wealth, compared to newcomers.

Today the *Cacique Mayor* is still a *primus inter pares*, but with a lot of competition from villagers who derived their power from brokerage relationships. Today, the best governor is the one who is most effective at linking up local needs to external resources, the one with the best contacts with NGOs and state institutions, the one who speaks both languages best and can manipulate meaning across both discourses, i.e. the one who can communicate with outsiders while remaining accessible to his own people (McDaniel, 2002). In short, the one who is the best broker. Today especially linkages to Cicol, and to a lesser but growing degree the municipality, are a resource in the village governance of food production. Those are the places to go to with requests for help for infrastructural works like roads or waterworks, around land issues and when some mishap befalls a family that is too large for a village to deal with. Villagers firmly believe that it is the responsibility of Cicol in particular to relay their requests to sources outside Lomerío.

The power in village decision-making is most clearly expressed in the role someone is granted in assemblies, in the taking of initiatives and in acting as a village spokesman toward important outsiders. It is clear that decision-making power is first of all dependent on one's relationship with Cicol or the municipality. In village assemblies villagers with strong relations with Cicol, such as (ex-) Cicol leaders and people who are or have been involved in Cicol-mediated projects speak most of the time, are best listened to, respected, take the lead in arguing the case and in decision-making. Others merely acquiesce or at least never openly contradict the brokers, but restrict themselves to humbly raised questions or remarks. They leave their contributions to the consideration of the assembly and do not persist unless their concerns are taken up by others.

Economic power is also important, but the remarks and actions of economically powerful people in the village assembly prove less decisive. There were two men in San Lorenzo that were relatively rich without important brokerage relations: one was a successful cattle breeder, the other the only mechanic in Lomerío. These men often spoke up at the meetings stressing their points of view, more often than not without brokerage relations. However, they never pressed for a decision to be reached but rather waited to see whether their points of view would be pursued by the ones with brokerage power.

The general picture is that contributions to the assemblies by people with little or no brokerage power only become relevant when taken up by others who know how to translate their concerns in something that makes sense in the outside world. It clearly requires brokerage power to speak, and be heard. Brokers rule, using their knowledge of 'how things go' in the outside world as a trump, using oratory skills that they have acquired in dealing with the outside world, and by backing each other up through a repetition of each other's deliberations.

5.4 Territorial governance

Cicol's activities are important for village subsistence farming in two ways. First, Cicol tries to secure the long-term security of access to natural resources as a means of production. Most of these natural resources needed for agricultural production, such as farm grounds, wood and water, are tied up in land. Second, Cicol tries to win development projects for the area.

Cicol's seven member board is elected every four years by representatives from the villages. These are village authorities such as the *Cacique Mayor*, the head of the school board and the representative of the women. Every week, on Monday, the board of Cicol holds an assembly in Puquio where all Lomerians with questions, requests or a general interest in territorial governance are invited. This assembly deals with all kinds of issues relevant for Lomerían, ranging from national politics to projects, to requests for help from villages or individual families.

On the one hand these assemblies are formal and bureaucratic, while on the other hand they represent a very open way of doing business. Discussions tend to develop in a very respectful and polite way. Topics of discussion are usually put forward with very formal, long introductions as to why somebody wants to burden *Cicol*. After often lengthy elaborations about the nature of one's question, many words of thanks follow, for the attention paid to the interests at hand and expressions of readiness to answer all further questions and doubts. The first reactions to these usually come from people outside the board with no formal authority, but who are regarded as competent or authoritative on other grounds. It is only at the end of the discussion that board members take the floor either to touch upon new points of view or facts or to sum up what has been said by way of conclusion.

As in the village assembly, powerful people in the meetings who support a certain decision, all take turns to explain why this is the right decision, going through all the arguments and positioning them in relation to arguments given by other powerful people. They will let discussions drool on endlessly, repeating their arguments over and over and dropping long, long silences to give 'everyone' the possibility to react. In the end these silences are often filled with counter-arguments or perspectives that offer other points of view. These are always extensively discussed and sometimes actually taken into account in the decision. At other times 'nothing' happens, silences remain silent. Assemblies can go on almost endlessly in this way. Much time is thus invested in consensus-building or at least in a ritual that suggests consensus-building.

Another interesting parallel between territorial governance by Cicol and the exchange of labour in family farming is the meal that people share after important Cicol assemblies. This meal is 'free', just like the one offered to conclude a minga. The one that is actually paying for the meal is often the party that has just used Cicol's brokerage power. This can be a development organisation or a village that invited Cicol to have a meeting in their village in order to accomplish something, for example to solve some land conflict.

The most important field of Cicol governance of natural resources is the issue of land titling. Historically, access to land has been very limited due to relationships of subjugation with outsiders such as land owners and mining and logging companies. At present, one of the strategies to secure access to the land is the fight for a communal land title for the Lomerían people. Before Cicol there was already a certain degree of co-operation between villages to secure land rights: to clear the borders between village lands, to fight large real-estate holders and extracting companies together and to organise politically as indigenous people in national demonstrations, for example in the typically Bolivian indigenous marches and road blockades. But with the arrival of Cicol this issue has been taken up seriously.

Cicol managed to become recognised by INRA as *the* representative organisation of the people of Lomerío in the procedure to get communal land titles to the territory. Even when it was highly doubtful for a long time whether this title would ever be fully granted, it provided Cicol with both a position which nobody interested in the land could afford not to reckon with and a strong legitimation in the governance of land in Lomerío. Since then all negotiations about access to land in this area have been taking place through the board of Cicol. Conflicts between villages, between villages and large real-estate holders and between villagers and extraction companies have been dealt with by granting Cicol the role of arbiter between Chiquitane groups or spokesman in conflicts with outsiders. When necessary Cicol can initiate court procedures against outsiders with the support of CEJIS. This is a lawyers' NGO that supports indigenous organisations in INRA procedures which provides Cicol with juridical knowledge.

Apart from the fight for land rights, Cicol works to gain development projects for the area. These may comprise infrastructural work on roads and communication: many villages received a shortwave radio and two of the 28 villages have a solar-powered satellite phone and water storage. Cicol also runs a project for an NGO that provides people with one cow per hectare of fenced-off land which will have to be paid back after five years. Other projects run on the introduction of new crops, such as pine-apple, setting up animal disease control, such as a vaccination programme against foot and mouth disease, or the introduction of new production methods, such as irrigated farming. Projects also include the foundation and organisation of co-operatives among villagers from different villages to produce cash crops, such as peanuts.

These projects bring new resources to the villagers, such as knowledge, infrastructural investments and physical resources, such as medicine, crop varieties and cattle. These are resources that Cicol does not have itself, but to which it is connected through its relationships with NGOs.

5.4.1 Brokerage

In Lomerío no authority has any other power than that provided by his relative position in webs of interdependencies. Authorities have almost no budget to dispense of or other physical power in terms of equipment, civil servants, policemen or jails whatsoever. Any activity that Cicol wants to carry out costs both money and manpower. NGOs are virtually the only source of income for the organisation and the villagers provide the manpower. Therefore, projects depend on the co-operation between the two and that is where Cicol comes in. Cicol functions as an intermediary between the Lomerían villagers and the NGOs. It is used by the villagers as an access point to get to the resources that NGOs can provide, their money, their knowledge, etc. At the same time it is used by the NGOs to get to the villagers: to their general willingness to co-operate in projects, to accept these strangers as teachers, to participate in their assemblies, to work in their projects, to invest their time in something else than their *chacos*.

Bierschenk's notion of development brokers as gatekeepers seems to suggest that access to the villagers would otherwise be easy and that it is restricted by the role of the development brokers (Bierschenk *et al.*, 2002). In the case of Lomerío it is the other way around. Cicol is more like a two-way bridge which facilitates the interplay between the Lomerío villagers and outsiders. This overcomes difficulties such as physical distances and language problems. But more subtle matters also play a role, such as how to handle established practices of communication and decision-making in the communities, which cannot easily be neglected. Cicol further cushions the general distrust of Lomerians towards strangers and their fear of them, both rooted in their history of subjugation.

Bierschenk *et al.* (2002) make a distinction between political and development brokerage, adding immediately that there can be a lot of straddling between the two. In the case of Lomerío it is very hard to make the distinction at all. This may be because the main reason why villagers contact NGOs or state institutions is for the acquisition of development funds for projects. After all, the main governance in Lomerío which comes from outside, i.e. from NGOs or state institutions, is in the form of projects. These projects almost invariably deal with Cicol to get access to the villagers needed to make a project successful.

NGOs need Cicol as a gateway, as a tool in their administration. In return villagers need Cicol to represent their wishes to the outside world, as a tool for problems that go beyond village resources.

As Bierschenk *et al.* (2002) point out, brokerage positions are always open to contestation by other players who want to occupy their position instead. Nevertheless in the case of Cicol there is a near monopoly in the important sector of land governance based on Cicol's connection to INRA's juridical procedure which allows for only one representative. As we have seen earlier conflicts over land have more often than not been dealt with by Cicol in its role as arbiter or spokesman, each time further legitimising Cicol's position as Lomerío representative. The INRA related contact with the CEJIS lawyers further strengthens Cicol's brokerage position vis-à-vis its constituents. This makes Cicol agents the only actors in Lomerío who can translate the juridical process to the Lomerían people and who can activate these lawyers as a resource for local actors in land conflicts in Lomerío.

Brokerage contestation does play a role in the relationship with APCOB which is Cicol's main gateway to development resources. Although alternative brokers until recently had not played a role of any significance, on a number of occasions APCOB refused to yield to demands by the Cicol agents and threatened to discontinue its co-operation with Cicol, thus implicitly opening up to new brokers. In recent years the municipal authorities, although established by Cicol itself and supported by Cicol to win legitimation through the elections in order to improve its own position, have become a serious competitor.

NGOs have started to make their support dependent on the joint approval of both Cicol and the municipality, thus bestowing a certain amount of brokerage power upon the municipal government, while at the same time allowing for some competition with Cicol. The result of this was that at the end of my field work there were signs that Cicol and the municipality would henceforth have to work together to accomplish what Cicol was previously able to accomplish alone, with the benefits of tax-based funding of the municipality.

6. Interdependency, discourse and power

In order to avoid repetition in this paper, I have already analysed the Lomerío case from a brokerage perspective as supported by Bierschenk *et al.* (2002) in the description of the case. In this analysis I have accounted for interdependencies in relationships and the role of brokerage in the bringing together of locally, culturally and socially distanced resources, such as in this case labour and money. In this section I will add to this an analysis of how both practices of family exchange and of village and territorial governance share a material condition in which relations of interdependency become discursively meaningful and thus further shape power relations in the actual access to food. For this I will relate the material conditions of subsistence farming and village and territorial governance to the legitimating discourses around these practices.

The dependencies found in village farming result in a culture of practices and legitimating discourses that constitute access to food security. The reciprocal interdependency relations in labour exchange are discussed in a discourse that emphasises the productivity and other positive aspects of these practices rather than the costs and problems associated with it. At village level governance comparable dependencies are a little different, but still result in quite reciprocal practices of governance with a legitimating discourse that is closely linked to the ones in family farming. At territorial level governance the modes of exchange are hardly reciprocal and much closer to clientelistic schemes. Nevertheless, the legitimating discourse still emphasises its reciprocal character.

The material conditions of agricultural production are such that people are very much dependent on each other for their subsistence farming. The needs of one family that requires relief by the help of others is discussed in a discourse on village labour exchanges that expresses honour and duty, and of a pleasure to help (Rozo

López, 2000). The fact that 'you depend on me' is not explicitly expressed, but hidden in other motivations. Not 'I'm helping you out,' but rather 'I am thankful to be working for you and to be treated by you as a guest, with food and drink'. Or even: 'I am working with you and afterwards we will feast together'. Who is benefiting where is not expressed as relevant, not made explicit. In the party of drinking and eating, debtor and creditor become equal by mutually complementing and celebrating each other and each other's work or food and drink and by treating each other to cups of *chicha*.

The discourse can be characterised as a mix of both obligation in honour of *comunario*-ship and voluntary, disinterested, selfless help and of the joy of working hard together and jointly celebrating the accomplishments. The reciprocity of family labour is thus expressed in certain regularities and expectations and institutionalised in the *minga* and the *cooperación*. Power relations nevertheless subtly unbalance the reciprocal character of these exchanges as has been shown. Another practice that expresses a positive meaning with regard to these interdependencies is the exchange of seeds between families. New *chacos* are to be sown by seeds from exchange with other families, otherwise it brings bad luck (Schwarz, 1994).

This symbolic aspect of reciprocity is carried over into the public sphere in the legitimating discourses that are used in village decision-making. Both the discourse and the material conditions in which practices of inter-family labour exchange are set, play an important role in the reciprocal character of village authorities. Although the character of these practices is slightly less reciprocal.

Village decision-making is not an exchange of resources in a personal sense, where action by an authority would be subject to personal remuneration, but in a more general way. This can be found in several instances: in the rotation of functions and in the multitude of authorities. Villagers become authorities to spend their time and energy on causes in line with village assembly's wishes. Or, more accurately, in line with what these wishes have become under the powerful influence of villagers with brokerage positions. In return, others will perform official authorities' duties for you, either at the same time as one of the other authorities or successively. Important to an understanding of reciprocity at this level is also the fact that the job of a formal village authority consists primarily of co-ordination. The job is about keeping people talking and trying to arrive at some sort of consensus, rather than decision-making or power-holding. The real power relations in the village are not directly linked to official positions, but rather to positions of brokerage.

The material conditions of village decision-making are also characterised by dependence on the help of others. In order for an authority to be effective he needs to spend his energy on co-ordination and listening and to leave decision-making to the village assembly. This is necessary because, as argued before, he totally depends on the villagers' labour for the execution of the relevant decisions.

The discourse is one of selflessness, an authority is doing the hard work for the good of all, and is doing so gracefully, because he wants to serve his people, because he perceives it as a duty for all to do so, each in turn, and because the villagers want him to do so, which they show in the elections and at the village assemblies where they back him up and advise him on how to act.

At the level of Cicol the reciprocity is not directly constituted in 'I help you and you help me' relationships, nor by a sense that everyone will in turn carry the burden of the job of authority for the good of the others. Cicol's job is much more one of top-down decision-making than that of village authorities. It requires more choices to set priorities and deal with the input by outsiders of new knowledge, new resources and new relationships. This on the one hand demands more than the following of sound traditions, and on the other hands provides authorities with some space to exercise power using management of meaning as a source of power. The fact that Cicol's agents serve a four-year term, rather than the common one-year term in villages also provides some more room for independent decision-making.

Although there is often no direct reciprocal link between Lomerían villagers and Cicol's actions and decision-making, the relationship between Cicol and its constituents is discursively constructed among Lomeríans as largely constituted on the basis of reciprocity. In this discourse the net of relationships that provides Cicol with its power basis is ultimately constructed to be dependent on a reciprocity in which villagers provide their labour, while Cicol in return represents the Lomeríans' wishes to the outside world. The power relationship between Cicol and it constituents is thus based on a combination of practices and legitimating discourses, emphasising on the one hand Cicol's representativeness and on the other hand the duty of villagers to provide their labour according to Cicol's wishes.

It is important to note that this discourse provides Cicol with an extra source of power in its brokerage relationship between villagers and NGOs. Cicol not only provides access to the villages which may otherwise be problematic, but also provides legitimation. For it is exactly the combined discourse of reciprocity of representativeness and village labour which makes Lomerío attractive for development organisations. The notion of representativeness is also part of discourses in the development business on democratisation and participation. Whereas the same is true for the provision of labour by the developing subject which is supposed to contribute to the sustainability of projects. This is why the reciprocity discourse at the level of subsistence farming that is replicated at village and territorial level governance, is not only relevant for the Chiquitane authorities but also for the governance practices of NGOs and state institutions. NGOs want to emphasise both sides of this reciprocal relationship in order to legitimate their choice of Lomerío as a location for their project and their co-operation with and trust in Cicol in doing so.

While Cicol's power vis-à-vis State and NGOs depends on the relationship between Cicol and its constituents, it also works the other way around. Its relationships with NGOs and the State provides Cicol with power over the villagers. For the provision of certain resources the Lomeríans depend on a good broker. Now that Cicol has managed to become the only viable broker, this makes the Lomeríans dependent on Cicol. This in turn provides the Cicol board with a certain amount of power over its constituents. Because Cicol has the strongest credentials for explaining what villagers need to do in order to acquire the wished resources, the board can use its power to manage meaning to fulfil the Lomeríans' wishes. Nevertheless, while managing meaning to this purpose, Cicol can at the same time use this to strengthen its own position.

Bierschenk *et al.* (2002) emphasise that this mutual benefit also includes a certain interest that brokers set out to gain. A broker helps actors on both sides of the divide but expects something in return. According to Bierschenk *et al.* the price is usually very vague and to be paid in some unclear future date: 'The broker thus takes a merchandise out of other people's expectations, anticipations and hopes' (Bierschenk *et al.*, 2002). In the case of Cicol these rates are extremely vague. The discourse, which is rooted in the general discourse on reciprocity, even denies that there is any payment. Nevertheless there are clearly some profits to be gained. These are to be found mainly in the field of personal evolvement and social status which can be used to influence future village and territorial governance, but there are also some cases of self-enrichment, although in public this is very much frowned upon.

This shows how governance power becomes a reality through a chain of interdependent relationships. This is not only true for Cicol, but also for village authorities, the municipality and the NGOs. All these authorities in the end depend directly on the approval of the villagers and of each other.

7. Conclusion

Food security depends on more than just the presence of food. It is embedded in local practices and discourses on the production and distribution of food. In this chapter I have shown how the resources needed for the production of food, especially labour from villagers and money from projects, are decisive for food security and thus the way local practices and discourses shape the access to these. Practices and discourses that crucially shape the access to these resources are found at the level of family exchange, and both in village and territorial governance.

I have focused on how governance of access to food production and food security in the indigenous area of Lomerío is influenced by the prevalence of dependencies in those relationships that are paramount in everyday agricultural production. These concern family food production and territorial governance which both need to be organised in material conditions of deficiency that lead to strong interdependencies. The absence of natural resources, other than land, forces families to rely on reciprocal forms of labour exchange. In the same way, this absence of natural resources forces Cicol to rely on its relational resources as a basis for its power.

In both localities of organisation the interdependency relationship and the harshness of the conditions which force people to co-operate is downplayed, while the voluntariness, honour, morality and the idea that the general good prevails over individual interests are emphasised. The way that discourse gives meaning to these conditions and the evolving governance functions to legitimate the power relations in this governance of access to food, mainly by downplaying the inequality involved.

The legitimating discourse around Cicol's brokerage position is twofold. Towards its people, Cicol's relations with the outside world are pictured as a powerful resource for its constituents. Powerful because of its special knowledge of outside actors, their programmes, their specific language and their way of dealing with things. At the same time Cicol is pictured as serving its people, being dependent on its people and not exercising power over its people. Thus downplaying the possibility for a Cicol agent to use its power for private purposes. Instead of this, again the voluntariness, honour, morality and the idea that the general good prevails over individual interests are emphasised. Towards the outside world the power of Cicol over its people is likewise understated. However this time Cicol is pictured as powerful and therefore useful, because of its knowledge of local needs and language and ways of dealing with things.

In short, the dependency of the Cicol board on its people is depicted as an honour to serve the general good on the one hand, while on the other hand this dependency is projected as representativeness. Cicol's power and usefulness on one side is constructed by emphasising its knowledge about and its relational resources on the other side.

Two types of resources are bottlenecks in the provision of food security: the labour from the villagers which is needed both in agricultural production and in development projects and the money from projects to make investments both in agricultural production as elsewhere. Labour is not only the major resource in inter-family village exchange; it is also the most important exchange value for brokers negotiating with the outside world. For it is through brokerage that local labour becomes available, which renders the external investments made by development organisations sensible. The scarcity of labour also means that development organisations need to compete not only among themselves to get access granted, but also with local agricultural production, because villagers choose where to contribute their labour.

An important factor in the success of development projects is therefore the management of meaning by brokers like Cicol and village authorities. It is their work that primarily shapes what people expect to gain in providing their labour in a project.

References

Benda-Beckmann, F., 1988. Between Kinship and the State: Social Security and Law in Developing Countries. Foris, Dordrecht, the Netherlands.

Benda-Beckmann, F. Von Benda-Beckmann, K. and Spiertz, H.J.L., 1997. Local law and customary practices in the study of water rights. In: Pradhan, R., Von Benda-Beckmann, F., Von Benda-Beckmann, K., Spiertz, H.J.L., Kadka, S.S., Haq, K.(eds.). Water Rights, Conflict and Policy. IIMI, Colombo, Sri Lanka, pp. 221-42.

Bierschenk, T., Chaveau, J.P., and De Sardan, O., 2002. Local development brokers in Africa: The rise of a new social category. Department of Anthropology and African Studies, Johannes Gutenberg Universitat, Mainz, Germany.

De Bakker, H.C.M., 2001. De cynische verkleuring van legitimiteit en acceptatie, Een rechtssociologische studie naar de regulering van seizoensarbeid in de aspergeteelt van Zuidoost-Nederland. Aksant, Amsterdam, the Netherlands.

Knogler, J., 1979. Relato sobre el país y la nación de los chiquitos an las Indias occidentales o América del sud y las Misiones en su territorio, redactada para un amigo. In: W. Hoffmann (ed.). Las Misiones Jesuíticas entre los chiquitanos, FECC.

Köhne-Hoegen, E.H., 2006. Boliviaanse contrasten. Trema: 344-348.

Leach, M., Mearns, R. and Scoones, I., 1999. Environmental Entitlements: Dynamics and Institutions in Community-based Natural Resource Management. World Development, 27: 225-247.

Long, N., 2001. Development Sociology. Routledge, London, UK.

McDaniel, J., 2002. Confronting the Structure of International Development: Political Agency and the Chiquitanos of Bolivia. Human Ecology 30: 369.

McDaniel, J.M., 2003. History and the Duality of Power in Community-based Forestry in Southeast Bolivia. Development and Change 34: 339-356.

Nuijten, M., 2003. Power, Community and the State: The Political Anthropology of Organisation in Mexico. Pluto Press, London, UK.

Nuijten, M., 2004. Governance in action. Some theoretical and practical reflections on a key concept. In: Kalb, D. (ed.). Globalization and development : themes and concepts in current research. Kluwer, Deventer, the Netherlands, pp. 103-130.

Nuijten, M., 2005. Power in practice: A force field approach to natural resource management. The Journal of Transdisciplinary Environmental Studies 4: 1-14.

Ribot, J.C. and Peluso, N.L., 2003. A theory of access. Rural sociology 68: 153-181.

Rozo López, B.E., 2000. Impacto diferencial de la economía de mercado en formas tradicionales de cooperación laboral: Estudio de las comunidades chiquitanas de San Lorenzo y El Cerrito de la zona de Lomerío. Departamento de Santa Cruz.' in Facultad e ciencias sociales, carrera de antropología y arqueología. Universidad Mayor de San Andrés, Bolivia.
Schwarz, B., 1994. Yabaicurr - Yabaitucurr - Chiyabaiturrup: estrategias neocoloniales de desarrollo versus territorialidad chiquitana. FIA. Semilla. CEBIAE, Bolivia.
Sen, A.K., 1981. Poverty and famines. Oxford University Press, New York, NY, USA.

Chapter 14
Feed security contested: soy expansion in the Amazon

Otto Hospes

1. Introduction

> 'Between 1993 and 2003, in developed countries the increase in cereal demand for feed will outstrip the increase in cereal demand for food in both absolute and relative terms. In the same period, developing countries' demand for cereals for animal feed is projected to double while demand for cereals for food for direct human consumption is projected to increase by 47 percent' (IFPRI, 1997: 11).

The key question is not 'can we feed the hungry?' but 'can we feed the animals?', stated Michel Keyzer of the Centre for World Food Studies (SOW, 2005) at an expert meeting on global action for food security.[559] At a global level, not only 850 million undernourished people have to be fed but also billions of pigs, chickens and cows. These converted cereals are to serve food preferences of (new) middle income classes, not the poor. Food experts have observed that high or increasing incomes go hand in hand with high or increasing shares of meat in the daily diet (FAO, 2003; IFPRI, 1997; Keyzer *et al.*, 2005).

The demand for feed for animals in a developing or emerging economy competes with the demand for food for the poor. In a globalising world, this feed-food competition has crossed national boundaries (Yotopoulos, 1985): 'Whether wheat produced in Australia will go to feed people in Bangladesh, pigs in the USSR or sheep to be exported to the EU becomes a question for the world market to determine' (Yotopoulos, 1985: 477). FAO's prospects for the world cereal balance for 2006/07 is a 'sharp drawdown in stocks' as a result of 'a recovery in feed use due to a rebound in poultry consumption' (FAO, 2006: 3). Increased consumption of chicken by (new) middle income classes in one part of a country or the world, may go hand in hand with persistent hunger and malnutrition in the same country or elsewhere. Though international human rights law emphasises that all human beings are equal, this suggests that some people are more equal than others (Sen, 1999).

---

[559] The expert meeting was held on 25 September 2006 in Amsterdam. The meeting was organised by the Foundation for Global Food Security and FAO Committee the Netherlands, in collaboration with Wageningen UR and Schuttelaar & Partners.

This chapter is about the recognition and regulation of far-reaching trade-offs of feeding animals to satisfy food preferences of the non-poor. The focus is on the purchase and use of soybean meal as a key ingredient in compound feed, next to cereals. Soy is the most traded oil-crop in the world. Brazil is the largest exporter of soy and the EU the largest importer of soy. The largest single EU country importing soy is the Netherlands. Unlike cereals, soy is not a food crop for the poor. Some may thus argue that soy production can be critical in replacing cereals as feed and herewith in addressing the food-feed competition in favour of poor people and their access to food. NGOs in both Europe and Brazil suggest, however, that this is to replace one problem with an even larger and more dramatic set of problems, of which food insecurity is just one. They report that soy expansion in Brazil to feed animals in the EU not only goes hand in hand with a decline in food cropping areas in Brazil but has also led to deforestation, loss of bio-diversity, human rights violations, intoxication and death, and spread of genetically modified soy in Brazil, also entering feed chains in the EU. These effects are what economists would call 'externalities', or in this case: the trade-offs or price of feed security that others have to pay somewhere at the very beginning of the soy chain.[560] This 'price transfer' incurs loss of other values (see Lang, 2006; Lang and Heasman, 2004), like protection of tropical forest and human rights.

To try and define feed security, one could simply replace 'people' with 'animals' in the FAO definition of food security: 'a situation that exists when all people, at all times, have physical, social and economic access to sufficient, safe and nutritious food that meets their dietary needs and food preferences for an active and healthy life' (FAO, 2003). However, animals do not buy feed, people do. More importantly, feed security is not about all people and not about consumers in the first place. Unlike food security, feed security refers to a more exclusive situation and is about agribusiness and livestock industry in the first place. This industry needs access to feeding stuff that is not only sufficient, safe and nutritious but also cheap. Economic and physical access to feed is a pre-condition for meat producers to meet the preference for meat of the non-poor. Unlike access to food for all, feed security is not a humanitarian concern and certainly not a human right but an outright business interest.[561]

[560] An external effect or externality is 'an effect of a purchase or use decision by one set of parties on others who did not have a choice and whose interests were not taken into account' (http://economics.about.com/cs/economicsglossary). External effects are 'costs and benefits caused by the activities of an industry which are not reflected in the price at which the product is sold or influence the quantities purchased' (www.indiainfoline.com/bisc/jmee.html).

[561] Borrowing from FAO's definition of food security, feed security could be defined as 'a situation that exists when all meat producers and processors, at all times, have physical and economic access to sufficient, safe and nutritious feed that meets their economic needs and enables them to meet the demand of meat for the non-poor'.

Feed security is about food security for the rich, that is, their access to food and food preferences. Such food security has been a largely ignored element of the international debate on food security that has been mainly focused on poor countries and poor people. This chapter is not about pitying the rich but reflecting on the possible negative effects of production of soy for the meat industry on human life conditions at the level of feed farming. These conditions include access to land and forest, biodiversity and respect for human rights, that together are critical for food security of small farmers, gatherers and indigenous people in rural areas.

NGOs want companies in both Brazil and the Netherlands to recognise negative environmental and social effects of soy expansion and to agree on regulation to reduce these effects. They want companies to recognise other values next to the value-for-money. The first objective of this chapter is to show how companies at different parts of the soy chain have responded to alarming reports of NGOs and how companies in one segment of the global soy chain have tried to define their responsibility for negative effects at the very beginning of this chain. Much of the debate on negative effects of soy and different values underlying this debate, has taken place in the form of international multi-stakeholder consultation, involving NGOs, companies and, to a limited extent, government officials. The second objective of this chapter is to assess the meaning of the instrument of multi-stakeholder consultation. Bringing together different and as many stakeholders as possible in meetings is generally considered to be a democratic, transparent and efficient instrument in which every opinion matters and can be accommodated to reach consensus. This chapter will show that the production and global trade in soy as a global commodity is very much subject to private regulation that is again constantly subject to negotiations between companies and NGOs on the truth about feed farming and its allegedly negative effects. These negotiations can hardly be qualified as democratic, transparent and efficient. For some opinions there is more room than for others. Some participants have turned their back on the 'participatory' meetings. Reaching consensus with many stakeholders proves to be a cumbersome and time-consuming process.

The structure of the remaining part of the chapter is as follows. First, some key characteristics will be given of soy and soy chain actors. Second, an overview and analysis will be given of the soy chain as a chain of dialogue and confrontation between companies and NGOs on the negative effects of soy cultivation and how to regulate these. For this purpose, three critical levels of interaction of NGOs and companies will be distinguished. In the next section, these three levels are taken together to describe a multi-level lobby of an international NGO and the response of Brazilian industry. The final section sums up the responses of companies to alarming reports of NGOs and reviews the meaning of the instrument of multi-stakeholder consultation.

2. Soy and soy chain actors

Soybeans have a remarkably high protein content with almost 80% of the bean processed into meal containing an average of 44-48% protein. Because soy is very protein-rich, it is a very suitable ingredient of animal feed (Dros, 2004: 7; Dros and Van Gelder, 2002; ISTA Mielke, 2005; Rabobank, 2006).

Soy is the single most important animal feed in the EU, accounting for 55% of protein-rich animal feed.[562] CIDSE (2005: 20) reports that,

> ...15 Member States of the EU produce 10.9 million tonnes of protein-rich material (PRM), in other words 22 percent of its protein requirements for animal feed. ... The remaining 78 percent, or 39.5 million tonnes, is imported. Soybean (in both bean and oil cake form) represents the overwhelming majority of European PRM imports: 33.8 million tonnes or 87 percent of vegetal proteins.

Europe is dependent on imports of soybean for its animal production, making it the world's largest importer of soy (both beans and oilcake). The increase in the EU's production of pigs and poultry, whose production doubled between 1970 and 2004, has contributed considerably to the increase of soy import. The BSE crisis and the ban on animal meal in cattle feed, which followed in 2000, led to an increased substitution by vegetable protein in animal feed and accentuated EU's dependence on imports of soy beans (CIDSE, 2005: 9, 21).

The production of soybeans outside the EU is critical for feed security in the EU. The USA has been the traditional supplier of soybean meal to the EU since the creation of the Common Agricultural Policy.[563] However, it is no longer the largest supplier of soy. Brazil took over the leading position of the USA as the world's biggest soy exporter in 2003. In 2004 Brazil supplied 63% of the EU soybean imports (Dros, 2004: 7-8, referring to ISTA Mielke, 2004).

Soy chains are truly global.[564] Soybeans are either exported as such or first crushed and then exported in the form of meal. The soy bean and soy meal chains connect producers, traders and crushers in Latin America and the USA with crushers, feed

[562] Available at: http://www.gmfreeireland.org/feed/index.php. See also Rabobank (2006).
[563] In its negotiations with the USA, the EU opted to promote and protect the production of cereals in exchange for which the USA could export cereal substitute products duty-free to the internal market of the EU, including soybean.
[564] Crushing of soybeans yields two products: soy meal and soy oil. Whilst soy oil is a by-product of soy meal production and is largely used for domestic consumption (Hin, 2002: 7), soy oil ranks as the most important edible oil in the world with a global market share of 23% (Dros and Van Gelder, 2002: i). Basically, the soy chain splits into two parts after crushing: one for meal and for oil.

industry, livestock industry, slaughter houses, retailers and consumers in Europe (ISTA, Mielke, 2005; see also Figure 1).

Soybean trading and crushing in the four South American soybean production countries (Brazil, Argentina, Paraguay and Bolivia) is dominated by four large international commodity trading companies, also called the ABCD of soy: Archer Daniels Midland (ADM), Bunge, Cargill and Dreyfus. The first three-mentioned are American trading companies. They are not only major players in Brazil but also control 80% of the European soybean crushing industry (Dros and Van Gelder 2002: iii).

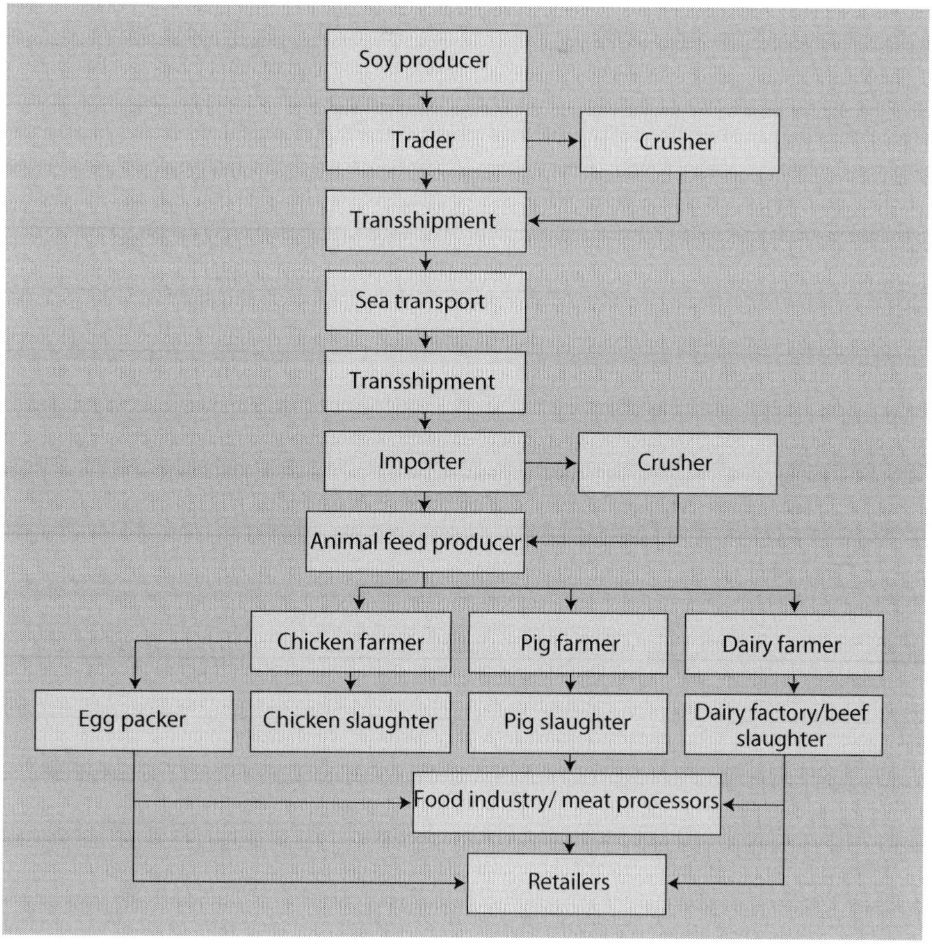

Figure 1. Soy chain from Brazilian producers to Dutch retailers. Adapted from: Dros and Van Gelder (2005).

3. Who is wrong, who is right?

NGOs, journalists and research institutes have each documented various negative effects of soy cultivation in the north-east and north of Brazil (Bickel, 2005; Bickel and Dros, 2003; Blankendaal, 2006; Dros and Gelder, 2005; Greenpeace, 2006a; Netherlands Soy Coalition, 2006). It is here where the soy frontier lies. Though the acreage of soybean planted areas is still relatively small in the north-east and north of Brazil, the increase in soybean planted areas has been highest in these regions in the period 1995-2004: 117% in the north-east and 767% in North (Dros, 2004: 21-23). Likewise, LEI reports most dramatic increases in soy areas in the northern states of Brazil: soy area increased five times in Piaui and seven times in Tocantins in the period 2000-2005 (Van Berkum *et al.*, 2006: 29). The actual and potential contribution of soy expansion to loss of tropical rainforest in the Amazon biome has formed a major trigger for these civil society-based organisations to collect data, to conduct lobbying and to inform a wider public about soy expansion and its effects. History gives them reason for concern.

Soy was introduced as a livestock feed crop in South Brazil in 1914. The dramatic increase in soy production in South Brazil in the 1960s and 1970s contributed to the near extinction of Atlantic Forest in this region. Throughout the 1980s and 1990s soy entered the Central West region. Between 1995 and 2004 the soy area in this region increased by 77%, herewith contributing to loss of natural vegetation in the Cerrado. This vast savannah originally covered 200 million hectares and is considered the most biologically diverse savannah of the world (Dros, 2004: 21-27).

Deforestation is not the only negative environmental effect that is recorded by Brazilian and international NGOs. Other reportedly negative environmental effects to which soy cultivation contributes directly or indirectly through deforestation, are loss of biodiversity, erosion, soil degradation, and environmental pollution and intoxication of people as a result of use of pesticides. A special issue is the use and spread of genetically modified soy in Brazil following the introduction of Roundup-Ready Soy by Monsanto in 1996. Experts estimate that about 30% of Brazil's soy is grown with GM seeds. The figure is thought to be closer to 90% in Brazil's southernmost state, where the seeds were first introduced in the 1990s after being smuggled in from Argentina (ICTSD, 2004). NGOs think that GM-soy is an environmental time-bomb: GM-soy undermines biodiversity and may form a threat to life as long as companies cannot prove that consumption of GM-soy is harmless to animals and human beings in the long run.

Last but not least, expansion of soy cultivation has also caused negative social effects. NGOs report that expansion of soy has reduced access to land, employment and food of people living at the soy frontier. Large soy farmers buy land from small food producers, or simply occupy land of people who do not have formal ownership titles. Such land acquisition or grabbing has not only resulted in more or less violent conflicts but also a reduction in food production for local markets. NGOs report that soy companies have built new storage and transshipment facilities at the soy frontier, gladly accepting the offer of tax exemption from local government but without providing the promised boost to local employment (Hospes, 2006). In 2004 the Brazilian government intervened in 236 cases of slavery involving 6,075 labourers. Anti-slavery campaigners think that '[t]he official estimate of the number of slaves is way off the mark. The real figure could be 250,000' (Greenpeace, 2006a: 31) because of the expansion of large-scale soy cultivation. The Netherlands Soy Coalition notices that 'hundreds of reports of slavery at soy farms are currently being investigated by the Brazilian Ministry of Labour' (2006: 26).

Some NGOs have qualified some or all of these negative social effects as violations of human rights (Netherlands Soy Coalition, 2006: 25; Bickel, 2005). Greenpeace has qualified the planting of 1.2 million hectares of soy in the Brazilian Amazon rainforest in 2004-2005, the use of slave labour to clear forest for agriculture, and greenhouses gas emissions resulting from clearing and burning the Amazon rainforest as a 'crime'. This environmental NGO has referred to the three US-based agricultural commodities giants ADM, Bunge and Cargill that reportedly invade the Amazon rainforest and spur the incursion of illegal farms as 'the criminals'. Noting that 80% of the world's soy production goes to the livestock industry, Greenpeace has qualified European agribusiness and livestock industry as 'partners in crime' (2006a: 5).

At different parts of the soy chain and in different ways, NGOs in Brazil and in the Netherlands have tried to reduce the negative effects of soy cultivation. Three critical points of interaction of NGOs and companies can be distinguished at which NGOs have tried to stop soy expansion or subject soy cultivation to some form of regulation: (1) the soy frontier in the north and north-west of Brazil; (2) hotels and conference rooms in Brazil, Paraguay and the Netherlands; (3) bilateral contacts of Dutch NGOs and Dutch companies in the Netherlands.

At the soy frontier, local NGOs have tried to gain public and government support for their actions and to bring companies to justice. In hotels and conference rooms Brazilian and Dutch NGOs have tried to agree with companies through a multi-stakeholder consultation on the effects of soy cultivation and criteria for responsible soy to be adopted by companies. These interactions were aimed at establishing soft law and some kind of self-regulation. Through bilateral contacts and targeted campaigns, Dutch NGOs have tried to spur *individual* Dutch companies to define their role and responsibility in the soy chain. Well-known retailers in Dutch society were selected as the 'most important lobby targets' of a letter campaign by the private sector working group of the Netherlands Soy Coalition. To influence public opinion and to address meat consumers in particular, at the same time public campaigns (including public action, media attention and quasi-tribunals) were directed at individual food producers and retailers. Herewith, these

NGOs exposed companies to a virtual but potentially very influential court: the consumer court. The following is an account of interactions between companies and NGOs at these three different interfaces.[565]

3.1 War at the soy frontier in Brazil

Soy companies and environmental NGOs in the north of Brazil are not on speaking terms. In fact, some kind of war is going on between Bunge or Cargill on the one hand and environmental NGOs on the other. After the arrival of Cargill in Santarem in Para state, the Movement to Defend the Amazon (FDA) has organised mass manifestations against Cargill and supported local campaigns of Greenpeace. In the same city cars carry a bumper sticker saying 'Greenpeace out'. A local manager of Cargill assured that employees were not allowed to put this sticker on their car and that the sticker was an own initiative of the city population. He also believes that the leader of the FDA, a Catholic priest, is spreading falsehoods. In the state of Piaui, the president of the environmental NGO Funaguas was physically threatened and sued in court after he had called Bunge responsible for the death of people due to chemicals used on large soy farms. Both soy companies and environmental NGOs try to convince the local population and local government at the soy frontier in the north of Brazil of their point of view, using their own studies and those of others for this purpose. It is not easy to distinguish facts from opinion in this battle. The line is very thin.

Researchers from the Catholic Pastoral Land Commission in Para (CPT-Para) which works with landless labourers, report that logging, cattle ranching and mining have substantially contributed to de-forestation of the Amazon and that soy cultivation is having a similar effect. The leader of the Movement to Defend the Amazon (FDA) believes that 'beautiful Santarem is being destroyed' with the arrival of Cargill and the expansion of soy cultivation. However, Cargill management in Santarem holds that deforestation is not taking place on a large scale in Santarem and Belterra districts as claimed by environmental NGOs: 'in fact, only one % of the total land is actually used and this could be very well increased to 10% without great losses of forests'. The president of Funaguas claims that large soy farmers and Bunge have greatly contributed to deforestation in the south of Piaui State. On the basis of satellite photos and on-site visits, he estimated deforestation rate in this state at 50% due to logging, production of charcoal and Bunge entering Piaui to buy soy.

The agricultural research corporation of the federal government EMBRAPA has concluded in a report that river basins have been contaminated as a result of chemicals used for soy cultivation. Cargill management in Santarem ensured,

[565] This account is based on an evaluation study on NGOs and the adoption of socially responsibility criteria for companies that purchase soy or soy-related products (Hospes, 2006).

however, that the company does not accept soy from areas that have been classified as 'permanent reservation areas', like riversides, lakes and steep areas. The official report of EMBRAPA was presented on TV but not during prime-time because 'local government was not happy with the report', said staff at CPT-Para. The president of Funaguas reported in a local newspaper that 16 farmers in the south of Piaui state died because they were intoxicated. This news was not to the liking of Bunge and the local government. The president was not allowed to return to the place again. In fact, he was physically threatened and hired body-guards for an 8-month period.

FDA states that the harbour of Santarem (where soy from in-land Mato Grosso is transferred into large ships) has been illegally constructed by Cargill as the company did not conduct an environmental impact study prior to building the harbour. Several federal prosecutors and a Brazilian university professor confirmed this matter and provided the evidence. CPT-Para reports that people from Mato Grosso state have come to Para state and occupied large pieces of federal state land for soy cultivation without an official ownership title. Cargill mentions that it has hired an environmental NGO (The Nature Conservancy, TNC) to monitor whether soy producers stick to Brazilian environmental law and do not use more than 20% of their plots in the Amazon for soy cultivation. Cargill management at Santarem admits that some 80-90% of their soy producers do not yet have a legal document, but added that this is not the fault of the producers but a government that is reluctant to issue title deeds for common land and for land larger than 100 hectares. To end 'this confusing situation', Cargill has hired an NGO (TNC) to organise the producers and to submit a joint application for ownership titles.

Next to these different opinions on 'facts', companies and NGOs hold diametrically opposed views on the relationships between soy expansion, development and human rights. For instance, environmental NGOs hold that large-scale soy cultivation drives small fishermen, small farmers and gatherers away from the Amazon, hereby violating their right to food, that is, the right to feed themselves. NGO leadership has experienced various attempts by companies to limit their possibilities to share their views with a wider public. Companies have frustrated local campaigns led by NGOs and shared their concerns with local authorities. NGO leaders also report physical threats. Some leaders were killed. The local manager of the Cargill factory in Santarem, however, believes that soy is synonymous with 'development' and that the population of Santarem 'has a right to development'. In his view, 'we cannot accept to freeze the Amazon economically'. According to the president of Funaguas, the investment plan of Bunge and expected job creation in Piaui state were just 'theory' and only meant to convince the government that tax exemption would be a fair gesture from the government.

Whereas local environmental NGOs prefer to meet companies in court because they do not believe that negotiations with them will be very helpful in addressing the problem, their legal action has not been very successful or effective. In co-operation with the federal prosecutor's office, FDA formulated a denouncement of the illegal construction by Cargill of the harbour in Santarem. Cargill lost the case in 2005 but the harbour is still in full operation. Due to the intervention of CPT-Para, three out of 15 people accused of murdering more than 1000 people were found guilty. Funaguas won a case against Bunge at state level but lost again in the appeal court at the federal level. The case of Funaguas against Bunge prompted this company to start one criminal and four civil cases against the president of Funaguas.

3.2 Multi-stakeholder consultation in hotels and conference rooms

In January 2004 nine Dutch NGOs – together forming the Netherlands Soy Coalition – organised a seminar in Amsterdam on 'Sustainable production of soy: a view on the future. A sense of urgency'. The aim was 'to initiate a dialogue between civil society organisations and companies from the soy trade and processing chain'. This dialogue 'should lead to the identification and development of concrete actions towards the sustainable production of soy'. The goal of the seminar was 'to inform the industry, banks and government about the impact of soy cultivation, the urgency to act and exchange thoughts on the role that should be played by Dutch actors in finding solutions'.

Just over one third of the 46 participants represented the Dutch food industry, feed industry, banks or government. They were shown a documentary portraying the negative effects of large-scale soy cultivation in the Cerrado of Brazil, including loss of biodiversity and forced migration of small farmers who are pushed off their lands and lose income and access to food. The audience was informed that the EU imports 40% of South American soy production and the Netherlands is the biggest single importer within the EU. The president of the Brazilian NGO Cebrac invited industry to consider and adopt social responsibility criteria that would be developed by a network of Brazilian NGOs and members of the academic world on the basis of a virtual debate, sponsored by Cordaid, Foundation DOEN and Solidaridad.

However, in those days Dutch industry did not really feel the urgency to act and exchange thoughts on the role that should be played by Dutch actors in finding solutions. The keynote speaker from the Product Board for Margarine, Fats and Oils (MVO) indicated that 'the share of the Dutch soy sector in South American soybeans will be lower in 2020 than the present 8%', giving way to export to Asia. He concluded that 'discussions about sustainable soy are only fruitful at an international level'. Questioning the relevance of having a discussion with the Dutch food or feed industry only, it was also indicated that 'it is of great importance that the large companies such as ADM, Cargill and Bunge become actively involved in the discussions'. In early 2004 the Dutch food industry thought that neither the cause of nor the solution to negative effects of soy cultivation was to be located or found in the Netherlands but elsewhere in the global soy chain.

The advice of Dutch industry to address the soy problem at an international level was taken on board by the Netherlands Soy Coalition. One of the members of this coalition, the Dutch private development organisation Cordaid, accepted the invitation of WWF and Unilever to join an organising committee to prepare an international conference on sustainable soy. For both WWF and Unilever, forest conversion due to (uncontrolled) soy expansion in Brazil was a major concern. To come to a different and more sustainable way of soy production, seeking 'the right balance between economic growth, environmental improvement and social equity', they believed it was necessary to involve as many different players as possible in the soy chain. A multi-stakeholder consultation or round table was seen as the mechanism to jointly produce solutions for sustainable soy production. The consultation was considered a process that would lead to a shared problem statement and shared view on how to solve problems.

March 2005 the first international conference of the Round Table on Sustainable Soy was held in Foz do Iguaçu, Brazil. More than 200 people from all over the world (and Latin America and Europe in particular) participated in this conference. Social NGOs, environmental NGOs, producer organisations and companies (including major players like Bunge and the Brazilian Association of Vegetable Oil Industries, ABIOVE) were represented at the meeting. However, the consultation did not lead to a shared problem statement but rather reproduced or even reinforced conflicting points of view of companies and NGOs on soy cultivation. Genetic modification of soy was an issue that created a huge divide (Quak, 2005).

Companies regretted that NGOs 'lacked understanding' and 'spread false messages' whereas NGOs accused companies of 'greenwashing': image building meant to obscure real problems related to soy cultivation. Companies did not accept or simply ignored 'social responsibility criteria for companies that purchase soy' that were formulated by a Brazilian network of NGOs on soy with support of members of the Netherlands Soy Coalition (Articulação Soja, 2004).[566] The representative of large soy producers in Brazil and member of the organising committee, Grupo Maggi, raised the rhetorical question 'how to expect a company to adopt social

[566] The three key criteria for realising an *immediate* reduction in negative impacts were: purchase of soy on land that was legally cleared before 31 December 2003, and in the case of the Amazon biome before October 1999; at least 20% of total annual purchases of soy must preferably come from family farms; only conventional soy (not genetically modified) or organic soy must be purchased. In addition, various *medium- and long-term* criteria were formulated, including the proposal that soy must be planted on fields no larger than 200 hectares. If a farm has contiguous fields over this size, the farmer must divide them into fields of no more than 200 hectares each, separated by at least a 50-metre strip of recovered native plant cover.

responsibility criteria if it is not allowed to participate in the discussion on these criteria'?

The real problem of the round table, however, was much more basic. During the consultation no consensus could be reached about the causes and scale of negative effects as identified by environmental and social NGOs. Companies emphasised that soy expansion cannot be considered as the one and only root cause of negative effects as reported or at least suggested by NGOs. They argued, for instance, that mining, logging and cattle farming have been main causes of deforestation. In addition, companies seriously wondered whether the negative effects were substantial indeed or rather of an incidental nature. Evidence of slavery, for instance, was found to be anecdotal. Companies felt that soy was demonised and too little attention was given by NGOs to the benefits of soy production, like export revenue and substantive income for the productive sector. The organising committee accepted the proposal of Cordaid, Cebrac, Greenpeace and other NGOs to rename the round table process and replace the concept of 'sustainable soy' with 'responsible soy'. These NGOs could not accept the idea that sustainable soy included large-scale and export-led soy cultivation, whereas the big companies had no problem with this at all and highlighted 'best practices' instead. The food and feed industry simply stuck to the concept of sustainable soy in their own writings. Companies and NGOs could only agree at the round table on a very short final, joint declaration, stating that 'soy production brings about social, economic, environmental and institutional benefits and problems'. The nature and scale of these problems were not specified, neither were possible solutions, let alone a timeframe for implementation of possible measures to reduce the negative effects of soy cultivation.

The first RTRS conference did not bring companies and NGOs closer together but rather pushed them further apart. For instance, Cordaid and its Brazilian partner organisations turned their back to the round table process. They concluded that there was too little prospect in the round table process of discussing their concerns about large-scale monoculture, slavery and genetically modified soy and too little support from the companies participating in the Organising Committee to change this.

After the disappointing first RTRS conference, membership of the Organising Committee changed as well. Representation from agri-business, food industry and banking sector was strengthened to include ABN-AMRO and ABIOVE next to Unilever, COOP Switzerland and Grupo Maggi. Membership also became more regional with the representation of NGOs from Paraguay (Guyra Paraguay) and Argentina (AAPRESID). NGOs from the soy frontier in Brazil and small family farmers producing non-GMO soy in Brazil were no longer represented in the committee once Cordaid and Fetraf-Sul had left.

Drawing lessons from the disappointing first conference at which companies and NGOs accused each other of not understanding reality, the new Organising Committee organised a *technical* meeting. The meeting was held in the office of ABN-AMRO in Sao Paulo in April 2006. The main thrust was to agree on the 'key impacts' of soy cultivation. Some 60 representatives of agri-business, international banks and NGOs together identified five environmental impacts of soy cultivation: habitat conversion and biodiversity loss; soil degradation and erosion; contamination and health effects of agrochemicals on man and environment; qualitative and quantitative hydrological changes; and infrastructure. In addition, four social impacts were identified: worker´s rights (violation of ILO standards); loss of livelihoods for small-scale land use systems; migration (rural to urban areas and rural to forest ecosystems); and land rights conflicts (illegal acquisition, land use rights violations and indigenous land rights). Genetic modification of soy was not listed as one of the environmental or social issues.

The key conclusions of the technical meeting do not mention the extent to which soy cultivation has led to biodiversity loss, contamination, violation of ILO standards, loss of livelihoods and land rights conflicts. With regard to the social impacts the Organising Committee noted that, 'While the presentations dealing with social issues contained a wealth of extremely useful first hand field experiences, there was a strong consensus among the participants on the need to strengthen future research efforts in the field of social impacts, given the lack of conclusive, comparable, verifiable and objective data upon which to build proactive policies, or develop reliable social performance indicators for responsible soy production' (available at: www.responsible.soy/eng). The technical meeting did not solve the major problem of the first Round Table but offered a checklist for future research and an agenda for further discussion on criteria and indicators.

At the second conference of the RTRS, held in August 2006 at the Yacht and Golf Club Paraguayo in Asunción, Paraguay, this agenda was accepted. Similar to the first RTRS conference, participants agreed that soy cultivation has caused problems but has also generated benefits. However, the latter was given a much more prominent place in the final declaration of the second conference than the first. The keynote and opening sentence of the declaration is that 'soy is a crop of key importance to world agricultural trade and makes an important contribution to the social and economic development of both producing and consuming nations'. A footnote to the declaration sums up the nine key impacts as identified at the technical meeting in Sao Paulo. The declaration does not specify the scale or intensity of these impacts. The participants of the RTRS conference committed themselves to the development and adoption of criteria within two years. In addition, the participants committed themselves to the development of verification and traceability systems in the same period. In spite of pressure from some NGOs, genetically modified soy and related issues (like health effects and intoxication due to pesticides or dependency) were not put on the future agenda of the Round Table. Likewise, no agreement was reached on short-term measures related to legal compliance, slavery and deforestation.

Three months after the second RTRS conference, the RTRS was established as an association under Swiss law. Participants of the RTRS conferences were invited to become a member of the association. From that moment, two institutional trajectories evolved within the RTRS. One was the continuation of international conferences during which participants (now members and non-members of the RTRS!) continued their debate on principles and criteria for sustainability.[567] The other was the establishment of the General Assembly as the highest decision-making body of the RTRS and consisting of members only.

The establishment of the RTRS as a member-based organisation in November 2006 represented a second major critical event. Like the first RTRS conference, the establishment of the RTRS as an association formed a kind of watershed or junction point. Those participants who did not feel at ease with the draft principles and criteria for sustainable soy and/or the way in which they were being discussed, were hesitant to become a member of the RTRS. Those who remained enthusiastic about the negotiation process from a principled and/or strategic perspective, enlisted as members. This way the route towards adoption of principles and criteria became more feasible and visible. A development group consisting of representatives from all three membership categories (producers, industry and finance, and civil society organisations) was formed. This group held five meetings in the period 2007-2009 and prepared the ground for adoption of principles and criteria at the third General Assembly in May 2009. The list of principles and criteria does not address the issue of genetic modification but comprehensively deals with both environmental and social concerns. Though the executive board and secretariat of the RTRS proudly announced the 'unanimous' adoption of the principles and criteria by the membership, there is one 'but': the principles and criteria still have to be field tested and the final responsible soy global standard is 'yet to be approved at a General Assembly' (RTRS, 2009).

3.3 Bilateral talks and public campaigns in the Netherlands

In August 2005, 22 food producers, food retailers, feed producers and banks in the Netherlands received a letter from the Netherlands Soy Coalition. With this letter the coalition wanted to call the attention of each and every individual company to problems related to soy production and to invite companies to discuss their role in achieving responsible trade of soy. The private sector working group of the Netherlands Soy Coalition had identified five companies as their 'most important target groups': Ahold, HEMA, Laurus, Nutreco and Unilever. It was expected that

[567] In April 2008 the third RTRS conference was held in Buenos Aires, Argentina, and 13 months later the fourth RTRS conference was held in Campinas, Brazil.

Ahold, Hema and Laurus would be willing to talk as they are well-known and reputable companies in the Netherlands, producing or selling meat (products) for which soy has been used as raw material (feed). It was expected that Nutreco and Unilever would also be willing to talk because members of the Netherlands Soy Coalition already had contacts with these companies.

Ahold took notice of the letter and background material but declined the invitation. From the brochures and information of the Netherlands Soy Coalition, the company concluded that the campaigns of the coalition were only focused on the Netherlands. Because Ahold believed that negative social and environmental effects of soy production can only be reduced on the basis of a global approach and forum, it considered a focus on the Netherlands inappropriate. The food retailer considered the international Round Table on Responsible Soy as the appropriate forum for discussion and indicated a willingness to stay informed about guidelines on soy through Swiss COOP and other European retailers.

HEMA also did not positively respond to the invitation of the Netherlands Soy Coalition to talk about what the retailer can do to address problems related to soy because it believed it could not do much about these problems. HEMA found it difficult to trace precisely where the soy used in HEMA sausages came from. According to HEMA, it is not possible to determine the contribution of HEMA sausages to problems related to soy cultivation. As a result, the company is reluctant to define its corporate social responsibility on soy. When Fair Food published an article in a university newspaper that contained critical information about HEMA, the retailer invited Fair Food to discuss the accusations. HEMA consulted its suppliers to see whether the accusations made by Fair Food could be verified. The company concluded that problems related to soy are beyond the reach of HEMA and situated in those parts of the soy chain on which HEMA cannot exert any influence.

The invitation to discuss soy reached Laurus at a time when it was preparing its transformation and mainstreaming into the company structure of Super de Boer. For that reason and given its involvement in several other committees, projects and activities related to corporate social responsibility, the company did not plan to take or join initiatives in the field of soy at short notice.

Both Nutreco and Unilever were unenthusiastic about bilateral talks with the Netherlands Soy Coalition because they considered international multi-stakeholder consultation as the tool by which to reach agreement on sustainable production of soy. They felt that bilateral talks between companies and NGOs in the Netherlands did not make sense and (wrongly) supported the idea that some form of regulation of the soy chain can be agreed upon by Dutch players only. Nutreco and Unilever considered the Round Table on Sustainable Soy to be the appropriate forum to generate criteria on production and purchase of soy.

Not only was the response of the most important target groups of the Netherlands Soy Coalition to the invitation to talk bilaterally rather negative, they also questioned views and assumptions underlying the main objective of the coalition to reduce the negative effects of soy cultivation. In their view, soy cultivation has historically been one of the drivers of deforestation but certainly not the only one. Though disqualifying uncontrolled soy expansion, Dutch multinational companies argued that loss of forest and other natural vegetation has always been part of agricultural development in Brazil. Finally, they emphasised that soy expansion can play an important role in addressing the increasing demand for food and feed of the still increasing world population.

Whereas the most important target groups of the Netherlands Soy Coalition did not respond very positively to the coalition's letter, it did not pass unnoticed by the Dutch food and feed industry. The letter campaign and questions of individual companies on what to do with it did trigger collective action at an industry level. In August 2005 the Product Board for Margarines, Fats and Oils (MVO) started a Working Group on Soy, consisting of companies in the Netherlands that process or purchase soy or soy-related products (like Unilever, Nutreco, Remia, ADM, Bunge and Cargill). The feed industry joined the initiative of the food industry: the Netherlands Association of Feed Companies (Nevedi) started to link its own internal discussion on soy with the MVO Working Group on Soy. The two industry associations together started to organise informal meetings with the NSC. The policy view of the two industry associations on how to address problems related to soy production reflects those of Nutreco and Unilever as individual members of the MVO working group on Soy: Nevedi and MVO both acknowledge problems related to soy production but consider multi-stakeholder consultation at the international level the best way to address problems and to identify criteria for the production and purchase of soy. Like Unilever, Nevedi prefers to address soy production and processing as a matter of sustainability rather than responsibility. June 2006 Nevedi drafted a fact sheet on 'soy and sustainability' in consultation with the MVO group.

Public action by Dutch NGOs triggered Dutch companies to give attention to the soy issue and to acknowledge problems related to soy production. These actions 'make us run', said the head of the bureau for quality and product integrity of a big Dutch retailer. A demonstration by Cordaid and Friends of the Earth in the harbour of Amsterdam that was given media attention formed a trigger for Unilever to recognise the negative effects of soy production. Similarly, negative news in the media prompted the Product Board for Fats, Oils and Margarines (MVO) to acknowledge problems related to soy production.

However, public action and bad publicity did not lead to a radical change in the view of the food and feed industry on how to reach agreements on sustainable or responsible soy. The complaint filed by Friends of the Earth against VION Food

Group at a human rights quasi-tribunal in Vienna for being responsible for the wrongs of soy production in Brazil offers a case in point. The complaint did not make VION which is the largest Dutch producer of pig meat, feel guilty or deficient.[568] The company believes that sustainable production of meat requires consultation and the concerted action of all stakeholders of an international feed and food chain. Because VION very much supports international dialogue on sustainable soy as part of the Round Table process, the company found the complaints and actions of Friends of the Earth misplaced. The announcement by Friends of the Earth of their intention to block the main factory in Boxtel (which was forbidden by a court of law) did not persuade VION to change its view.

4. Multi-level lobbying and a moratorium on the trade of soy from the Amazon

During multi-stakeholder consultations and at the soy frontier in Brazil, large soy companies have used different strategies to counter, disqualify or otherwise neutralise problem statements, public actions and accusations from Brazilian and foreign NGOs. These included non-reply, counter-lobby directed at local authorities and physical threats at the soy frontier and agenda-setting, issue squeezing and participation in the organising committee of the international Round Table. Given these various strategies and moves at both the soy frontier and multi-stakeholder consultations in Brazil, the moratorium in July 2006 of ABIOVE and ANEC came as somewhat of a surprise.[569] These two associations and their member companies declared a moratorium of two years on trade of soy that will be planted as of October 2006 coming from deforested areas within the Amazon biome.

Much of this can be seen as the success of a multi-level lobby by Greenpeace. Governments were not involved at all. Greenpeace conducted public campaigns at the soy frontier in Brazil and targeted public action in Europe at McDonald's and other food companies. Instead of investing a lot of time and energy in dialogue at the Round Table, this NGO adopted a confrontational approach. Greenpeace collected on-the-ground information in collaboration with local environmental NGOs and confronted the European food industry and a wider audience with the report 'Eating Up The Amazon' (Greenpeace, 2006a). This report and targeted action by Greenpeace either convinced European food producers and retailers that soy expansion contributes to deforestation of the Amazon and/or made them realise that their reputation and market share may be harmed if no action was

[568] The complaint was also filed against other Dutch chain actors: feed producers Nutreco and Provimi, retailers Albert Heijn and Laurus, and Rabobank. Friends of the Earth used the report 'Van oerwoud to kippenbout: effecten van sojateelt voor veevoer op mens en natuur in het Amazone – a ketenstudie' (Dros and Van Gelder, 2005) as evidence.

[569] ABIOVE is the Brazilian Association of Vegetable Oil Industries. This association has 12 member companies (including ADM, Bunge, Cargill and Dreyfus) that together account for about 72% of soybean processing volume in Brazil. ANEC is the National Association of Grain Exporters.

undertaken. McDonald's, El Corte Ingles, Waitrose, Asdao, Ritter-Sport and Tegut agreed to form an alliance with Greenpeace with a view to demanding responsible soy from their suppliers. Together they made a proposal to the Brazilian soy industry to accept a moratorium on the trade of soy from the Amazon. Greenpeace had proposed a moratorium of five years but ABIOVE and ANEC decided on two years only. The two industry associations committed themselves to co-operate with the Brazilian government during the moratorium to prepare an effective mapping and monitoring system for the Amazon Biome and to develop strategies to encourage and move soy producers to comply with the Brazilian Forest Code.[570]

The moratorium was declared less than two months before the second RTRS conference held in August 2006. The moratorium strengthened the internal division among Brazilian NGOs on whether or not to seek dialogue with companies. Some NGOs were happily surprised whereas others remained suspicious about the moratorium. The declaration does not include an explicit acknowledgement of deforestation or other negative effects as a result of soy expansion. The intention to encourage soy producers to comply with the Brazilian Forest Code is an implicit recognition of illegal deforestation but does not mention whether illegal deforestation happens incidentally or is a widespread practice. The preparation of an effective mapping and monitoring system does not involve NGOs. And why two years only?

Due to the appreciation of the Real and a decline in the world market price of soy at the Chicago futures market since 2004, soy cultivation has turned into a less profitable business (LEI, 2006: 33-34, Blankendaal, 2006: 18). In March 2006 prospects still looked very grim (LEI, 2006: 34). These economic developments made soy expansion a risky investment in 2006-2007. Under such market conditions, a moratorium is not such a bad idea, but just for two years as the economic climate and profitability prospects may have improved after this period.

The moratorium is a rather unorthodox application of the PPP principle as one of the leading principles of corporate social responsibility: the moratorium is to give up the value-for-money (Profit) in favour of other values (Planet and People), yet at a time when it was very unprofitable to invest in soy; the moratorium is to save an ecologically important part of the Planet (the Amazon) but on a temporary basis; the moratorium takes concerns of People (who live in the Amazon or work for environmental NGOs) seriously and has created goodwill but has also strengthened the internal division among Brazilian NGOs as the opponents of companies in their conflict at the soy frontier, in court rooms and at round tables.

[570] According to the Brazilian Forest Code, producers must set aside 80% of their landholdings in forested areas to be allowed to cultivate the remaining 20% (Cargill, 2006; Dros, 2004: 26).

In 2008 the moratorium was extended by one year. More importantly, the Brazilian Ministry of Environment was invited to join the Soybean Working Group. This group consisted of representatives from the business sector and civil society.[571] It was established with a view to defining the *modus operandi* for the moratorium. The membership of the government means that the soy moratorium has changed from a private-private initiative into a private-public partnership. Greenpeace loudly applauded the rapprochement of the Government of Brazil as this environmental NGO had earlier stated that, 'the moratorium stays until proper procedures for legality and governance are in place and until there is an agreement with the Brazilian Government and key stakeholders on long term protection for the Amazon rainforest. (Greenpeace, 2006b). At this point, Greenpeace assumes that bringing the state back in will help to overcome one of the major problems that kick-started the moratorium as a form of private governance: the lack of capacity of the government to enforce its own environmental laws and to stop deforestation in the Amazon. The question is whether the soy moratorium as a form of private-public governance can provide an enlightening example to the Brazilian government, or whether the rapprochement of the government will lead to politicisation or bureaucratisation of the soy moratorium. In 2009 the moratorium was again extended by one year.

5. Conclusions

In response to the alarming reports of NGOs on the negative environmental and social effects of soy cultivation in Brazil, companies in Brazil and the Netherlands first questioned whether there was reason for alarm at all. Soy companies in Brazil accused environmental NGOs of spreading false messages and not understanding reality. Up until today and throughout the soy chain, trading companies do not agree or at least have serious questions about the scale and intensity of negative effects as reported by NGOs. These companies wonder whether soy planting has indeed led to substantial loss of rainforest and whether there is comprehensive proof of widespread slavery at the soy frontier. Companies have also questioned to what extent negative effects (like deforestation) as identified by NGOs can be attributed to soy expansion and soy producers in the first place. Moreover, commodity traders assure that they do not buy soy from producers that violate Brazilian environmental law, Brazilian anti-slavery law or international ILO standards. Greenpeace has called ADM, Bunge and Cargill 'the criminals' and the livestock industry 'the partners in crime' (Greenpeace 2006a: 5). Companies, however, wonder whether any crime was committed in the first place.

[571] The Soya Working Group includes soy traders such as Bunge, Cargill, ADM and Amaggi, as well as non-governmental organisations including Greenpeace, Conservation International, TNC, IPAM and WWF.

The second and more offensive reply of companies in response to alarming reports and accusations of NGOs has been to emphasise the benefits or positive effects of soy cultivation. This is not only about generation of employment, income, agricultural trade and export revenues. In response to the Greenpeace Report 'Eating Up the Amazon', Cargill declared in a press release that, 'Economic development is the long-term solution to protecting both the Amazon's peoples and the environment' (May, 2006). A local manager of Cargill explained that his company has built a harbour in Santarem and has financed soy producers in the Amazon because 'the population of Santarem has a right to development'. In the name of human rights and development, Cargill has invested in soy expansion in the Amazon. Cargill does not consider itself a criminal but a hero. Companies in the Netherlands and Brazil point a critical finger at environmental NGOs, like Friends of the Earth and Greenpeace. These NGOs want to hold them accountable for deforestation and human rights violations in Brazil but 'cannot be held accountable themselves' for misleading campaigns, according to these companies. They believe that NGOs spend money from individual donors on lobbying against companies with incorrect or biased information.

A third response by food and feed companies to the alarming reports and concrete proposals of NGOs has been to refer to the international Round Table process. Companies in the Netherlands were rather hesitant in the early phase of this process to take pro-active or unilateral steps. They argued that guidelines and criteria had to be developed on the basis of a multi-stakeholder consultation involving all chain actors. In theory, this multi-stakeholder consultation is based on a modern 'triangular dynamic' (Lang, 2006) of companies, civil society and state to determine criteria on responsible soy. In practice, Dutch and Brazilian governments have played a very minor role in the consultation process. They were not represented in the Organising Committee of the Round Table process. The Brazilian minister for Agriculture explained at the first Round Table on soy how economically important this crop is. The Dutch Minister for Development Co-operation very much wants the private sector and NGOs, not the ministry itself, to discuss at a Round Table how to make soy production sustainable. As a consequence, the Round Table process has adopted a duality model with companies and NGOs to negotiate what the problem is and to agree on how to make soy production and processing either more sustainable or responsible. The first RTRS conference was not very successful in generating consensus between these two parties on problems related to soy cultivation, let alone social responsibility criteria. NGOs that were very much concerned about deforestation in the Amazon, slavery conditions and the spread of genetically modified soy turned their back on the Round Table, hereby further eroding the meaning of multi-stakeholder consultation. At the second RTRS conference, no agreement was reached on the scale and intensity of problems related to soy cultivation. Nine environmental and social problems formed a checklist for future research and an agenda for further discussion on criteria and indicators. The third and fourth RTRS conference took place after the establishment of the RTRS as a foundation in 2006. As a club organisation, the RTRS prepared draft principles and criteria for responsible soy. These principles were adopted for field testing in 2009, and are still awaiting approval at a General Assembly.

In theory, corporate social responsibility and criteria for responsible soy can be very well defined on the basis of multi-stakeholder consultation. Such soft law and social regulation can be used to hold companies accountable for the negative effects of soy cultivation. However, if multi-stakeholder consultation does not allow for negotiation between companies, NGOs and states on a more or less equal footing to define soft law and regulation of the soy chain, this consultation comes close to a one-man show or window-dressing. Under such conditions, private companies define the rules of the game and the course of multi-stakeholder consultation. Multinational companies are keen not only on controlling the trade of soy but also that of images on sustainable or responsible production of soy. The two-fold challenge of soy as a global commodity for key corporate players in the soy chain is to secure production in the South and to protect their reputation in the North.

References

AIDEnvironment, 2005. Fact sheet Soy. Basic facts on soy production, trade and social and environmental impacts. Amsterdam, the Netherlands.

Articulação Soja. 2004. Criteria for Corporative Responsibility of Soy Buyer Enterprises. Cebrac, Brazil.

Bickel, U., 2005. Human rights violations and environmental destruction through soybean production in Brazil. Working Paper.

Bickel, U. and Dros, J.M., 2003. The impacts of soy bean cultivation on Brazilian ecosystems. Three cases studies illustrating the social and environmental impacts of soy production in the states of Amazonas, Mato Grosso and Piaui in Northern Brazil in the 1998-2003 period. WWF Forest. Conversion Initiative, Zurich, Zwitserland.

Blankendaal. S., 2006. De schaduwzijde van soja. Onze Wereld, October: 14-19.

Cargill, 2006. Cargill's view on the Greenpeace Report 'Eating Up the Amazon'. Press release.

CIDSE, 2005. An alternative vision to today´s soy market: the prospects for mobilizing European organizations. Background paper.

Dros, J.M., 2004. Managing the soy boom. Analysis of expansion of soy bean production in Argentina, Bolivia, Brazil and Paraguay in the 1998-2003 period. Paper commissioned by World Wide Fund.

Dros, J.M. and Van Gelder, J.W., 2003. Corporate actors in the South American soy production chain. A research paper prepared for World Wide Fund for Nature Switzerland.

Dros, M. and Van Gelder, J.W., 2005. Van oerwoud tot kippenbout: effecten van sojateelt voor veevoer op mens en natuur in het Amazonegebied – een ketenstudie. Commissioned by Milieudefensie and Cordaid, Den Haag, the Netherlands.

Dutch Soy Coalition, 2004. Sustainable production of soy: a view on the future. A sense of urgency'. Seminar report, Amsterdam, the Netherlands.

FAO (Food and Agriculture Organisation) Economic and Social Department, 2003. Trade reforms and food security: conceptualizing the linkages. FAO, Rome, Italy.

FAO (Food and Agriculture Organisation) Economic and Social Department, 2006. Crop prospects and food situation no.2. FAO, Rome, Italy.

Greenpeace, 2004. GM and dairy cow feed: Steps to a GM-free future for the UK dairy industry. A study on the short and long term alternatives to GM (soy) feed components for the British dairy sector, London, UK.

Greenpeace, 2006a. Eating up the Amazone. Amsterdam: Greenpeace International, Amsterdam, the Netherlands. Available at: www.greenpeace.org/international/news/McVictory-200706. Accessed March 19, 2010.

Greenpeace, 2006b. Victory as fast food giant pledges to help protect the Amazon. Downloaded from www.greenpeace.org/international/news/McVictory-200706 on March 19th of 2010.

Hin, C.J.A., 2002. Marktpersptieven voor maatschappelijk verantwoord geproduceerde soja. CLM Onderzoek en Advies, Utrecht, the Netherlands.

Hospes, O., 2006. An evaluation study of Cordaid and the adoption of social responsibility criteria for soy production by companies. Policy and Operations Evaluation Department, Dutch Ministry of Foreign Affairs, Den Haag, the Netherlands.

International Centre for Trade and Sustainable Development, 2004. Newsletter. Bridges trade Biores: 19.

IFPRI (International Food Policy Research Institute), 1997. The world food situation: recent developments, emerging issues and long-term prospects. Food Policy Report 2020 vision.

ISTA Mielke, 2004. Oil world annual. Hamburg, Germany.

ISTA Mielke, 2005. Commodity analysis soy complex and palm oil. Hamburg, Germany.

Keyzer, M.A., Merbis, M.B. Pavel, I.F.P.W and Wesenbeeck, C.F.A., 2005. Diet shifts towards meat and the effects on cereal use: can we feed the animals in 2030? Ecological Economics 55: 187-202.

Lang, T. and Heasman, M., 2004. Food wars: the global battle for mouths, minds and markets. Earthscan, London, UK.

Lang, T., 2006. Social and ethical aspects of food policy: some thoughts on the transition from a 'value-for-money' society to a 'values for money'. Paper presented at workshop on Ethics on your plate: dining and discussing on ethical issues concerning food, agriculture, nature, environment, landscape and animals. November 15, 2006, Wageningen, the Netherlands.

Netherlands Soy Coalition. 2006. Soja doorgelicht. De schaduwzijde van een wonderboon.

Quak, E., 2005. Soja: splijtzwam voor maatschappelijke organisaties. Vice Versa 39(3).

Rabobank, 2006. Rabo ag focus: soybeans. The Netherlands.

Round Table on Responsible Soy, 2009. RTRS P&C Implementation Manual for Producers. Available at: www.responsiblesoy.org. Accessed March 19, 2010.

Sen, A., 1999. Development as freedom. Oxford University Press, London, UK.

SOW Centre for World Food Studies/VU Amsterdam, 2005. Can we feed the animals? SOW-VU brief, available at: www.sow.vu.nl. Accessed March, 2010.

Van Berkum, S., Roza, P. and Pronk, B., 2006. Sojahandel- en ketenrelaties; Sojaketens in Brazilië, Argentinië en Nederland. LEI, Den Haag, the Netherlands.

Yotopoulos, P.A., 1985. Middle-Income Classes and Food Crises: The 'New' Food-Feed Competition. Economic Development and Cultural Change 33: 463-483.

Keyword index

F

G

R

S

Z

Printed in the United States
by Baker & Taylor Publisher Services